Understanding Relativity

A Simplified Approach to Einstein's Theories

LEO SARTORI

UNIVERSITY OF NEBRASKA—LINCOLN

University of California Press

BERKELEY   LOS ANGELES   LONDON

University of California Press
Berkeley and Los Angeles, California

University of California Press
London, England

Copyright © 1996 by The Regents of the University of California

Library of Congress Cataloging-in-Publication Data

Sartori, Leo.
    Understanding relativity : a simplified approach to Einstein's
theories / Leo Sartori.
        p.   cm.
    Includes bibliographical references and index.
    ISBN 0-520-07986-8 (c). — ISBN 0-520-20029-2 (p)
    1. Relativity (Physics)   I. Title.
    QC173.55.S367   1996
    530.1'1—dc20   94-49358

Printed in the United States of America

 2 3 4 5 6 7 8 9

The paper used in this publication meets the minimum requirements of
American National Standard for Information Sciences—Permanence of
Paper for Printed Library Materials, ANSI Z39.48-1984 ∞

To Eva, Anne, and Jenny

Contents

Preface

Albert Einstein's special theory of relativity, one of the supreme achievements of the human intellect, is more accessible to a nonspecialist audience than is commonly believed. Over the past twenty years I have taught relativity with some success to a diverse group of students at the University of Nebraska, including some with little or no previous physics and with limited mathematical preparation. This book is intended to serve as a text for such a course, as well as to guide the reader who wishes to study the subject independently.

Relativity is a challenge, but the challenge is in the ideas, not in the mathematics. The reader should not be put off by the fairly large number of equations; nearly all of them involve nothing more than simple algebra. Some of the more complex mathematical sections can be omitted without disrupting the development.

Although this book is on an elementary level, it is by no means a watered-down version of relativity. The logical arguments are presented in a rigorous manner, and the major conceptual difficulties are addressed. For example, I discuss in detail the connection between the possible reversal in the time order of events, predicted by relativity, and the logical requirement of causality (the cause must precede the effect). It is my hope that even sophisticated readers will find the presentation stimulating.

The emphasis throughout is on the concepts and the logical structure of the theory. To that end, introduction of the Lorentz transformation is delayed until chapter 4. In chapter 3, the relativity of simultaneity and the time dilation and length contraction effects are deduced directly from Einstein's two postulates and are analyzed exhaustively. These profound conclusions concerning the nature of space and time constitute the heart of the theory. The reader who masters the content of chapter 3 will understand relativity.

An entire chapter is devoted to the paradoxes of relativity, including the famous twin paradox as well as several based on the length contraction effect. In addition to their intrinsic interest, the paradoxes are an effective teaching tool and provide a convenient way to introduce some important and subtle ideas.

A novel feature of this book is the use of the Loedel diagram, a geometrical representation of relativity that has received little attention in the literature. The Loedel diagram exhibits the space-time consequences of the theory in a visual way that many students find helpful and does not share the major drawback of the better-known Minkowski diagram, namely, the need to employ different scales for the length and time axes. Because some instructors may prefer not to delve into the Loedel diagram, however, it is set off at the end of chapter 5 and can safely be omitted. I use it to help clarify the resolution of the paradoxes in chapter 6—but only as a supplement to the more conventional explanations.

Chapter 7 deals with the application of relativity to atomic energy and particle physics. I present Einstein's first derivation of the mass-energy relation $E = mc^2$ and use the same thought experiment to derive the expression for relativistic momentum.

The last two chapters are an introduction to general relativity and to cosmology. Those subjects cannot be treated at an elementary level in the same depth and with the same rigor as can the special theory, but most students are eager to learn something about curved space and the expanding universe. These two chapters should be considered optional. They fit comfortably into a full-semester course such as the one I teach, but an instructor who wants to devote only part of a semester to relativity may choose to restrict the syllabus to the special theory.

I have included more historical material than is found in most treatments of the subject. Two mysteries in particular have fascinated me for a long time: the role of the Michelson-Morley experiment in the genesis of relativity and the contributions of Lorentz and Poincaré and their influence on Einstein. Detailed discussion of these topics is found in sections 3.2 and 4.8. Galileo's views on relativity are discussed in section 1.3.

Notes: Relativity makes frequent reference to observers, who come in two genders; rather than repeatedly employing the construction "he or she," I have assigned genders at random. Instructors may obtain answers to the problems by writing to the author, Department of Physics and Astronomy, University of Nebraska, Lincoln 68588.

Acknowledgments

In writing and revising the book I have benefited from many discussions with my colleagues Paul Finkler, C. E. Jones, and William Campbell. Each of them read several chapters critically and made useful suggestions, as did Anne Sartori and Jonathan Parker. Roger Buchholz skillfully produced the illustrations.

I should like to acknowledge my debt to several works in the published literature. Abraham Pais's scientific biography of Einstein, *Subtle Is the Lord* (New York: Oxford University Press, 1982), provided much useful material on Einstein's development of both the special and the general theory as well as on the work of Lorentz and Poincaré. Arthur I. Miller's *Albert Einstein's Special Theory of Relativity* (Reading: Addison-Wesley, 1981) was another useful historical source. Gerald Holton's publications are the definitive word on the influence of the Michelson-Morley experiment on Einstein's thinking.

Edward Harrison's excellent book, *Cosmology: The Science of the Universe* (Cambridge: Cambridge University Press, 1981) clarified for me many of the subtleties of that subject, including the important distinction between the cosmological and the Doppler interpretations of the redshifts, a subject not well treated in most books. I learned a great deal about the Loedel diagram from Albert Shadowitz's *Special Relativity* (Philadelphia: W. B. Saunders, 1968).

Finally, the well-timed publication of volume 2 of *The Collected Papers of Albert Einstein* (Princeton: Princeton University Press, 1989) gave me access to the text (as well as valuable English translations) of all of Einstein's relevant publications.

1   Galilean Relativity

1.1. RELATIVITY AND COMMON SENSE

A child walks along the floor of a moving train. Passengers on the train measure the child's speed and find it to be 1 meter per second. When ground-based observers measure the speed of the same child, they obtain a different value; observers on an airplane flying overhead obtain still another. Each set of observers obtains a different value when measuring the same physical quantity. Finding the relation between those values is a typical problem in relativity.

There is nothing at all startling about these observations; relativity was not invented by Albert Einstein. Einstein's work did, however, drastically change the way such phenomena are understood; the term "relativity" as used today generally refers to Einstein's theory.

The study of relativity began with the work of Galileo Galilei around 1630; Isaac Newton also made important contributions. The ideas described in this chapter, universally accepted until 1900, are known as "Galilean relativity."

Galilean relativity is fully consistent with the intuitive notions that we call "common sense."[1] In the example above, if the train moves at 30 meters per second (m/sec) in the same direction as the child, common sense suggests that ground-based observers should find the child's speed to be 31 m/sec; Galilean relativity gives precisely that value. Einstein's theory, as we shall see, gives a different result.

In the case of the child, the difference between the two theories is minute. The speed measured by ground observers according to Einstein's

1. According to Einstein, common sense is "that layer of prejudice laid down in the mind prior to the age of eighteen."

1

relativity differs from the Galilean value 31 m/sec only in the fourteenth decimal place; no measurement could possibly detect such a tiny difference. This result is characteristic of Einsteinian relativity: its predictions are indistinguishable from those of Galilean relativity whenever the observers, as well as all objects under observation, move slowly relative to one another. That realm is generally called the nonrelativistic limit, although Galilean or Newtonian limit would be a more apt designation. "Slowly" here means at a speed much less than the speed of light.

The speed of light plays a central role in Einstein's theory; whenever any speed in the problem approaches that value, Einsteinian relativity departs dramatically from that of Galileo and Newton. Because the speed of light is so great, however, most commonly observed phenomena are adequately described by Galilean relativity.

The "special" theory of relativity, which is the principal subject of this book, is restricted to observers who move *uniformly*, that is, at constant speed in the same direction. If observers move with changing speeds, or along curved paths, the problem of relating their measurements is much more complicated. Einstein addressed that problem as well, in his "general" theory of relativity. Because the general theory involves quite advanced mathematics, I can give only a descriptive treatment in chapter 8. The special theory, in contrast, requires only elementary algebra and geometry and can be presented with full rigor.

Many of the conclusions of special relativity run counter to our intuition concerning the nature of space and time. Before Einstein, no one doubted that time is absolute. Newton put it as follows in his *Principia:* "Absolute, true, and mathematical time, of itself and from its own nature, flows equably without relation to anything external."

Special relativity obliges us to abandon the absolute nature of time. We shall see, for example, that the time order of two events can depend on the relative motion of the observers who view them. One set of observers may find that a certain event *A* occurred before another event *B*, whereas according to a second set of observers, who are moving relative to the first, *B* occurred before *A*. This result is surely difficult to accept.

In some cases, a reversal of time ordering would be truly bizarre. Suppose that at event *A* a moth lands on the windshield of a moving car; the car clock reads 12:00. At event *B* another moth lands; the car clock now reads 12:05. For the driver of the car, the order of those events is a direct sensory experience: she can see both events happen right in front of her and can assert with confidence that *A* happened first. If observers on the ground were to claim that event *B* happened first, they would be denying that sensory experience; moreover, the car clock would according to them be running backward! (It would read 12:05 before it reads 12:00.)

As we shall see, special relativity implies that moving clocks run slow. That is itself a strange result, but clocks running backward would be too much to swallow. No such disaster arises, however. In the case of the moths, event *A* happens first according to all observers. A reversal of time ordering can occur only for events spaced so far apart that no single observer (and no single clock) can be present at both. The order of such events is not a direct sensory experience for anyone; it can be determined only by comparing the readings of two *distinct* clocks, one present at event *A* and the other present at *B*. If two sets of observers disagree on the order of those events, no one's sensory experience is contradicted and no one sees any clock running backward. The proof of this assertion, given in chapter 5, depends on the fact that nothing can travel faster than light, one of the important consequences of special relativity.

A logical requirement of any theory is *causality*. If event *A* is the cause of event *B*, *A* must occur before *B*: the cause must precede the effect. We will see in chapter 5 that special relativity is consistent with the causality requirement. Whenever a cause-and-effect relation exists between two events, their time order is absolute: all observers agree on which one happened first.

Figure 1.1 shows a hypothetical experiment to illustrate the relativistic reversal of time ordering. Event *A* takes place in San Francisco and event *B* in New York. According to clocks at rest at those locations, *A* occurs before *B*. The same events are monitored by observers on spaceships moving from west to east at equal speeds; one ship is over San Francisco when event *A* occurs, and the other is over New York when event *B* occurs. Special relativity predicts that if the ships are moving fast enough, their clocks can show event *B* happening before *A*. Notice that no single clock is present at both events; the relevant times in the problem are recorded by four distinct clocks, two on the ground and two on the spaceships.

I hasten to add that no such experiment has ever been performed. The fastest available rockets travel a few kilometers per second, only about one hundred thousandth the speed of light. At that speed, the events of figure 1.1 would have to be separated in time by less than a millionth of a second if a reversal of time order were to be detectable. Moreover, the speeds of the two spaceships would have to be equal to within a very small tolerance. The experiment is just too hard to carry out. But we can be confident that if faster rockets were available and if other technical requirements were met, the effect could be detected.

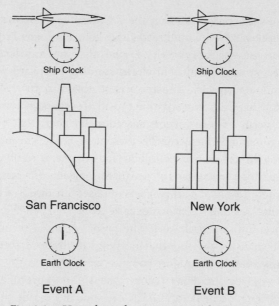

Fig. 1.1. Hypothetical experiment to demonstrate
the reversal of time ordering predicted by special
relativity. Event *A* occurs in San Francisco, event *B*
in New York. Each event is detected by two sets
of observers—one set fixed on earth and the other
located on spaceships flying at equal (constant)
speeds. Each set of observers measures the times of
the two events on its own clocks, which have been
previously synchronized. According to earth clocks,
event *A* happens before *B*, whereas according to
spaceship clocks, *B* happens before *A*. The time intervals shown on the clocks are much exaggerated.

The evidence that confirms special relativity comes principally from
atomic and subatomic physics. In many experiments particles move at
speeds close to that of light, and the effects of special relativity are dramatic. Particles are created and annihilated in accord with the famous Einstein relation $E = mc^2$. No understanding of such phenomena, or of the
kinematics of high-energy particle reactions, would be possible without
relativity. Thus Einstein's theory is confirmed daily in every high-energy
physics laboratory. Particle reactions are not within the realm of everyday
experience, however; in the latter realm, everything moves fairly slowly[2]

2. An obvious exception is light itself, which *is* part of everyday experience. Although any phenomenon that involves light is intrinsically relativistic, most opti-

and relativistic effects are not manifested. If the speed of light were much smaller, the effects of special relativity would be more prominent and our intuition concerning the nature of time would be quite different.

The preceding discussion is intended to provide a taste of what is to come and to encourage the reader to approach relativity with an open mind. I am not suggesting that any conclusion contrary to one's intuition be accepted uncritically, even though the context may be restricted to unfamiliar phenomena. On the contrary, any such conclusion must be vigorously challenged. Before abandoning ideas that appear to be self-evident, one must be satisfied that the experimental evidence is sound and the logical arguments are compelling.

1.2. Events, Observers, and Frames of Reference

I begin by defining some important terms. In relativity an *event* is any occurrence with which a definite time and a definite location are associated; it is an idealization in the sense that any actual event is bound to have a finite extent both in time and in space.

A *frame of reference* consists of an array of observers, all at rest relative to one another, stationed at regular intervals throughout space. A rectangular coordinate system moves with the observers, so that the x, y, and z coordinates of each observer are constant in time. The observers carry clocks that are synchronized: each clock has the same reading at the same time.[3]

Each observer records all events that occur at her location. Each event has four coordinates: three space coordinates and a time. By definition, the space coordinates are the coordinates of the observer who detected the event and the time of the event is the reading of her clock when it occurs.

A second frame of reference consists of another array of observers, all at rest relative to one another and all moving at the same velocity relative to the first set. They have their own coordinate system and their own (synchronized) clocks, and they also record the coordinates of events. The coordinates of a given event in two frames of reference are, in general,

cal phenomena can be explained without invoking the specific value of the speed of light. For example, refraction (the bending of a light ray when it crosses the boundary between air and glass) depends only on the ratio of the speeds of light in the two media. Hence a nonrelativistic theory of refraction is quite adequate.

Effects of relativity are manifested in experiments that depend on the time required for light to traverse a specified path, such as the Michelson-Morley experiment, discussed in detail in chapter 2.

3. The synchronization of clocks is discussed in detail in chapter 3.

different. The central problem of relativity is just to determine the relation between the two sets of coordinates; this turns out to be not so simple a matter as it first appears.

Throughout this book, whenever observations in two frames of reference are being compared, one frame will be called S and the other S'. Coordinates measured in frame S' will be designated by primed symbols, and those measured in frame S will be designated by unprimed symbols. Events will be labeled E_1, E_2, E_3, and so on. Thus, x_1', y_1', z_1', and t_1' denote the coordinates of event E_1 measured in frame S'; x_2, y_2, z_2, and t_2 denote the coordinates of event E_2 measured in frame S, and so on.

As an illustration, let us return to the problem of the child walking on a train. Figure 1.2 shows the child's motion as seen in two frames of reference, one fixed on the train (sketches [a] and [b]) and one fixed on the ground (sketches [c] and [d].) S is the ground frame and S' the train frame. The two sets of axes are parallel to one another. The train's motion as seen from the ground is taken to be in the x direction and the floor of the car is in the x-y plane. Since the child has no motion in the z direction, the figure has been simplified by omitting the z and z' axes.

In figure 1.2a, the child is just passing a train observer labeled H'; this is event E_1. The space coordinates of E_1 in S' are $x_1'=2$, $y_1'=1$, $z_1'=0$; its time coordinate t_1' is the reading of the clock held by H' as the child passes her. Some time later, as shown in figure 1.2b, the child passes a second train observer, labeled J'; this is event E_2. The space coordinates of E_2 are $x_2'=3$, $y_2'=4$, $z_2'=0$; its time coordinate t_2' is the reading of the clock held by J'.

Figure 1.2c shows event E_1 as seen in the ground frame. The child is just passing ground observer B. The space coordinates of E_1 in S are $x_1=2$, $y_1=1$, $z_1=0$; its time coordinate is read off B's clock. Figures 1.2a and 1.2c should be thought of as being superposed: the positions of ground observer B, train observer H', and the child all coincide when E_1 occurs.

Figure 1.2d similarly shows E_2 as seen in frame S; the child is now passing ground observer Q. The space coordinates of E_2 in frame S are $x_2=5$, $y_2=4$, $z_2=0$. The positions of Q, J', and the child all coincide at E_2. Notice that B and H', whose positions coincided at E_1, no longer coincide at E_2. As seen from the ground, all the train observers have moved to the right during the interval between the two events. (As seen from the train, all the ground observers have moved an equal distance to the left.)

Inspection of the figures reveals that the length of the child's path measured in the ground frame is greater than that measured in the train frame. The child's speed in the ground frame is correspondingly greater

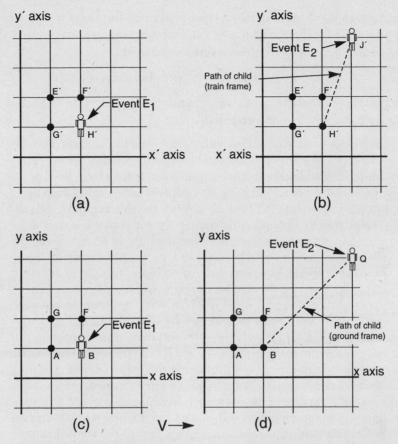

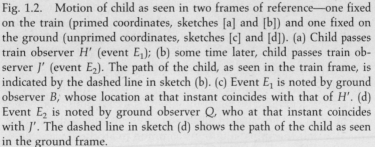

Fig. 1.2.   Motion of child as seen in two frames of reference—one fixed on the train (primed coordinates, sketches [a] and [b]) and one fixed on the ground (unprimed coordinates, sketches [c] and [d]). (a) Child passes train observer H' (event E_1); (b) some time later, child passes train observer J' (event E_2). The path of the child, as seen in the train frame, is indicated by the dashed line in sketch (b). (c) Event E_1 is noted by ground observer B, whose location at that instant coincides with that of H'. (d) Event E_2 is noted by ground observer Q, who at that instant coincides with J'. The dashed line in sketch (d) shows the path of the child as seen in the ground frame.

(provided the elapsed time is the same in both frames, which is true in Galilean relativity).

The notion of a frame of reference as an (essentially infinite) array of observers is not intended to be a literal description of how measurements are carried out. It would be impractical, to say the least, to station observers throughout all space in the manner prescribed. But there is no reason in principle why that could not be done. In what follows, every event is assumed to be monitored by observers on the scene.

1.3. The Principle of Relativity and Inertial Frames

The principle of relativity was first enunciated by Galileo in 1632. Galileo's argument is clear and graphically put.

> *Salviatus:* Shut yourself up with some friend in the main cabin below decks on some large ship and have with you there some flies, butterflies, and other small flying animals. Have a large bowl of water with some fish in it; hang up a bottle which empties drop by drop into a wide vessel beneath it. With the ship standing still, observe carefully how the little animals fly with equal speed to all sides of the cabin. The fish swim indifferently in all directions; the drops fall into the vessel beneath; and, in throwing something toward your friend, you need throw it no more strongly in one direction than another, the distances being equal; jumping with your feet together, you pass equal spaces in every direction. When you have observed all these things carefully (though there is no doubt that when the ship is standing still everything must happen in this way), have the ship proceed with any speed you like, so long as the motion is uniform and not fluctuating this way and that. You will discover not the least change in all the effects named, nor could you tell from any of them whether the ship was moving or standing still. In jumping, you will pass on the floor the same spaces as before, nor will you make larger jumps toward the stern than toward the prow, . . . despite the fact that during the time that you are in the air the floor under you will be going in a direction opposite to your jump. . . . Finally the butterflies and flies will continue their flights indifferently toward every side, nor will it ever happen that they are concentrated toward the stern, as if tired out from keeping up with the course of the ship, from which they will have been separated during long intervals by keeping themselves in the air. . . . The cause of all these correspondences of effects is the fact that the ships' motion is common to all the things contained in it.[4]

4. Galileo Galilei, *Dialogue Concerning the Two Chief World Systems*, translated by Stillman Drake (Berkeley and Los Angeles: University of California Press, 1953), 186–187. It is not clear whether Galileo ever actually performed the ship experiments. Following the quoted speech, Sagredo says, "Although it did not occur to me to put these observations to the test when I was voyaging, I am sure that they would take place in the ways you describe" (*Dialogue*, 188). This suggests that Galileo had not done the experiment. But see the remarks below.

Galileo is asserting, in effect, that *the laws of nature are the same in any two frames of reference that move uniformly with respect to one another.* If identical experiments are carried out by two sets of observers, with identical initial conditions, all the results will be the same. It follows that there is no way to determine by means of experiments carried out in a given frame of reference whether the frame is at rest or is moving uniformly. Only the relative velocity between frames can be measured. This set of assertions is called the principle of relativity.

Galileo's motivation was to refute Aristotle's argument that the earth must be standing still. If the earth were moving, Aristotle had claimed, a stone dropped from the top of a tower would not land at its base, since the earth would have moved while the stone was falling. Galileo argues that the earth plays a role entirely analogous to that of the ship in his example; just as a stone dropped from the top of a mast lands at its foot whether the ship is moving or at rest, so does one dropped from a tower on earth. And just as observations carried out within the ship cannot be used to decide whether the ship is standing still or moving uniformly, so the observed motion of objects on earth implies nothing about the motion of the earth other than that it is (approximately) uniform.

Although Galileo may not have carried out all the ship experiments, he definitely performed the falling rock experiment as well as many others on falling bodies. In a famous letter replying to Francesco Ingoli, who had attacked his views and sided with Aristotle, Galileo says, "whereas I have made the experiment, and even before that, natural reason had firmly persuaded me that the effect had to happen in the way that it indeed does."[5]

Several remarks are in order concerning Galileo's principle of relativity. First, the observations on which the principle was based were necessarily limited to quite slow speeds. Perhaps if the ship were moving very rapidly, shipborne observers might detect unusual effects that would enable them to conclude that their ship was indeed in motion. If that were to happen, the relativity principle would be only approximately valid. The laws of nature might be (very nearly) the same in two frames of reference

5. *The Galileo Affair,* editor and translator Maurice A Finocchiaro (Berkeley and Los Angeles: University of California Press, 1989), 184. The motion of an object dropped from a moving vehicle had been debated long before Galileo. Tycho Brahe, as late as 1595, still sided with Aristotle, but Thomas Digges gave a correct analysis in his book, *A Perfitt Description of the Celestial Orbes,* published in 1576. Giordano Bruno also studied the problem and came to the correct conclusion. According to Drake, Galileo probably knew about Bruno's work although he did not refer to it. (Bruno had been burned at the stake as a heretic.)

that move slowly relative to one another but quite different in two frames whose relative velocity is great. Galileo's observations obviously could not exclude such a possibility, and even today the direct evidence from physics in moving laboratories is limited to fairly low velocities. Indirect evidence, however, strongly supports the hypothesis that the relativity principle holds for any speed.

Galileo's experiments all deal with phenomena in what is nowadays called mechanics; on the basis of those experiments, therefore, one can conclude only that a principle of relativity applies to the laws of mechanics. Perhaps other experiments, involving different phenomena, can distinguish among frames.

Nineteenth-century physicists believed that electromagnetic and optical phenomena provide just such a distinction. According to the view prevalent during that period, there exists a unique frame of reference in which the laws of electromagnetism take a particularly simple form. If that were so, the principle of relativity would not apply to electromagnetic phenomena: the results of some experiments would depend on the observer's motion relative to the special frame.

Many experiments were performed with the aim of determining the earth's motion relative to the special frame, but they all failed to detect any effect of that assumed motion. The most important was the Michelson-Morley experiment, described in chapter 2.

For Einstein, it was aesthetically unsatisfying that a principle of relativity should hold for one set of phenomena (mechanics) but not for another (electromagnetism.) He postulated that Galileo's principle applies to *all* the laws of nature; this generalization forms the basis for special relativity.

The relativity principle has an important philosophical implication. If there is no way to distinguish between a state of rest and a state of uniform motion, absolute rest has no meaning. Observers in any frame are free to take their own frame as the standard of rest. Shore-based observers watching Galileo's ship are convinced that they are at rest and the ship is in motion, but observers on the ship are equally entitled to regard themselves as being at rest while the shore along with everything on it moves. The question, Which observers are *really* at rest? has no meaning if there is no conceivable experiment that could answer it. (According to observers in an airplane flying overhead, both shore observers and ship observers are in motion.)

In sum, the principle of relativity denies the possibility of absolute rest (or of absolute motion). Motion can be defined only relative to a specific frame of reference, and among uniformly moving frames strict democracy prevails: any frame is just as good as any other. Any reference to a body "at rest" should be understood to mean "at rest in a frame of reference fixed on the earth" (or in some other specified frame).

The restriction to uniform motion is essential to the relativity principle. The laws of nature are *not* the same in all frames of reference.[6] As Galileo fully realized, accelerated motion is readily distinguished from uniform motion. If a ship moves jerkily or changes direction abruptly, things behave strangely: suspended ropes do not hang vertically, a cake of ice placed on a level floor slides away for no apparent reason, and the flight pattern of Galileo's butterflies appears quite different than it does when the ship is moving uniformly. Any of these effects tells the observers that their frame is accelerated.

The distinguishing feature of uniformly moving frames is that in any such frame the *law of inertia* holds: a body subject to no external forces remains at rest if initially at rest, or if initially in motion, it continues to move with constant speed in the same direction. In an accelerated frame, the law of inertia does not hold. Instead bodies seem to be subjected to peculiar forces for which no agent can be identified. Those forces, called "inertial forces," have observable consequences.

Frames of reference in which the law of inertia holds are called *inertial frames;*[7] all others are noninertial. In terms of this nomenclature, we can rephrase Galileo's principle of relativity as follows:

> If S is an inertial frame and S' is any other frame that moves uniformly with respect to S, then S' is also an inertial frame. All the laws of mechanics are the same in S' as in S, and no (mechanical) experiment can distinguish S' from S.

The discussion here will be confined almost entirely to inertial frames.

Observers in a given frame can determine whether their frame is inertial by carrying out experiments to test whether the law of inertia holds. A frame of reference fixed on earth satisfies the criterion fairly closely; for most purposes such a frame can be regarded as inertial. Because of the earth's rotation, however, an earthbound frame is not strictly inertial.

Even a frame of reference fixed at the pole, which does not partake of the earth's rotation, is not strictly inertial because the earth is moving in

6. See, however, the discussion of the principle of equivalence in chapter 8.
7. Einstein called them "Galilean frames."

a curved orbit around the sun. And the sun is itself in orbit about the center of the galaxy. An inertial frame is an idealization in the sense that no experiment can assure us that our frame is strictly inertial, that is, that a body subject to no forces does not experience some tiny acceleration.

1.4. The Galilean Transformation

An event E occurs at time t at the point x, y, z, as measured in some inertial frame S. What are the coordinates (x', y', z', t') of E in another inertial frame, S', that moves at velocity V relative to S? The answer that any physicist would have given to this question before 1905 is the Galilean transformation, derived here. The derivation is straightforward and the results appear almost self-evident. As we shall see, however, special relativity gives a different answer.

For convenience, let the two sets of axes be parallel to one another, with their relative motion in the x (or x') direction (fig. 1.3). At some instant the origins O and O' coincide and all three pairs of axes are mo-

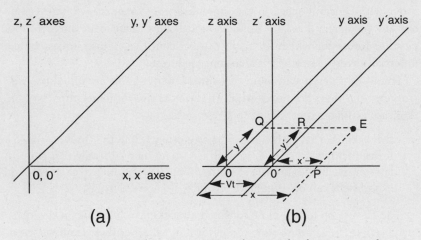

(a)   (b)

Fig. 1.3. Coordinates of an event in two frames of reference, according to Galilean relativity. The primed coordinate system is moving from left to right, as seen by observers in the unprimed system. (a) Primed and unprimed axes momentarily coincide. The clocks of all observers are arbitrarily set to zero at this instant. (b) The state of affairs at some later time, t. O', the origin of the primed system, has moved a distance Vt down the x axis. The x and x' axes still coincide. The event in question occurs at the point labeled E. The space coordinates of the event in both frames are indicated. As the figure shows, y and y' are equal, but x' is less than x. (For simplicity, the z coordinate of the event is assumed to be zero.)

mentarily superposed (fig. 1.3a). At that moment[8] all observers in both frames synchronize their clocks by setting all their readings to zero.

The relation between the times t and t' can be written directly. Since time is absolute, we have simply

$$t' = t \qquad (1.1a)$$

The spatial coordinates of E in S' can be taken from the figure. At time t the two origins are separated by the distance Vt. Hence the relation between x and x' is

$$x' = x - Vt \qquad (1.1b)$$

The y and z coordinates of E are the same in both frames:

$$y' = y \qquad (1.1c)$$
$$z' = z \qquad (1.1d)$$

Equations (1.1a–d) constitute the Galilean transformation.

Inverse Transformation

Suppose we are given the coordinates of an event in S' and want to find its coordinates in S. Solving equations (1.1a–d) for the unprimed coordinates in terms of the primed ones, we get

$$t = t' \qquad (1.2a)$$
$$x = x' + Vt' \qquad (1.2b)$$
$$y = y' \qquad (1.2c)$$
$$z = z' \qquad (1.2d)$$

which is the desired inverse transformation.

If primed and unprimed coordinates are interchanged and V is changed to $-V$, equations (1.1a–d) turn into equations (1.2a–d), and vice versa. This result is a logical necessity. It cannot matter which reference frame we choose to label S and which S'; the same transformation law must apply. But V is defined as the velocity of S' relative to S. If we interchange the labels, the magnitude of the relative velocity is unchanged but its sign is reversed. (If ground observers see a train moving from left to right at a given speed, train observers must see the ground moving from right to left at the same speed.)

8. Because time is absolute in Galilean relativity, phrases like "at that moment" and "t seconds later" have the same meaning in both frames. When we come to special relativity, we shall have to exercise great care in using such language.

Invariance of Distance

Suppose train and ground observers wish to measure the distance between two telephone poles situated alongside the track. The S positions of the poles are independent of time:

$$x_1 = A \qquad x_2 = B \tag{1.3}$$

and the distance $x_2 - x_1$ between them is just $B - A$.

The equations of motion of the poles in S' are obtained by applying equation (1.2b) to both x_1 and x_2 in (1.3). The result is

$$x_1' = A - Vt' \tag{1.4a}$$
$$x_2' = B - Vt' \tag{1.4b}$$

Both these equations describe bodies moving from right to left at speed V, as they must.

Suppose train observers measure the position of pole #1 at time t_1' and that of pole #2 at time t_2'. The difference between the two readings is

$$x_2' - x_1' = B - A - V(t_2' - t_1') \tag{1.5}$$

Inasmuch as the poles are moving in S', the two position measurements must be made at the same time if their difference is to yield the correct distance between the poles. With $t_2' = t_1'$, equation (1.5) gives $x_2' - x_1' = B - A$, the same as the result obtained in frame S.

This discussion introduces the important concept of *invariance*. A quantity is said to be invariant if it has the same value in all frames of reference. I have shown that the spatial separation between two events that occur at the same time is invariant in Galilean relativity. The analogous statement in special relativity is not true.

Transformation of Velocity

Suppose a body moves in the x direction at velocity v, as measured in the ground frame S.[9] If the body sets out from the origin at $t = 0$, its position at time t is

$$x = vt \tag{1.6}$$

Using equations (1.2a,b) to express x and t in terms of x' and t', we obtain

9. Throughout this book, velocities of objects are denoted by lowercase letters. The velocity of one frame relative to another is denoted by a capital letter, usually V.

$$x' + Vt' = vt'$$

or

$$x' = (v - V)t' \tag{1.7}$$

Equation (1.7), like (1.6), expresses motion at constant velocity; the magnitude of the velocity, which we may call v', is

$$v' = v - V \tag{1.8a}$$

The inverse transformation is obviously

$$v = v' + V \tag{1.8b}$$

With $v' = 1$ m/sec and $V = 30$ m/sec, equation (1.8b) gives $v = 31$ m/sec, the value cited earlier as the "commonsense" result.

If the motion is not confined to the x direction, we can write instead of equation (1.6)

$$x = v_x t \tag{1.9a}$$
$$y = v_y t \tag{1.9b}$$
$$z = v_z t \tag{1.9c}$$

where v_x, v_y, and v_z denote the three components of velocity in S.

When we transform to S' coordinates as before, the x equation reproduces the result expressed in equation (1.8a), with a subscript x on v and v':

$$v_x' = v_x - V \tag{1.10a}$$

Since $y = y'$ and $t = t'$, equation (1.9b) becomes

$$y' = v_y t'$$

which implies that

$$v_y' = v_y \tag{1.10b}$$

Similarly, we find that

$$v_z' = v_z \tag{1.10c}$$

Only the component of velocity in the direction of the relative motion between frames changes when we change frames.

In deriving equation (1.8) we assumed that the velocity of the body in question was constant. If the velocity is changing, the result still holds provided v and v' refer to the *instantaneous* values of velocity (measured at the same time, of course). This is readily shown with the help of the calculus; one simply differentiates equation (1.1) with respect to time. The same result can be derived by purely algebraic methods.

Combination of Galilean Transformations

Suppose two trains travel along the same track, one at velocity V and the other at velocity U relative to the ground. Let x', y', and z' and x'', y'', and z'' denote coordinates in frames of reference attached to the first and second train, respectively. The transformation from (x, y, z, t) to (x', y', z', t') is given by equation (1.1); that from (x, y, z, t) to (x'', y'', z'', t'') must be given by a similar set of equations, with U in place of V:

$$x'' = x - Ut$$
$$y'' = y \qquad z'' = z \qquad (1.11)$$
$$t'' = t$$

What about the transformation from the coordinates of the first train to those of the second? Eliminating (x, y, z, t) from equations (1.1) and (1.11), we find directly

$$x'' = x' - (U - V)t'$$
$$y'' = y' \qquad z'' = z' \qquad (1.12)$$
$$t'' = t'$$

Equation (1.12) describes another Galilean transformation, with relative velocity $U - V$. This is just the velocity of the second train as measured by observers on the first.

Acceleration

Finally, we examine the transformation properties of acceleration, the rate of change of velocity. This can be done without any equations.

According to equation (1.8), the velocities of a moving body in S and S' always differ by the same amount, V. If the velocity measured in S changes from v_1 to v_2 during some time interval, the velocity measured in S' changes from $v_1 - V$ to $v_2 - V$; the increment in velocity in S' is $v_2 - v_1$, the same as in S. Since acceleration is defined as change in velocity per unit time, it has the same value in both frames. Letting a and a' denote the accelerations measured in the two frames, we have simply

$$a' = a \qquad (1.13)$$

Acceleration is invariant in Galilean relativity. We shall see in chapter 4 that in special relativity it transforms in a much more complicated manner.

1.5. STELLAR ABERRATION

An interesting application of the velocity transformation law is provided by stellar aberration, the change in the apparent direction of a star caused by the earth's motion around the sun.[10] A similar effect can be detected when driving through a rainstorm: raindrops falling vertically appear to be moving obliquely.

Let S be a frame of reference in which the sun is at rest and the earth's orbit is in the $x - y$ plane. Suppose the orbital velocity V points in the x direction.

Consider a star that is located on the z axis and is not moving relative to the sun (fig. 1.4a). The analysis is simplest for this special case, although a similar result applies to any star.

The velocity components in frame S of a light ray that reaches earth from the star are

$$v_x = 0 \tag{1.14a}$$
$$v_y = 0 \tag{1.14b}$$
$$v_z = -c \tag{1.14c}$$

If the earth were not moving, a telescope pointed in the z direction would receive light from the star.

We want to find the direction of the light ray in S', the earth's rest frame, which moves at velocity V relative to S. The Galilean velocity transformation, equation (1.10), gives the velocity components in S':

$$v'_x = v_x - V = -V \tag{1.15a}$$
$$v'_y = v_y = 0 \tag{1.15b}$$
$$v'_z = v_z = -c \tag{1.15c}$$

Figure 1.4b shows the direction of the light ray in frame S'. The "apparent" direction of the star (the direction in which our telescope must be pointed) differs from its "true" direction by a small angle called the *aberration angle*, α. For the special case under consideration, the aberration angle is determined by the trigonometric relation

$$\tan \alpha = \frac{v'_x}{v'_z} = \frac{V}{c} \tag{1.16}$$

10. Aberration is not to be confused with stellar parallax, which is due to the changing *position* of the earth as it traverses its orbit. Unlike aberration, parallax depends on the distance of the star. Even for the nearest stars, the parallax angle is much smaller than the aberration angle.

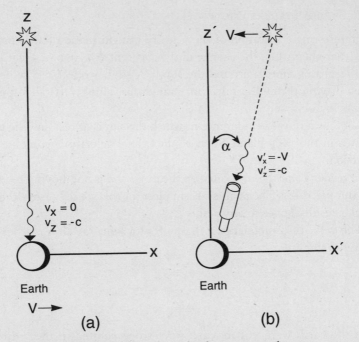

Fig. 1.4.   Effect of the earth's orbital motion on the apparent
position of a star. Sketch (a) is drawn in a frame of reference, S,
in which the star is at rest on the z axis and the earth is moving
in the x direction at velocity V. A light ray from the star, moving
in the negative z direction, reaches earth. Sketch (b) shows the
same ray in frame S', in which the earth is at rest and the star
is moving. S' moves at velocity V relative to S. The velocity
components of the star are given by eq. (1.14) in S and by (1.15)
in S'. To see the star, an astronomer must point his telescope at
an angle, α, given by eq. (1.16); this effect is called aberration.

The value of V is known to be 30 km/sec. Equation (1.16) therefore
gives $\tan \alpha = 10^{-4}$ or $\alpha = 20''$ of arc. Although this is a very small angle,
it is readily measurable with a good telescope.

If the earth's motion were uniform, the aberration effect would be undetectable since the "true" direction of the star is unknown. But because
the direction of the earth's orbital velocity changes regularly, the aberration effect likewise changes. Six months after the situation shown in the
figure, the earth's velocity in frame S will have reversed its direction and
in S' the star will appear to be on the other side of the z axis. Over the
course of a year, the apparent position of the star traces out a circle whose
radius is about 20 seconds of arc; for a star in an arbitrary direction, the

path is an ellipse. The effect was detected and explained correctly by James Bradley in 1725.[11]

The aberration formula can also be derived by analyzing the situation from the outset in frame S. Between the time the light ray enters the telescope and the time it reaches the eyepiece, the telescope has moved. Unless the telescope is tilted, therefore, the light ray will run into the side of the instrument. The result obtained by such an argument is, of course, the same as equation (1.16). The advantage of the derivation given here is that it is readily generalized when we analyze the problem from the point of view of special relativity in chapter 4.

1.6. The Covariance of Physical Laws

The principle of relativity, set forth in section 1.3, asserts that the laws of physics are the same in any two inertial frames of reference. This principle can be rephrased as a requirement on the mathematical properties of physical laws.

A typical physical law expresses a mathematical relation between quantities like velocity, acceleration, and force. All these quantities must be measured in some frame of reference, say, S. Suppose we transform to a second frame S' using the Galilean transformation. If the primed quantities are related in exactly the same way as the corresponding unprimed ones, the relation is said to be *covariant* under the transformation. The principle of relativity demands that all the laws of mechanics be covariant under a Galilean transformation.

Consider the basic law of mechanics, Newton's second law:

$$F = ma \tag{1.17}$$

Here m is the mass of a body, F is the total external force acting on it, and a is its acceleration, all measured in some frame S.

Let F' and a' be the force and acceleration measured in some other inertial frame S'. If the second law is covariant, the relation

$$F' = ma' \tag{1.18}$$

must follow from (1.17).

To be perfectly general, we should have written equation (1.18) as

$$F' = m'a' \tag{1.19}$$

11. Bradley used his theory of aberration to deduce a quite accurate value for the speed of light.

with m' the mass appropriate to frame S'. In classical mechanics, however, mass is considered an intrinsic property of a body and must be invariant. Hence we can put $m' = m$.

We showed earlier (eq. [1.13]) that acceleration is invariant under a Galilean transformation: $a' = a$. Hence equation (1.18) follows from (1.17) if and only if the force F is also invariant,[12] that is, if

$$F' = F \qquad (1.20)$$

One has to investigate case by case whether the forces that exist in nature satisfy the invariance requirement (eq. [1.20]). For example, the gravitational attraction between two bodies is given by Newton's law of gravity:

$$F = G \frac{m_1 m_2}{r^2} \qquad (1.21)$$

Here m_1 and m_2 are the masses of the bodies, r is the distance between them, and G is a constant. We have already shown that distance is invariant under a Galilean transformation. Hence the gravitational force is indeed invariant and the law of gravity is consistent with the principle of relativity.

The most important example of a force law that is not covariant in Galilean relativity is electromagnetism. The laws of electricity and magnetism were codified in the late nineteenth century in a system known as Maxwell's equations. If one assumes that these equations hold in some frame S and makes a Galilean transformation to another frame S', the equations in S' do not have the same form: Maxwell's equations are *not* covariant under a Galilean transformation. This was very troubling to Einstein and motivated his quest for a new theory.

There were three logical possibilities:

(i)   The relativity principle does not apply to electromagnetism; Maxwell's equations are valid only in one special frame of reference.

(ii)   The relativity principle does apply to electromagnetism, but Maxwell's equations are only approximately correct; they must be replaced by a more general set of equations that are strictly covariant.

---

12. If every term in a relation is invariant, as in the present example, the relation is obviously covariant. This is not a necessary condition, however. Covariance requires only that both sides of the equation transform in the same way. Suppose, for example, that under some hypothetical transformation, $a' = 2a$ and $F' = 2F$. Eq. (1.18) would be satisfied and the law would be covariant.

(iii) The relativity principle applies universally and Maxwell's equations are exact, but the Galilean transformation is wrong.

Alternative (i) was the one preferred by nineteenth-century physicists; the ether frame, discussed in chapter 2, was postulated to be the special frame in which Maxwell's equations hold. Einstein rejected that view and boldly asserted that alternative (iii), which a priori seems the least plausible, is in fact correct. This assumption led him to special relativity.

1.7. THE CONSERVATION OF MOMENTUM

As a final application of Galilean relativity, we examine the law of conservation of momentum and show that it is covariant.

Newton defined momentum (which he called "quantity of motion") as the product of mass and velocity. The modern symbol for momentum is p; thus $p = mv$.

Momentum is a vector quantity: it has direction as well as magnitude. By definition, the momentum of a body points in the same direction as its velocity. The components of momentum along a given set of axes are

$$p_x = mv_x \qquad p_y = mv_y \qquad p_z = mv_z \qquad (1.22)$$

Any of these components can be positive or negative.

An important law in classical mechanics is the conservation of momentum: the total momentum of an isolated system remains constant. The law is a consequence of Newton's second and third laws.

The most common application of the conservation of momentum is to problems that involve collisions. When two bodies collide, the momentum of each body changes as a consequence of the forces exerted on it by the other. The *total* momentum of the system, however, is the same after the collision as before. If the collision involves motion in more than one dimension, each component of momentum is separately conserved.

Figure 1.5 illustrates the conservation of momentum in collisions between bodies of equal mass m. In each case body A, moving at velocity v in the x direction, collides with body B, which is initially at rest. The total momentum of the system before the collision is mv. In collision (a), body A comes to rest and B moves at velocity v after the collision in the direction of A's initial motion. The final momentum is mv, in accord with the conservation law.

Figure 1.5b shows a different possible outcome: A and B stick together, and both move at velocity $v/2$ after the collision. The final momentum of each body is $mv/2$, and the total is again mv; momentum is conserved.

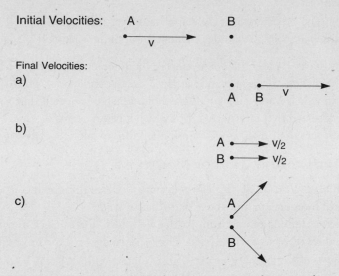

Fig. 1.5. The conservation of momentum according to Galilean relativity. Shown are three collisions, in each of which momentum is conserved. The initial state in each collision is that shown in the top sketch: body *A*, moving in the *x* direction, collides with body *B*, initially at rest. The bodies have equal masses. After collision (a), *A* is at rest and *B* moves with *A*'s original velocity, *v*. After collision (b), each body moves with velocity *v*/2 in the *x* direction. After collision (c), each body has some *y* motion as well as some *x* motion. The *y* components of velocity (and of momentum) are equal and opposite; thus the total *y* momentum is zero.

Figure 1.5c shows yet another possible outcome. In this case, each body has some *y* velocity and therefore some *y* momentum after the collision. Since the initial *y* momentum was zero, the total *y* momentum after the collision must also be zero. The *y* momenta of the emerging bodies must have equal magnitudes and opposite signs; since the masses are equal, the *y* velocities must likewise be equal and opposite.

As these examples demonstrate, conservation of momentum does not determine the outcome of a collision. If we know only the initial velocities of the colliding bodies, we cannot predict whether the final velocities are those shown in (a), in (b), in (c), or something different still.[13] The outcome depends on data that have not been specified, such as the elastic

13. The conservation law (see eq. [1.23]) is one equation with two unknowns (the two final velocities). It has an infinite number of solutions.

properties of the bodies and whether the collision is head-on or at a glancing angle. A head-on collision between two billiard balls would lead to outcome (a), whereas a glancing collision between the same billiard balls could lead to outcome (c). Colliding lumps of putty are likely to stick together, outcome (b).

Consider a collision between two bodies, A and B, whose motion is confined to one dimension. Conservation of momentum is expressed by the relation

$$m_A v_A + m_B v_B = m_C v_C + m_D v_D \tag{1.23}$$

where C and D refer to the bodies that emerge from the collision. C and D might be the same as A and B, but they might be different. Such "rearrangement" collisions are of particular interest in nuclear physics.

To prove that momentum conservation is a covariant law, we assume that equation (1.23) holds in some inertial frame S and show that the same relation holds also in any other inertial frame S'.

The velocity of each body in S' is related to its velocity in S by the Galilean velocity transformation, equation (1.7):

$$v_A = v_A' + V, \quad v_B = v_B' + V, \quad \text{and so on} \tag{1.24}$$

where, as usual, V is the speed of frame S' relative to S.

Using equation (1.24) we can express each term in equation (1.23) in terms of velocities measured in S'. The result is

$$m_A(v_A' + V) + m_B(v_B' + V) = m_C(v_C' + V) + m_D(v_D' + V) \tag{1.25}$$

Expanding and grouping terms, we obtain

$$m_A v_A' + m_B v_B' = m_C v_C' + m_D v_D' + V(m_C + m_D - m_A - m_B) \tag{1.26}$$

Equation (1.26) represents momentum conservation in frame S', provided the last term on the right side vanishes. Since V is not zero, this requires that

$$m_C + m_D = m_A + m_B \tag{1.27}$$

Condition (1.27), which must be satisfied if momentum conservation is to be a covariant law, expresses the conservation of mass. In classical mechanics, mass can be neither created nor destroyed; equation (1.27) must be valid. If, as a result of a collision, the colliding bodies exchange mass or even break up into many fragments, the total mass of the emerging bodies must be exactly the same as the total mass of the bodies that collided. As we shall see, that statement is not true in special relativity.

The proof of covariance is readily extended to collisions in more than one dimension. Each component of momentum can be treated separately. The preceding analysis shows that if x momentum is conserved in S, it is conserved as well in S'. (V, the relative velocity of the two frames, is assumed to be in the x direction.) But according to equation (1.10), the y velocity of each body is the same in S' as in S. Since momentum is velocity times mass, it follows that the y momentum of each body is likewise the same in both frames. Hence if y momentum is conserved in frame S, it is likewise conserved in S'. The same is true of z momentum.

In summary, the law of conservation of momentum, with momentum defined as mass times velocity, is covariant under a Galilean transformation, provided that mass is conserved. We shall see in chapter 7 that in special relativity, momentum must be redefined if the conservation law is to be covariant.

PROBLEMS

1.1. A train moves at a constant speed. A stone on the train is released from rest.

(a) Using the principle of relativity, describe the motion of the stone as seen by observers on the train.

(b) Using the Galilean transformation, describe the motion of the stone as seen by observers on the ground. Draw a sketch.

1.2. This problem deals quantitatively with the experiment of problem 1.1. Let S denote the ground frame of reference and S' the train's rest frame. Let the speed of the train, as measured by ground observers, be 30 m/sec in the x direction, and suppose the stone is released at $t' = 0$ at the point $x' = y' = 0$, $z' = 7.2$ m.

(a) Write the equations that describe the stone's motion in frame S'. That is, give x', y', and z' as functions of t'. (Note: A body starting from rest and moving with constant acceleration g travels a distance $\frac{1}{2} gt^2$ in time t. Gravity produces a constant acceleration whose magnitude is approximately 10 m/sec/sec.)

(b) Use the Galilean transformation to write the equations that describe the position of the stone in frame S. Plot the stone's position at intervals of 0.2 sec, and sketch the curve that describes its trajectory in frame S. What curve is this?

(c) The velocity acquired by a body starting from rest with acceleration g is gt. Write the equations that describe the three components of the stone's velocity in S', and use the Galilean velocity transformation to find the velocity components in S.

(d) Find the magnitude of the stone's speed at $t = 1$ sec in each frame.

1.3. A jetliner has an air speed of 500 mph. A 200-mph wind is blowing from west to east.

(a) The pilot heads due north. In what direction does the plane fly, and what is its ground speed? (Hint: Define a frame of reference S' that moves with the wind. In S' there is no wind; hence the plane always moves in the direction it is headed, at 500 mph. Use the velocity transformation to find the components of the plane's velocity in the ground frame.)

(b) In what direction should the pilot head in order to fly due north? What is the plane's ground speed in this case?

1.4. A river is 20 m wide; a 1 m/sec current flows downstream. Two swimmers, A and B, arrange a race. A is to swim to a point 20 m downstream and back while B swims straight across the river and back. Each can swim at 2 m/sec in still water.

(a) In what direction should B head in order to swim straight across? Illustrate with a sketch. (See the hint for the preceding problem.)

(b) Who wins the race, and by how much time?

1.5. An elastic collision is one in which kinetic energy as well as momentum is conserved, that is, the total kinetic energy after the collision is equal to the total initial kinetic energy. The Newtonian definition of kinetic energy is $K = \frac{1}{2} mv^2$.

Consider the collision shown in fig. 1.5a. The mass of each body is 2 kg; the initial velocity of body A is 0.6 m/sec. In frame S, the frame in which the figure is drawn, the collision is obviously elastic. (The initial kinetic energy of B and the final kinetic energy of A are both zero; in the collision A's momentum and kinetic energy are simply transferred to B.)

Analyze the same collision in a frame S' that moves to the right at 0.2 m/sec relative to S. Find the kinetic energy of each body before and after the collision and verify that the collision as seen in S' is elastic.

1.6. The object of this problem is to investigate whether, as suggested by the result of problem 1.5, the definition of an elastic collision is invariant under a Galilean transformation.

Consider the general one-dimensional collision discussed in section 1.7, in which bodies A and B collide and bodies C and D emerge. (C and D might be the same as A and B, or they might be different.) Assume that momentum is conserved, that is, eq. (1.23) is satisfied.

(a) Write the equation that expresses the conservation of kinetic energy in frame S. Now transform to a frame S' that moves at velocity V relative to S. Show that kinetic energy is conserved in S', provided mass conservation is satisfied.

(b) Suppose the collision is inelastic: the total kinetic energy of C and D in frame S differs from the total kinetic energy of A and B by an amount Q. Is the value of Q invariant? Justify your answer.

1.7. Raindrops are falling vertically at 2 m/sec. A person is running horizontally at 3 m/sec. At what angle to the vertical should she hold her umbrella for maximum effectiveness? (Consider the path of the raindrops in the runner's rest frame.)

1.8. Analyze the stellar aberration effect for a star that lies in the plane of the earth's orbit. How does the magnitude of the aberration angle vary as the earth traverses its orbit?

2   The Michelson-Morley Experiment

2.1. THE ETHER

The Michelson-Morley experiment occupies a special niche in the pantheon of relativity. Contrary to many accounts, the experiment did not strongly influence Einstein's discovery of special relativity.[1] It nonetheless provides strong experimental underpinning for the theory and was instrumental in promoting its widespread acceptance. Albert A. Michelson's experiment was rooted in late-nineteenth-century ideas concerning the nature of light. I begin therefore with a brief exposition of those ideas.

That the speed of light is very great had been known for a long time. Galileo had tried to measure it but did not succeed. The first determination of the speed of light was obtained in 1676 by Ole Römer from his observations of the eclipses of one of Jupiter's moons. Because the earth-Jupiter distance changes, the interval between successive eclipses varies; the variation measures the time required for light to travel the additional distance. Römer's result for the speed of light, 2.2×10^8 m/sec, was about 25 percent low.[2] The best modern value is 2.998×10^8 m/sec.

During the eighteenth century a lively controversy raged over the question, Does light consist of tiny particles or is it a wave phenomenon? Expert opinion was divided; Newton, for example, favored the particle hypothesis. Many properties of light, such as reflection and refraction, can be explained in either view.

Strong evidence in favor of the wave hypothesis was provided by experiments performed by Thomas Young, Augustin-Jean Fresnel, and oth-

1. See section 3.2 for a detailed historical discussion.
2. The principal source of error in Römer's determination was uncertainty in the distance between earth and Jupiter.

ers, which showed that light exhibits interference and diffraction. These characteristic wavelike phenomena are well-nigh impossible to explain on the basis of a particle description.

Although the wave character of light seemed firmly established by the early 1800s, the nature of the waves was not at all clear. The model favored at first was that light waves, like all other known waves, are a mechanical oscillation of some material medium. (A sound wave in air, for example, consists of longitudinal vibrations of air molecules.) The light medium was called the "luminiferous ether"; I shall refer to it simply as the ether.

The ether, if it exists, has quite unusual properties. It must pervade all space, even where no matter is present. (Unlike sound, light propagates readily through the best vacuum.) It must be extremely tenuous, inasmuch as the earth and all other astronomical bodies pass through ether-filled space with no detectable loss of speed. Finally, the ether must be capable of vibrating at extremely high frequencies. (The frequency of visible light is more than 10^{14} cycles per second, much higher than that of any known mechanical oscillation.)

Important progress took place when James Clerk Maxwell showed that the equations of electricity and magnetism have solutions that consist of traveling waves, whose speed can be calculated in terms of known constants. The calculated speed of those electromagnetic waves turned out to be almost exactly equal to the measured speed of light. This was convincing evidence that light is in fact an electromagnetic phenomenon.

After Maxwell's work, the mechanical model of light was abandoned. No material substance vibrates when an electromagnetic wave propagates; the oscillation is in the magnitudes of the electric and magnetic fields, which are only mathematical quantities.

If no mechanical oscillation takes place, no medium is required. Most physicists were nonetheless unwilling to accept the notion that electromagnetic disturbances can propagate through an absolute vacuum. The ether thus lived on, viewed now as a medium that somehow "supports" the oscillations associated with the propagation of light even though it does not itself vibrate. The nature of that medium became even more mystifying.

2.2. PRELUDE TO MICHELSON-MORLEY

Numerous attempts were made during the late nineteenth century to confirm the existence of the ether. The Michelson-Morley experiment is the best known of those attempts. The experiment is described in the next section; here I indicate its basic idea by sketching an analogous experiment using water waves. The discussion is entirely within the framework of Galilean relativity.

In a wave phenomenon that involves a medium, the rest frame of the medium is a unique frame of reference. Observers in any frame can carry out experiments to determine their velocity relative to the medium.

Suppose a ship is at rest in still water. Observers on the ship measure the speed of water waves moving in various directions. Because the medium is isotropic (the water looks the same in all directions), the measured speeds must all be equal.[3] Let c denote that common speed.

An identical experiment carried out on a moving ship has a quite different outcome: the wave speed in that case varies with direction. A wave traveling in the same direction as the ship moves more slowly than one traveling in the opposite direction. In fact, if the ship's speed is c, a wave traveling in the same direction as the ship does not appear to move at all.

The speeds of a given wave in the rest frame of the water and in the rest frame of the ship are related by the Galilean velocity transformation, equation (1.8). The fact that one is dealing with a wave instead of a material object does not affect the validity of the simple argument by which the transformation equations were derived.

By measuring the speeds of water waves in all directions, then, shipborne observers can determine the velocity of their ship relative to the water. The direction in which waves travel slowest must be the ship's heading, and the magnitude of the minimum speed is $c - V$, where V is the speed of the ship. If all waves are found to travel at the same speed, the ship must be at rest relative to the water.

The same argument can be applied to the propagation of light, with the ether in place of the water and the earth playing the part of the ship. In the rest frame of the ether light travels at the same speed c in all directions, whereas in the earth frame the speed of light should vary with direction; the magnitude of the variation depends on V, which now denotes the speed of the earth relative to the ether. Michelson proposed to determine the value of V by detecting the difference in travel times of light rays traversing a given distance in different directions.

One does not have to believe in the ether to conclude that the speed of light measured on earth should vary with direction. The Galilean velocity

---

3. Any motion of the source does not affect the speed of the waves, which is determined by the forces between adjacent water molecules.

transformation implies that light can travel at the same speed in all directions only in one reference frame; we may call that the "isotropic frame" if we are not committed to the existence of an ether. Unless the earth happens to be at rest in the isotropic frame (which is highly improbable a priori), the speed of light in the earth's frame must depend on its direction.

2.3. THE EXPERIMENT

Michelson's experiment differed in one important respect from the hypothetical water wave experiment described in the preceding section. Waves in water travel only a few meters per second; a ship can easily move as fast as (or even faster than) the waves. The speed of water waves measured on a moving ship can therefore vary by a substantial fraction. In the case of light, however, the ratio V/c was expected to be very small. The effect that Michelson was trying to detect was a very weak one.

A lower bound on the value of V/c can be deduced from the earth's orbital velocity around the sun, $\mathbf{V}_{orb}$, whose magnitude is known to be about 30 km/sec. $\mathbf{V}$, the earth's velocity relative to the ether, is the (vector) sum of $\mathbf{V}_{orb}$ and the sun's velocity relative to the ether, $\mathbf{V}_s$; the latter is completely unknown.

If $\mathbf{V}_{orb}$ and $\mathbf{V}_s$ happen to point in opposite directions, their sum can be much smaller in magnitude than either one. Since the direction of $\mathbf{V}_{orb}$ reverses every six months as the earth traverses its orbit, however, such cancellation cannot persist. No matter what the magnitude of $\mathbf{V}_s$, V must sometimes be at least 30 km/sec. The minimum value of V/c over the course of a year is therefore 10^{-4}.

To be sure, $\mathbf{V}_s$ might happen to be very large, in which case V/c would be much greater than 10^{-4}. But Michelson could not count on such good fortune; he had to be prepared to detect a variation of only one part in ten thousand in the speed of light. Michelson's great achievement was to construct a device, called an interferometer, sensitive enough to detect so small a variation.

Figure 2.1 is a schematic view of Michelson's interferometer. Light from a source, S, strikes a glass plate, P, inclined at 45°. The plate is silvered so that it reflects about half the light incident on it and transmits the rest. The transmitted ray strikes mirror A and is reflected back to P, where it is again partially reflected and finally reaches observation point C. (The transmitted part of this ray is of no interest and is not shown.)

The reflected portion of ray SP strikes a second mirror, B, which reflects it back to P. The transmitted part of this ray reaches the observation

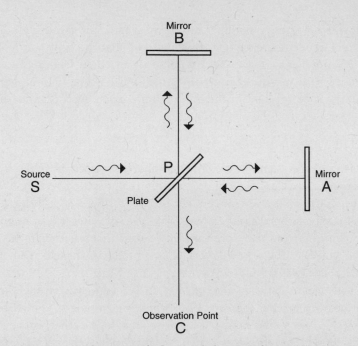

Fig. 2.1. Schematic of the Michelson-Morley experiment. Light from source S strikes the half-silvered mirror P, where part is reflected, reaching mirror B; the rest is transmitted, reaching mirror A. The two rays return to P, where each is again partially transmitted and partially reflected. Some of the light that reaches a detector located at C has followed path $SPAPC$, while the rest has followed path $SPBPC$. The two components "interfere" and produce a pattern of light and dark fringes at C.

point together with the ray that has traveled along the other path. The "arms" PA and PB are of equal length, L, and the entire apparatus is mounted on a rotatable platform.

If the earth is moving relative to the ether, as Michelson believed, the speed of light is different along each arm of the interferometer and the travel times for the paths $SPAPC$ and $SPBPC$ differ slightly. Since the first segment, SP, and the last, PC, are common to both paths, the difference in travel times is determined by segments PAP and PBP.

Assume for the moment that SPA is aligned in the direction of the earth's motion through the ether, which we take as the x direction. The speed of ray PA is then $c - V$, while that of ray AP is $c + V$; the travel times for the two segments are

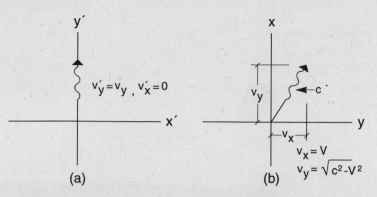

Fig. 2.2. (a) A light ray moves in the y direction in the earth frame, S'. (b) In the ether frame, S, the ray has both an x and a y component of velocity. The y components of velocity are the same in both frames; the x components are related by eq. (1.10a). Since $v_x' = 0$, (1.10a) gives $v_x = V$.

$$t_{PA} = \frac{L}{c-V} \quad t_{AP} = \frac{L}{c+V}$$

and the time for the round-trip PAP, which we call T_1, is

$$T_1 = \frac{L}{c-V} + \frac{L}{c+V} = \frac{2Lc}{c^2-V^2} = \frac{2L}{c(1-\beta^2)} \tag{2.1}$$

I have defined

$$\beta = V/c \tag{2.2}$$

This is standard notation in relativity.

The speeds of rays PB and BP can be found with the help of the Galilean velocity transformation. Let S be the ether frame and S' the earth's rest frame. Figure 2.2a shows the direction of ray PB in S'. By hypothesis, this ray is traveling in the y' direction and v_x' must be 0. According to equation (1.8b), then,

$$v_x = v_x' + V = V \tag{2.3}$$

Figure 2.2b shows the direction of the same ray in frame S. Since in this frame all rays travel at speed c, we can write

$$c^2 = v_x{}^2 + v_y{}^2 = V^2 + v_y{}^2$$

which gives

$$v_y = \sqrt{c^2 - V^2}$$

Finally, using equation (1.10b),

$$v'_y = v_y = \sqrt{c^2 - V^2} \tag{2.4}$$

Since $v'_x = 0$, (2.4) is the desired speed of ray PB in the earth frame. The speed of ray BP is clearly the same. Hence

$$t_{PB} = t_{BP} = \frac{L}{\sqrt{c^2 - V^2}} = \frac{L}{c\sqrt{1 - \beta^2}} \tag{2.5}$$

and the time, T_2, for path PBP is

$$T_2 = t_{PB} + t_{BP} = \frac{2L}{c\sqrt{1 - \beta^2}} \tag{2.6}$$

Since β is less than 1, $\sqrt{1 - \beta^2}$ is greater than $1 - \beta^2$ and T_2 is less than T_1. (For $\beta = 0$, both T_1 and T_2 are equal to $2L/c$.)

It is instructive to carry out the same calculation in the ether frame, S, in which the entire apparatus is moving to the right with velocity V. In this frame all rays travel at the same speed, c, but the paths of the rays differ in length. The travel time of each ray must of course be the same in S as in S'.

Figure 2.3a shows the path of ray PA in frame S. The ray travels a distance $L + Vt_{PA}$, where Vt_{PA} is the distance the apparatus moves during the time, t_{PA}, that the ray is in transit. Since the speed of the ray is c, we can write

$$L + Vt_{PA} = ct_{PA}$$

Solving this simple algebraic equation for t_{PA}, we find

$$t_{PA} = \frac{L}{c - V}$$

which is the same as the travel time for this ray calculated in the earth frame.

The analysis of the return leg AP is similar. The length of its path is $L - Vt_{AP}$ (fig. 2.3b). We have then

$$L - Vt_{AP} = ct_{AP}$$

whose solution is

$$t_{AP} = \frac{L}{c + V}$$

again the same as the result calculated in the earth frame.

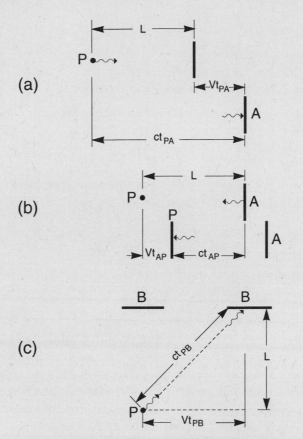

Fig. 2.3. Paths of the rays in Michelson's experiment as seen in the ether frame. (a) Ray *PA*, which travels in the $+x$ direction, travels a distance greater than *L*. The difference is the distance the apparatus has moved during the time interval t_{PA} or Vt_{PA}. (b) Ray *AP*, which moves in the $-x$ direction, travels a distance less than *L* by the amount Vt_{AP}. (c) Ray *PB* moves diagonally in this frame. The distance it travels is the hypotenuse of a right triangle whose legs are *L* and Vt_{PB}. All three rays travel at speed *c*.

Figure 2.3c shows the path of ray *PB* in frame *S*. *PB* is the hypotenuse of a right triangle whose legs are *L* and Vt_{PB}. Hence we can write

$$(ct_{PB})^2 = L^2 + (Vt_{PB})^2$$

whose solution is

$$t_{PB} = \frac{L}{\sqrt{c^2 - V^2}}$$

once more the same as the earth frame result.

In either frame, then, the difference in travel times is

$$T_1 - T_2 = \text{(time for path } SPAPC) - \text{(time for path } SPBPC)$$

$$= \frac{2L}{c}\left(\frac{1}{1 - \beta^2} - \frac{1}{\sqrt{1 - \beta^2}}\right) \tag{2.7}$$

Since β is expected to be small, we can write an approximate form for equation (2.7). The binomial expansion (see the appendix to chap. 4) gives

$$\frac{1}{1 - \beta^2} \approx 1 + \beta^2 \qquad \frac{1}{\sqrt{1 - \beta^2}} \approx 1 + \frac{1}{2}\beta^2$$

and (2.7) becomes

$$T_1 - T_2 \approx \frac{L}{c}\beta^2 \tag{2.8}$$

Notice that in equation (2.8), from which the unknown V is to be calculated, only the square of the small quantity β appears. Since the square of a number much smaller than 1 is smaller still, the expected effect is tiny. With L a few meters and $\beta = 10^{-4}$, $T_1 - T_2$ is only about about 10^{-16} sec, much too small to measure even with present-day equipment. Michelson's interferometer was, however, capable of detecting *changes* in $T_1 - T_2$ of that order of magnitude; that was sufficient for his purposes.

As the name suggests, the operation of an interferometer is based on the "interference" between two or more waves. Figure 2.4 shows two examples of rays that start at the same point and are reunited after traversing different paths. In case (a), the paths contain equal numbers of waves. Hence a crest of one wave arrives at the observation point O_1 together with a crest of the other and the two waves reinforce one another. This is called *constructive* interference; it occurs whenever the numbers of waves in the paths differ by an integer.

In case (b), the lower path contains half a wave more than the upper one. A crest of one wave arrives at O_2 together with a trough of the other and the waves cancel each other. This is called *destructive* interference. Cancellation occurs whenever the numbers of waves in the two paths differ by an integer plus a half. This can be caused either by a difference in the lengths of the paths or by a difference in the speeds of the rays (or by a combination of both factors).

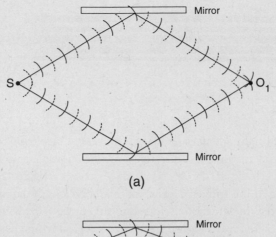

(a)

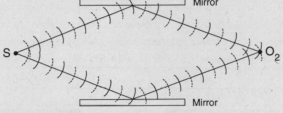

(b)

Fig. 2.4. Wave interference. Waves start at source point S and arrive at two adjacent observation points, O_1 (a) and O_2 (b). Crests of each wave are shown by solid arcs, troughs by dashed arcs. In (a), the total paths of the two rays are of equal length. Two crests are shown arriving together at the observation point (constructive interference). In (b), the path of the lower ray is half a wavelength longer than that of the upper ray. Hence a trough of the lower ray arrives together with a crest of the upper one, creating destructive interference. In the interferometer, a bright fringe would be observed at O_1 and a dark fringe at O_2.

In Michelson's experiment, rays $SPAPC$ and $SPBPC$ interfere when they reach the observation point. Because light rays are not geometrical lines but have a finite width, light arrives not just at the single point marked C but over a finite region in the vicinity of C. For every point in that region, the lengths of the paths followed by the two rays differ slightly. At some points constructive interference takes place, while at others the interference is destructive. The observer sees a pattern of

Fig. 2.5. Fringe pattern from a repeat of the Michelson-Morley experiment carried out by Georg Joos in 1930.

closely spaced bright and dark bands called *interference fringes*. Adjacent bright fringes correspond to a difference of one additional wave between the two paths.

The spacing of the fringes depends on the wavelength of the light and on the geometry. In Michelson's interferometer, the fringes were a fraction of a millimeter apart. Figure 2.5 shows a typical fringe pattern.

If, for any reason, the difference between the numbers of waves in the two paths changes, the fringe pattern shifts. A change of one wave causes the pattern to shift by one full fringe. A shift of a fraction of a fringe is readily detectable; the interferometer is therefore sensitive to a change in the light paths as small as a fraction of a wave.

The high sensitivity can be demonstrated by inserting a thin glass plate in the path of one of the rays. Since light travels more slowly in glass than in air, the travel time of the ray that passes through the glass increases by an amount that depends on the thickness of the glass and on its index of refraction. The fringe pattern is observed to shift by just the predicted amount.

The calculation thus far is based on the assumption that ray PA travels in the direction of **V**, the earth's velocity relative to the ether. Since the direction of **V** was unknown, Michelson had no way of knowing how the arms of his instrument were oriented relative to that direction. He therefore observed the behavior of the fringes as he slowly rotated the apparatus. At some time, arm PAP must be aligned in the direction of **V**.[4] At that instant T_1 is given by equation (2.2), T_2 by equation (2.6), and $T_1 - T_2$ by equation (2.8).

After an additional rotation of 90°, the two arms will have changed places: arm PB is now aligned in the direction of **V**, and arm PA is perpendicular to that direction. At that instant T_1 is given by equation (2.6) and T_2 by equation (2.2). Hence $T_1 - T_2$ has the magnitude expressed in equa-

4. This statement is correct only if the direction of **V** is in the plane of the interferometer. In general, PAP will at some time be aligned with the projection of **V** on that plane.

tion (2.8), but its sign is negative. One expects, therefore, that as the interferometer is rotated the fringe pattern will shift back and forth, with a maximum shift given by twice the magnitude of (2.8). By measuring the maximum shift, Michelson expected to determine the value of V.

In Michelson's first experiment, performed in 1881, the arms of the interferometer were 1.2 meters long. With $\beta = 10^{-4}$, equation (2.8) gives $T_1 - T_2 = 4 \times 10^{-17}$ sec. The maximum change in $T_1 - T_2$ is twice this amount, or 8×10^{-17} sec. During that time light travels 2.4×10^{-8} m. Since the wavelength of visible light is about 5×10^{-7} m, the expected maximum shift in the fringe pattern was about one-twentieth of a fringe.

Michelson, a skilled observer, was confident that he could detect even so small a shift. To his surprise, he found no shift whatever in the fringe pattern as he rotated his interferometer. He concluded that the velocity of the earth relative to the ether is zero and interpreted the result as supporting George Stokes's ether drag theory. Such an interpretation is, however, untenable. (See the discussion of ether drag in sec. 2.4.)

The result of the 1881 experiment was considered inconclusive because the predicted fringe shift was so small.[5] Because of the importance of the problem, Michelson was urged to refine the apparatus and improve its sensitivity; that task occupied him for several years. In 1887, in collaboration with Edward Morley, he performed a much-improved version of the experiment, in which multiple reflections increased the effective path length to about 11 meters. Figure 2.6, taken from Michelson's paper,[6] shows the 1887 interferometer.

The amplitude of the expected variation in the 1887 experiment was about half a fringe, far more than the sensitivity of the instrument. But amazingly, there was still no detectable effect. The fringe pattern exhibited no measurable shift as the interferometer was rotated, implying that the earth must be at rest relative to the ether. Michelson concluded that the upper limit on the value of V was about 1 km/sec.

As noted earlier, V could be that small if the motion of the sun relative to the ether happens to cancel the earth's orbital velocity. But even if such an improbable coincidence were to occur at some particular time, it could not persist. Six months later, when the direction of the earth's orbital

5. Because of an error in Michelson's theoretical analysis, his calculated value of the expected shift was too small by a factor of two. The error was pointed out by Hendrik Lorentz and independently by Alfred Potier.
6. A. A. Michelson and E. W. Morley, "On the Relative Motion of the Earth and the Luminiferous Ether," *American Journal of Science* 34 (1887):333–345.

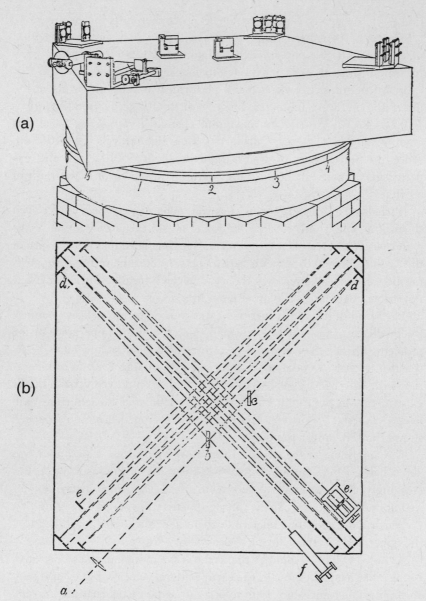

Fig. 2.6. Michelson's interferometer. Sketch (a) shows the apparatus; (b)
shows rays being reflected from the array of mirrors. The multiple reflection
effectively increases the path of the rays and would have increased the magnitude of any fringe shift. These figures are taken from Michelson's paper: A. A.
Michelson and E. W. Morley, "On the Relative Motion of the Earth and the
Luminiferous Ether," *American Journal of Science* 34 (1887):333–345.

velocity had reversed, V would have been about 60 km/sec and a large fringe shift should have been detected.

In his paper, Michelson promised to repeat the experiment at three-month intervals to eliminate the possibility of an accidental cancellation. He never did so, but others did and obtained the same result. No matter what time of year the experiment was done, no measurable shift in the fringe pattern was detected.

All conceivable sources of experimental error were carefully investigated; none was found. During the following half-century, the experiment was repeated many times by different investigators, with improved technique and variations in the procedure. No reliable experiment has ever yielded a positive result; in fact, the upper limit on the value of V consistent with the data was reduced even further. The experimental result had to be accepted as correct.

Although Michelson's experiment is now regarded as one of the most important in the history of physics, he himself considered it a failure. He had not achieved his objective—to measure the speed of the earth relative to the ether.

2.4. Attempts to Salvage the Ether

Taken at face value, the Michelson-Morley result implies that the speed of light in the earth's frame is always the same along both arms of the interferometer, no matter how the instrument is oriented. But from the point of view of Galilean relativity and the ether hypothesis, the speed of light is independent of direction only when the earth is at rest relative to the ether. How can the earth always be at rest relative to the ether when we know that the earth's velocity is continually changing? This was the enigma posed by Michelson's experiment.

Einstein's answer is simple and yet revolutionary. He *postulates* that the speed of light is independent of direction in *any* inertial frame. Michelson's result follows directly, but Einstein's postulate is clearly incompatible with the Galilean velocity transformation. In the context of conventional nineteenth-century physics, it makes no sense.

In the next chapter we shall see how Einstein's postulate leads to special relativity. Here I describe three attempts to explain the null result of the Michelson-Morley experiment within the framework of nineteenth-century physics: ether drag, emission theories, and the FitzGerald-Lorentz contraction. Although all three attempts were unsuccessful, one turned out to be on the right track.

Ether Drag

If the moving earth drags along the ether in its immediate vicinity, earth and ether are always at rest with respect to one another. The velocity of light is then always independent of direction, and the Michelson-Morley result follows.

The possibility that some ether might be dragged by moving bodies had been considered by many investigators during the nineteenth century. It is not an unreasonable hypothesis; material fluids such as air and water are, after all, dragged along to some extent when bodies move through them. Why should not the same thing happen to the ether?

Sir Oliver Lodge tried to confirm the existence of ether drag by passing beams of light near the edge of rapidly spinning metal spheres. He found no measurable effect. His result was inconclusive, however. The amount of ether dragged might depend on the size of the object; if so, the effect might be undetectable in the case of a small sphere and yet be substantial for the large earth.

Several versions of the ether drag hypothesis had been put forward. A theory proposed by Stokes assumed total drag: a moving body imparts its full velocity to the ether in its immediate vicinity; the amount of drag diminishes with distance. Such a model accounts for Michelson's failure to detect any fringe shift. Michelson in fact at first interpreted the result of his experiment as confirmation of Stokes's theory.

Although the assumption of full ether drag accounts for the result of Michelson-Morley, it runs into difficulty with other optical phenomena, notably stellar aberration. If the ether that surrounds the earth were fully dragged, no aberration effect would be detected. The direction of a light ray in the earth's frame would be unaffected by the earth's motion, and the apparent position of any star would be the same as in a reference frame in which the earth is moving.[7] The observations on stellar aberration therefore render Stokes's theory (or any other in which the ether is fully dragged) untenable.

Other theories assumed that the ether is only partially dragged. They defined an ether drag coefficient, which measures the extent to which the ether shares the velocity of a body moving through it. In Fresnel's theory, the ether drag coefficient inside a moving body is $1 - 1/n^2$, where n is the index of refraction of the material in question; the ether outside the body remains stationary. This was the favored theory because it explains the

7. See D. Bohm, *The Special Theory of Relativity* (New York: W. A. Benjamin, 1965), for a detailed discussion and a useful acoustic analogy.

result of Hippolyte Fizeau's experiment on the speed of light in moving water. (See sec. 4.5.) Fresnel's theory is consistent with the aberration data because for air, whose index of refraction is very nearly unity, it gives a drag coefficient nearly zero. However, Fresnel's theory cannot account for the result of Michelson-Morley.

In summary, no ether drag theory can explain both stellar aberration and the null result of the Michelson-Morley experiment.

Emission Theories

A very different hypothesis was put forward by Walter Ritz and others. They proposed that the speed of light is c relative to the source of the light instead of relative to the ether.[8] This is admittedly strange behavior for waves; it is more characteristic of particles. However, Ritz managed to construct an "emission theory" in which electromagnetic waves behave in this peculiar fashion. Einstein himself, before he developed special relativity, apparently leaned toward the emission theory.

An emission theory readily explains the result of the Michelson-Morley experiment. Inasmuch as the light source in the experiment was always at rest with respect to the interferometer, the speed of light is always the same and no change in the fringe pattern is to be expected as the interferometer is rotated.

The emission theory has been directly disproven in an experiment by T. Alvager et al. that detected the high-frequency radiation (gamma rays) emitted in the decay of rapidly moving neutral particles called pions.[9] If the speed of light were c relative to the source, then (according to Galilean relativity) the laboratory speed of a gamma ray emitted in the direction of the pion's velocity should be greater than c while that of a gamma ray emitted in the opposite direction should be less than c. No such difference in the speeds was observed.

Another disproof of the emission theory was provided by experiments of the Michelson-Morley type performed with light from extraterrestrial sources. The emission theory predicts a number of complicated effects. For example, because the sun is rotating, light originating from different parts of the sun's disk should travel at different speeds; this would complicate the interference pattern. The experiments were performed in 1924 by

8. One has to make an additional assumption about the velocity of a light ray reflected from a moving mirror. In Ritz's theory, the velocity of the reflected ray is c relative to the mirror.
9. T. Alvager, F. J. M. Farley, J. Kjellman, and I. Wallin, "Test of the Second Postulate of Relativity in the GeV Region," *Physics Letters* 12 (1964):260–262.

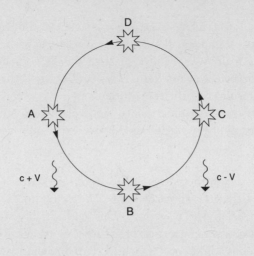

Fig. 2.7.   Light reaches earth from a member of
a binary star system, which moves in a circular
orbit. When the star is at point A, it is moving
toward earth; when it is at point C, it is moving
away from earth. According to Ritz's emission
theory, the light from A should travel at speed
$c + V$ and the light from C at speed $c - V$, where
V is the star's orbital velocity.

R. Tomaschek using starlight and in 1925 by Dayton Miller using sunlight; in each case, none of the effects predicted by the emission theory
was observed.

At the turn of the century, neither of the aforementioned pieces of
evidence was available. However, a fairly convincing argument against the
emission theory was put forward by Willem de Sitter, based on the properties of binary stars.

Binary stars orbit around their center of mass; if the stars have equal
mass, the center of the orbit is halfway between them. Consider that simple case and assume that the orbit is circular, with the earth in the plane
of the orbit (fig. 2.7). The apparent separation of the stars as seen from
earth oscillates, reaching a maximum when the stars are at positions A
and C. One quarter of a period later, when the stars are at positions B and
D, they eclipse one another and only one image is visible. If the rotation

is in the sense indicated in the diagram, each star is approaching earth as it passes position A and receding as it passes position C.

According to the emission theory, the light emitted by either star in the direction of earth when the star is at A travels at speed $c + V$ relative to us, while that emitted at C travels at speed $c - V$; V here denotes the orbital speed of the star. The difference in travel times of light reaching earth from the two positions gives rise to peculiarities in the observed motion. The predicted effect is not small; the difference between the two travel times is given by

$$t_1 - t_2 = \text{(time for light to reach earth from position } C) -$$
$$\text{(time for light to reach earth from position } A)$$
$$= \frac{2LV}{c^2 - V^2}$$

where L is the (mean) distance of the stars from earth. Since V is much smaller than c, the above expression is approximately

$$t_1 - t_2 \approx 2LV/c^2 \tag{2.9}$$

L/c, the mean travel time, can be a thousand years or more. Even if V/c is as small as 10^{-4}, therefore, the time difference $t_1 - t_2$ can be several months or more. For close binaries this can exceed the period of the stars' orbital motion. If that were the case, the light emitted from the later position, A, traveling faster, would reach earth *before* that emitted from C. During part of its orbit the star would appear to be moving backward! Numerous peculiar effects of this nature would be expected. On some occasions two images of the same star would be observed. The fact that no such effects are detected conclusively rules out the emission theories, or so it was thought at the time.[10]

The FitzGerald-Lorentz Contraction

Finally, we examine the most interesting of the attempts to reconcile the Michelson-Morley result with conventional theory—the proposal, first put forward by George FitzGerald, that the dimensions of material bodies change when they move through the ether. Lorentz had the same idea independently a short time later and developed it in detail.[11]

10. The validity of de Sitter's argument was challenged many years later by J. G. Fox, "Experimental Evidence for the Second Postulate of Special Relativity," *American Journal of Physics* 30 (1962):297–300.
11. FitzGerald discussed the contraction hypothesis often in his lectures but published only a brief qualitative note on the subject.

Lorentz and FitzGerald postulated that when a body moves through the ether, its dimension *in the direction of motion* is contracted by the factor

$$1/\sqrt{1-\frac{V^2}{c^2}} \tag{2.10}$$

The transverse dimensions remain unchanged.

Repeating the analysis of section 2.3, we can verify that the contraction hypothesis accounts for the result of the experiment. If the length of the longitudinal interferometer arm PAP is reduced by the factor expressed in equation (2.10), T_2 is reduced by the same factor and becomes equal to T_1. Hence no change in the fringe pattern is to be expected as the interferometer is rotated.[12]

As Lorentz himself pointed out, the contraction hypothesis is hard to test directly. The effect is very weak: the diameter of the earth contracts by only 6 centimeters as a result of its assumed motion through the ether. Moreover, any meterstick used to measure lengths itself shrinks; hence the contraction effect is not manifested in a simple length measurement. To detect the effect, one would have to use a meterstick that moves at a speed different from that of the object being measured; that is not an easy experiment to carry out.

Lorentz proposed a physical explanation for the contraction in terms of changes in the intermolecular forces when a body is in motion; his mechanism is described in section 4.9. He admitted, however, that his contraction proposal was essentially ad hoc.

According to Lorentz, the null result of Michelson-Morley represents the cancellation of two effects. The contraction of the longitudinal arm of the interferometer compensates for the differences in the speed of light along the various paths. The correct explanation, provided by special relativity, requires no such fortuitous cancellation. As we shall see, special relativity does predict a length contraction but only in a reference frame in which the object is moving. In the earth's frame, Michelson's apparatus is at rest and is not contracted.

A slightly modified version of the Michelson-Morley experiment, carried out by Roy Kennedy and Edward Thorndike many years later, discriminated between Lorentz's contraction hypothesis and Einstein's the-

12. The same conclusion follows if the transverse dimensions *increase* by the factor $1/\sqrt{1-V^2/c^2}$ while the longitudinal dimension remains unchanged. Lorentz quickly decided that the contraction hypothesis is more plausible.

ory; the results favored Einstein. Kennedy and Thorndike used an interferometer with arms of unequal lengths, L_1 and L_2. Repeating the derivation of section 2.3, we find for the difference in travel times instead of equation (2.7), the expression

$$T_1 - T_2 = \frac{2}{c}\left(\frac{L_1}{1-\beta^2} - \frac{L_2}{\sqrt{1-\beta^2}}\right). \qquad (2.11)$$

Under the contraction hypothesis, T_1 is reduced by the factor $\sqrt{1-\beta^2}$. Hence the time difference $T_1 - T_2$ becomes

$$T_1 - T_2 = \frac{2(L_1-L_2)}{c\sqrt{1-\beta^2}} \approx \frac{2(L_1-L_2)}{c}\left(1+\frac{1}{2}\beta^2\right) \qquad (2.12)$$

When the apparatus is rotated, the value of $T_1 - T_2$ does not change; hence in the standard Michelson-Morley experiment, no shift in the fringe pattern is to be expected. If β were to change, however, the value of $T_1 - T_2$ would change and the fringe pattern ought to shift.

To observe the effect, one has to keep the fringes under observation until the earth's velocity relative to the ether has changed appreciably. A small diurnal change in β is caused by the earth's rotation about its own axis; the rotational speed alternately adds to and subtracts from the orbital speed. Since the rotational speed even at the equator is only about 0.5 km/sec, however, this effect is very small.

A much larger effect is caused by the earth's orbital motion, which, as discussed earlier, causes V to change by about 60 km/sec during the course of a year. In 1932, Kennedy and Thorndike performed the experiment and managed to keep the fringes under continuous observation for several months. According to the ether theory, even with a correction for the FitzGerald-Lorentz contraction, a shift in the fringe pattern should have been detected; no such effect was found. In special relativity, as we shall see, the time difference is always $2(L_1-L_2)/c$, and no effect is to be expected.

Other experiments related to the motion of the earth through the ether and the contraction hypothesis were performed during the first years of the twentieth century. Several were based on the property of many solids and liquids that they become doubly refracting under strain: the indices of refraction for the two possible polarizations of light differ slightly. Since the FitzGerald contraction can be viewed as a strain, it should give rise to a double refraction whose magnitude changes as the apparatus is rotated. The magnitude of the effect should be proportional to $(V/c)^2$. In

1902, Lord Rayleigh looked for the effect and failed to find it. Dewitt Bristol Brace at the University of Nebraska repeated the experiment in 1904 with greatly increased accuracy but also obtained a null result.[13]

The sensitivity of Brace's experiment was such that he could have detected a difference between the two refractive indices as small as 10^{-13}; an effect at least twenty times greater was expected. Brace argued that the absence of double refraction was evidence against Lorentz's contraction hypothesis. However, an explanation for the null result was proposed by Lorentz and also by Joseph Larmor in the same year.

Another interesting experiment was carried out in 1903 by Frederick Trouton and H. R. Noble, who suspended a parallel-plate condenser from a string. For fairly complicated reasons, there should be a torque on the condenser that depends on the angle between the condenser plates and the direction of the earth's velocity through the ether. The predicted effect was of the order $(V/c)^2$. The outcome of this experiment too was negative: no torque was detected.

PROBLEMS

2.1. Galileo proposed a simple way to measure the speed of light. He stationed himself on a hilltop with a lantern and an assistant with an identical lantern on another hilltop. Both lanterns were initially lit. The assistant was instructed to shut off his own lantern as soon as he saw Galileo's go out. The time delay between the instant when Galileo shut off his own lantern and the instant when he saw the assistant's lantern go out would measure the time required for light to travel back and forth between the two hilltops.

Galileo's method failed because the times in question were too short to measure. Assuming that the distance between hilltops was 10 km and that Galileo could detect a time delay of 0.1 sec, what is the highest value for c that he could have detected? With the actual value of c, how far apart would the hilltops have to be in order to permit a determination? Neglect the human reaction time, which would also contribute an error.

2.2. A Michelson interferometer is operated with light of wavelength 5×10^{-4} mm. By how many fringes should the interference pattern shift when a glass plate 1 mm thick is inserted in the path of one of the beams? The index of refraction n of the glass is 1.2. (The speed of light in a transparent medium is c/n.)

2.3. This problem deals with de Sitter's objection to Ritz's emission theory, in which the speed of light is assumed to be c relative to its source. Refer to fig. 2.7. Let T be the period of the star's circular orbit, L its mean distance from earth, and V its orbital velocity. (L is much greater than the radius of the orbit, so the light

13. D. B. Brace, "On Double Refraction in Matter Moving through the Aether," *Philosophical Magazine* 7 (1904):317–328.

rays from any point on the orbit may be assumed to be parallel.) Neglect any relative motion between earth and the center of the star's orbit and assume that the emission theory is valid.

(a) Find the difference between the arrival times of the light emitted when the star is at position C and that emitted when it is at position A, at a time T/2 later. Your answer should be expressed in terms of V, c, L, and T.

(b) Find the relation between V, L, and T if light rays emitted by the star at A and C arrive at earth at the same time. To an observer on earth, the star appears to be at both places at once.

(c) Let $V = 100$ km/sec and $T = 10^6$ sec (about 10 days). For what value of L does the star appear to be at two places at once? Express your answer in light-years.

2.4. In the derivation of the expected time difference in Michelson's experiment in section 2.3, arm PA of the interferometer was assumed to be aligned in the direction of the earth's motion through the ether. Since that direction is totally unknown, the assumption is not justified. The object of this problem is to derive a more general result.

Assume that arm PA makes an angle ϕ with the direction of **V**, the earth's velocity relative to the ether. Repeat the derivation and obtain a general expression for the quantity $T_2 - T_1$. (For $\phi = 0$, the result should reduce to eq. [2.7].) What happens when ϕ is 90°?

3 The Postulates of Relativity and Their Implications

3.1. THE POSTULATES

In 1905 Albert Einstein, a twenty-six-year-old technical expert third class at the Swiss patent office in Bern, published three monumental papers in the *Annalen der Physik*. One of those papers set forth the theory now known as special relativity.[1] The theory is based on two postulates, which I paraphrase as follows:

Postulate 1 (Principle of Relativity): The laws of nature are the same in all inertial frames.

Postulate 2 (Constancy of the Velocity of Light): The speed of light in empty space is an absolute constant of nature and is independent of the motion of the emitting body.

All of special relativity follows by logical deduction from these two postulates. The only other assumption required is that space is homogeneous and isotropic, that is, no region of space is intrinsically different from any other and there are no preferred directions in space.

As described in chapters 1 and 2, a principle of relativity had been propounded in the seventeenth century by Galileo but for a long time was believed to apply only to the laws of mechanics. Einstein's first postulate

1. "On the Electrodynamics of Moving Bodies," *Annalen der Physik* 17 (1905):891–921. For a long time the only English translation available was in *The Principle of Relativity: A Collection of Original Memoirs*, first published by Methuen in 1923 and reprinted by Dover Books in 1952. That translation has been criticized as containing many inaccuracies. New translations are found in Arthur I. Miller, *Albert Einstein's Special Theory of Relativity* (Reading, Mass.: Addison-Wesley, 1981), and in the second volume of *The Collected Papers of Albert Einstein*, ed. John Stachel (Princeton: Princeton University Press, 1989).

is simply the extension of Galileo's principle to encompass *all* the laws of physics, including those of electromagnetism and optics.[2] Such a generalization had great intuitive appeal for Einstein. In his popular exposition of the theory, he says "that a principle of such broad generality should hold with exactness in one domain of phenomena, and yet should be invalid for another, is a priori not very probable."[3] The quest for generality and for unifying principles guided Einstein in all his research.

The second postulate is the revolutionary part of special relativity. As Einstein says in the introduction to the 1905 paper, the second postulate is "only apparently" irreconcilable with the first. What he means is that the two postulates are irreconcilable only if one insists on retaining the Galilean transformation, which implies that the speed of light (like that of anything else) should be different when measured by two sets of observers in relative motion.

The second postulate implies that the Galilean transformation, self-evident though it may appear, must be rejected. Its replacement, the Lorentz transformation, is derived in chapter 4. The principal conceptual consequences of special relativity can, however, be demonstrated directly from Einstein's two postulates without employing the Lorentz transformation. That is the task of this chapter.

As noted in chapter 2, the second postulate provides a simple explanation for the null result of the Michelson-Morley experiment: if the speed of light is an absolute constant, the light travel times along the arms of the interferometer are always equal, no matter how the instrument is oriented. Hence, no fringe shift is to be expected as the interferometer is rotated. The Michelson-Morley experiment therefore provides strong (though indirect) support for the second postulate. (See, however, the historical remarks in sec. 3.2.)

Direct experimental confirmation of the second postulate came only many years afterward. The most convincing data were provided by Alvager's experiment on the decay of neutral pions, cited in section 2.4 in connection with Ritz's emission theory. In that experiment, pions moving at $0.9998c$ were observed to decay into two photons (light pulses). One photon was emitted in nearly the forward direction (the direction of the

2. In his 1905 paper, Einstein states the relativity principle as follows: "In all coordinate systems in which the mechanical equations are valid, also the same electrodynamic and optical laws are valid" (p. 140 of translation in *Collected Papers*).

3. A. Einstein, *Relativity: The Special and the General Theory*, 15th ed. (New York: Crown Publishers, 1952), 14. The first edition was published in 1916.

decaying pion) and the other in nearly the opposite direction. According to Galilean relativity, the forward-moving photon should travel at speed $1.9998c$ and the other at only $0.0002c$. Instead, both photons were observed to travel at speed c to within about three parts in 10^5.

Demise of the Ether

In nineteenth-century electromagnetism, the ether played a central role. It provided a unique frame of reference in which Maxwell's equations hold and the speed of light is the same in all directions. According to Einstein's postulates, *every* inertial frame has that property. There remains no role for the ether to play in the description of natural phenomena; it has become superfluous.

Many physicists, including both Lorentz and Michelson, were reluctant to abandon the ether. Although Lorentz quickly accepted Einstein's relativity, he continued to maintain that a medium of some kind is needed as the carrier of the electromagnetic field. In his *Theory of Electrons*, published in 1909, he said, "I cannot but regard the ether, which can be the seat of an electromagnetic field with its energy and its vibrations, as endowed with a certain degree of substantiality, however different it may be from all ordinary matter."[4] Michelson expressed similar views. References to the ether continued to appear in the literature long after relativity had gained general acceptance. Gradually, however, the ether faded from discussion.

Special relativity dispenses also with the notions of absolute rest and absolute motion. So long as an ether was believed to exist, its rest frame provided a standard with respect to which absolute motion might be defined. A body could be said to be in a state of absolute rest if it was at rest relative to the ether. With the demise of the ether, all inertial frames are completely equivalent. There is no way to define absolute rest or absolute motion.

The first postulate implies that no experiment can have a result that favors one inertial frame over another. Nature imposes strict democracy among inertial observers; this rule dictates the outcome of many experiments. For example, consider two uniformly moving trains that approach each other on a straight track. The trains are equipped with identical speed-measuring devices, such as the radar employed by traffic patrols.

4. H. A. Lorentz, *The Theory of Electrons*, 2d ed. (New York: Dover Publications, 1952), 230. First published in 1909.

Observers on each train aim their device at the other train and measure its speed of approach. How are the results of the two measurements related?

In a world governed by Galilean relativity, it is easy to show that the two measured speeds must be equal. Suppose one train moves at 30 m/sec and the other at 40 m/sec relative to the ground. The distance between the trains diminishes by 70 m each second, and both speed indicators must read 70 m/sec.

In special relativity, this simple argument is not valid. As we shall presently discover, space and time have many unexpected properties. Observers on each train find that the other train is contracted and its clocks run slow. Hence it is not at all obvious that the two speed measurements should yield identical results. That outcome is demanded by the first postulate, however, for any other would violate the requirement of equality among inertial observers. If the two radars were to register unequal readings, which speed should be the greater? There is nothing in the problem to distinguish between the two trains, other than that they are moving in opposite directions; the isotropy of space assures us that this cannot make any difference. The only outcome consistent with Einstein's first postulate is that the two measured speeds are equal.

3.2. The Role of the Michelson-Morley Experiment in the Genesis of Relativity

Special relativity appears to be a classic case of a theory constructed expressly to explain a puzzling experimental finding. The second postulate, the heart of the theory, seems to be based directly on the result of the Michelson-Morley experiment. This indeed is the way the story is presented in much of the literature. Robert Millikan, in an article written in honor of Einstein's seventieth birthday, put it as follows:

> That unreasonable, apparently inexplicable experimental fact [the result of Michelson-Morley] was very bothersome to 19th-century physics, and so for almost twenty years physicists wandered in the wilderness in the disheartening effort to make it seem reasonable. Then Einstein called out to us all, "Let us merely accept this as an established experimental fact and from there proceed to work out its inevitable consequences," and he went at that task himself with an energy and a capacity which very few people on earth possess. Thus was born the special theory of relativity.[5]

5. R. A. Millikan, "Albert Einstein on His Seventieth Birthday," *Reviews of Modern Physics* 21 (1949):343.

Millikan's account paints a dramatic (and credible) picture. Yet in his own writings Einstein suggests that Michelson's experiment played at most a minor role in the genesis of special relativity. The 1905 paper makes no mention of the experiment, although Einstein does refer to "unsuccessful efforts to discover any motion of the earth relative to the 'light medium,' " without identifying those efforts.[6] The Michelson-Morley experiment was only one of them.

There is some question as to whether Einstein even knew about the experiment when he wrote his paper. In an interview conducted in 1950, Einstein told Robert Shankland that he became aware of the Michelson-Morley result through the writings of Lorentz, but *only after 1905.* "Otherwise," he said, "I would have mentioned it in my paper." He added that the experimental results that had influenced him most were the observations on stellar aberration and Fizeau's experiment on the speed of light in moving water. "They were enough," he said.[7]

When Shankland raised the question again two years later, Einstein gave a different response. "This is not so easy," he said. "I am not sure when I first heard of the Michelson experiment. I was not conscious that it had influenced me directly during the seven years that relativity had been my life." He added that in the years 1905–1909 he thought a great deal about Michelson's result. He then realized that he had also been conscious of the result before 1905, partly from the papers of Lorentz and more because he had "simply assumed this result of Michelson to be true."[8]

Abraham Pais, who knew Einstein well and wrote his scientific biography, is certain that Einstein did know about the Michelson experiment before 1905.[9] He points out that Einstein was in his seventies and not in good health when he talked to Shankland; at the first interview he probably did not remember that Michelson's experiment was discussed in Lorentz's 1895 monograph, which he had definitely read before 1905 (see sec. 4.8).

Even if Einstein was aware of Michelson-Morley's result, we must accept his assertion that it was not a major motivating factor in the genesis of relativity. He repeatedly uses terms like "negligible," "indirect," and

6. The 1905 paper contains no references whatever. Einstein's first reference to the Michelson-Morley experiment was in a review article he published in 1907.
7. R. S. Shankland, "Conversations with Albert Einstein," *American Journal of Physics* 31 (1963):47–57.
8. Ibid.
9. A. Pais, *Subtle Is the Lord* (Oxford: Oxford University Press, 1982), 116.

"not decisive" to describe the influence of Michelson's experiment on his thinking.

In his penetrating analysis of the issue, Gerald Holton concludes that "the role of the Michelson experiment in the genesis of Einstein's theory appears to have been so small and indirect that one may speculate that it would have made no difference to Einstein's work if the experiment had never been made at all."[10] In light of this assessment, Einstein's achievement looms all the more remarkable.

If he was not influenced by Michelson's experiment, how did Einstein arrive at the second postulate? That is the intriguing question. Einstein's paper provides little guidance. In it he presents the second postulate with no explanatory remarks or motivation, as though it were a commonly accepted proposition instead of a daring departure from conventional notions. An illuminating passage is found in the autobiographical notes written by Einstein in 1949.

> By and by I despaired of the possibility of discovering the true laws by means of constructive efforts based on known facts. The longer and the more despairingly I tried, the more I came to the conviction that only the discovery of a universal formal principle could lead us to assured results. . . . After ten years of reflection such a principle resulted from a paradox upon which I had already hit at the age of sixteen: If I pursue a beam of light with the velocity c, I should observe such a beam as a spatially oscillatory electromagnetic field at rest. However, there seems to be no such thing, whether on the basis of experience or according to Maxwell's equations. From the very beginning it appeared to me intuitively clear that, judged from the standpoint of such an observer, everything would have to happen according to the same laws as for an observer who, relative to the earth, was at rest. For how, otherwise, should the first observer know, i.e., be able to determine, that he is in a state of fast uniform motion?[11]

The seed of the theory of relativity had evidently been planted when Einstein was only sixteen years old! The idea that light has the same speed in all inertial frames, so difficult for an ordinary mind to grasp, was a

10. G. Holton, "Einstein, Michelson, and the Crucial Experiment," a chapter in his *Thematic Origins of Scientific Thought* (Cambridge: Harvard University Press, 1988). See also G. Holton, "Einstein and the Crucial Experiment," *American Journal of Physics* 37 (1968):968–982.
11. "Autobiographical Notes," in *Albert Einstein: Philosopher-Scientist*, ed. P. A. Schilpp (Evanston: Harper Torchbook, 1949), 53.

quite natural one for Einstein. He was prepared to accept it even without strong experimental evidence.

In the years following 1905, the postulates of relativity have been confirmed by ample experimental evidence, including refined versions of the Michelson-Morley experiment. But the genesis of the theory, apparently, lay in Einstein's inspired intuition.

3.3. THE RELATIVITY OF TIME: SIMULTANEITY

The most profound conceptual implication of special relativity is the change it has brought about in our perception of the nature of time. Relativity requires us to reject the notion of absolute time, which was taken for granted by Newton and by all the thinkers who followed him. (Newton's definition of absolute time was quoted in chap. 1.)

In a world in which time is absolute, the following proposition is surely valid: *if observers in different inertial frames measure the time interval between two events, using identically constructed clocks, the measured time intervals will in every instance be equal.* The proposition is (at least in principle) subject to experimental test. Although our intuition strongly suggests that the result of all such tests must be positive, it is conceivable that the measured time intervals might sometimes turn out to be different. In that case, absolute time would have to be abandoned. Einstein's postulates predict just such an outcome, and the prediction is confirmed (albeit indirectly) by experimental evidence. All the major conceptual consequences of special relativity can be related to the relativity of time.

I introduce the relativity of time by analyzing the concept of simultaneity. If time is absolute, two events that occur at the same time according to one set of observers must be simultaneous as well for any other set. The second postulate leads inescapably to a contrary conclusion: other observers find that the events are *not* simultaneous. This result suffices to establish the relativity of time.

One bit of reassurance can be offered the reader. If two events occur simultaneously *at the same place*, a single observer can directly experience both: she can see them happen together. Such events are simultaneous in any frame even according to special relativity. In fact, since an event is characterized by its space and time coordinates, two events that occur at the same time *and* at the same place can be regarded as parts of a single event.[12]

12. The Lorentz transformation, derived in chapter 4, is a prescription for calculating the coordinates of an event in frame S', given its coordinates in S. Each S' coordinate can depend only on the four numbers x, y, z, and t. If all four of those

If, however, two events take place at *different* locations in a particular frame of reference, no single observer in that frame can experience both. Determination of simultaneity in such a case is a complicated procedure, as has already been discussed in chapter 1. Two observers are required, one at the location of each event; each observer records the reading of a clock on the scene, and the two readings are subsequently compared. If the readings are the same (and if the clocks are properly synchronized), the observers conclude that the events in question were simultaneous.

After all that exchange of information, however, simultaneity at a distance must be regarded as an inference rather than a sensory observation. Special relativity predicts that if observers belonging to a different inertial frame record the times of the same two events on their (separated) clocks, those clock readings will be unequal. That prediction, although counter-intuitive, does not contradict any sensory experience.

The experiment shown in figures 3.1 and 3.2 demonstrates the relativity of simultaneity. A train moves with constant speed V relative to the ground. One frame of reference, S, is fixed on the ground while another frame, S', moves with the train. F', R', and M' are train observers situated at the front, rear, and midpoint of the car. At some instant, M' flashes a light. Call this event E_1, and call events E_2 and E_3 the arrivals of the light flashes at the rear and the front platforms, respectively.

We analyze the experiment first in frame S' (fig. 3.1). Let t_1', t_2', and t_3' be the times of the three events as recorded by train clocks located at the scene of each event. Since F' and R' are at equal distances from M', the light flashes, traveling at the same speed, take equal times to arrive. Hence

$$t_2' = t_3' \qquad (3.1)$$

In the train frame of reference, then, E_2 and E_3 are simultaneous.

Figure 3.2 shows the same set of events as seen in the ground-based frame of reference. Ground observer M witnesses the emission of the light flashes, and ground observers R and F witness their arrivals at the rear and front of the car. While the backward-moving flash was in transit, the train moved ahead. The distance MR traversed by that flash is therefore *less* than half the length of the car (fig. 3.2b). A similar argument shows that the distance MF traversed by the forward-moving flash is *more* than half the length of the car (fig. 3.2c). Hence MF is longer than MR.

---

numbers are the same for event B as for event A, then no matter what form the transformation law may take, the result for B must be the same as for A.

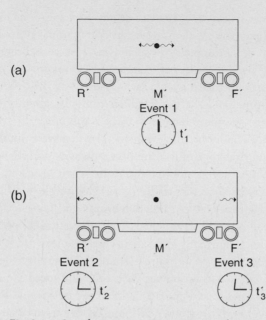

Fig. 3.1. Simultaneity experiment as seen by train observers. (a) Light flashes leave M' (event 1); (b) light flashes arrive at R' (event 2) and at F' (event 3). Since the paths of the two rays are of equal length, their arrivals are simultaneous: $t_2' = t_3'$.

According to Einstein's second postulate, both flashes travel at the same speed c in the ground frame as well as on the train; consequently, ground observers must see the backward-moving flash arrive first. In the ground frame, then, events E_2 and E_3 are *not* simultaneous. If t_1, t_2, and t_3 denote the times of the three events measured by ground clocks, we infer that

$$t_2 < t_3 \tag{3.2}$$

The second postulate is directly responsible for this perplexing result. It is easily verified that in Galilean relativity E_2 and E_3 are simultaneous in both frames. If the light flashes travel at the same speed c according to train observers, then in the ground frame the forward-moving one travels at speed $c + V$ and the backward-moving one at speed $c - V$. The faster speed of the forward-moving flash exactly compensates for the greater distance it has to travel (see problem 3.1) and the arrivals of the two flashes are simultaneous just as they are in the train frame. The hypothesis that the two light flashes travel at the same speed c in *both* frames

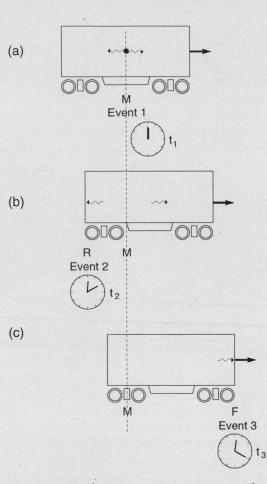

Fig. 3.2. Simultaneity experiment as seen by ground observers. (a) Light flashes leave *M* (event 1); (b) light flash arrives at rear of train (event 2); (c) light flash arrives at front of train (event 3). Event 2 occurs *before* event 3. Note that in sketch (b), the forward-moving ray is still en route.

leads to the discomforting conclusion that their arrivals are simultaneous in one frame but not in the other.

Notice we have not assumed that ground observers agree with those on the train as to the length of the car; that assumption would be wrong, as we shall presently see. (Ground observers see the train as contracted.) Both sets of observers do have to agree that the flashes start out from the midpoint of the car. That is assured by symmetry considerations: since the front and rear portions of the car are identical and move with the same speed, there is no reason why either should appear longer than the other in any inertial frame.

The magnitude of the difference between the arrival times of the two flashes in the ground frame depends on the length of the car and on its speed. For reasonable values of length and speed, the difference turns out to be extremely small—too small to be detected even with sensitive modern equipment. (See problem 3.2.) For that reason, the simultaneity experiment has never been performed. It is, however, conceptually straightforward. We shall describe many such "thought experiments"; even though they have never been performed, their outcome is unambiguously predicted by special relativity and their analysis is helpful in clarifying the subtleties of the theory.

If, instead of being at the midpoint of the car, M' were a short distance ahead of the midpoint when he flashed his light, the *forward*-moving flash would travel a shorter distance than the other and would therefore arrive first according to train observers. Provided the displacement from the midpoint were small enough, ground observers would still see the *backward*-moving flash arrive first. The order of E_2 and E_3 would thus be reversed in the two frames. This is the phenomenon discussed in chapter 1 to illustrate the strange predictions of special relativity.

Causality

If a cause-and-effect relationship exists between two events, a reversal of their time ordering is clearly unacceptable. In our light-flash experiment, event E_1 is clearly the cause of event E_2. A theory that predicts that any observers would see those two events in the opposite order, that is, see the arrival of a light flash before its emission, would be bizarre indeed. The principle of causality—that the cause must precede the effect—is not one we are prepared to abandon. Fortunately, special relativity does not require us to do so. I shall prove in chapter 5 that if two events are causally related, the cause precedes the effect in any frame of reference. Conversely, if the order of some pair of events is different in two frames of reference, neither one can be the cause of the other.[13] Special relativity is thus fully consistent with the causality principle.

According to the first postulate, any inertial frame of reference is as good as any other. If observers in one frame see a given pair of events as

13. In the light-flash experiment, events E_2 and E_3 are not causally related: neither is the cause of the other.

simultaneous while those in another frame see them as separated in time, it is fruitless to inquire whether the two events are "really" simultaneous; there is no operational way to answer the question. We conclude, rather, that simultaneity is not absolute but can hold only in a particular frame of reference. This conclusion, a direct consequence of the second postulate, is an integral part of special relativity.

A few remarks may help the reader come to grips with a conclusion that seems so contrary to common sense. First, as already emphasized, absolute simultaneity is being abandoned only in the case of spatially separated events, for which it is an inference rather than a direct observation. Moreover, the magnitude of the effect is substantial only when the relative velocity of the observers involved is close to the speed of light or when the events are separated by a huge distance. It is therefore not observable in everyday experience.

Finally, I would call attention to the following analogy. Two events that coincide in space (i.e., that occur at the same location) in one frame of reference do not coincide in any other frame if the events occur at different times. Suppose a ball is thrown straight up inside a moving train. According to observers on the train, the ball lands, some time later, in the same place it started. Ground observers, however, see the ball follow a parabolic path and land at some point down the track. This result is not at all surprising (it is equally true in Galilean relativity), but it is analogous to the earlier result that is so contrary to our intuition, except that the roles of space and time are interchanged:

Statement A: Events that happen at the *same place* but at *different times* according to one set of observers happen at different places according to a second set of observers moving relative to the first.

Statement B: Events that happen at the *same time* but at *different places* according to one set of observers happen at different times according to another set of observers moving relative to the first.

The symmetry between the two statements is striking, yet the first seems plausible while the second is perplexing. We are accustomed to thinking of space and time as entirely different entities; in the pre-Einsteinian formulation of the laws of physics, space and time take very different roles.

Special relativity, in contrast, treats space and time on a much more symmetrical basis. We shall find that except for minus signs here and there, all the equations of the theory are unaltered when space and time are interchanged. This symmetry endows the theory with a certain elegance. It does not, of course, prove that the theory is right; a theory can be elegant and yet not conform to nature. Still, theories that are simple and elegant have an undeniable appeal.

Although the discussion of simultaneity has been based on the properties of light, the conclusions are in fact quite general and apply to *any* pair of events. Consider two events, E_1 and E_2, that occur simultaneously in some frame of reference at positions P and Q. Imagine that halfway between P and Q, light flashes were emitted at some earlier time; the flashes arrive at P and Q simultaneously.

If the emissions were timed so that one flash arrives at P at the time of E_1, the other flash arrives at Q at the time of E_2. The results of this section imply that in any other frame of reference, the arrivals of the light flashes at P and Q are not simultaneous. However, since E_1 and the arrival of the first flash at P coincide in both time and space, those two events must be simultaneous in all frames. Likewise, E_2 and the arrival of the second flash at Q must be simultaneous in all frames. Since the two light flashes arrive at different times in the second frame, the same must be true of events E_1 and E_2. We have proved that if *any* two events occur simultaneously at different locations in some frame S, those events will be observed to occur at different times in any frame moving relative to S.

3.4. The Synchronization of Clocks

In the preceding section, the simultaneity of separated events was defined in terms of clock readings: the time of every event is measured by a clock at the site of the event, and two events are simultaneous if the clock readings are equal. Implicit in this definition is the assumption that the clocks are properly synchronized; this is a far from trivial condition when the clocks are separated. In fact, we seem to be confronted with a serious logical problem. The assertion that clocks A and B are synchronized is equivalent to saying that the events "clock A reads 6:00" and "clock B reads 6:00" are simultaneous. But if simultaneity is defined in terms of clock readings, how can the simultaneity of the clock readings themselves be verified? The argument appears to be circular.

A possible procedure would be to synchronize the clocks while they are together and then move them slowly to their final locations. That procedure is not satisfactory because we cannot be sure that motion does not affect synchronization. One would like to be able to test directly whether two separated clocks are synchronized.

Einstein's second postulate provides the solution to the problem. Two separated clocks can be synchronized by exchanging light signals between them, taking advantage of the fact that all light signals travel at the same

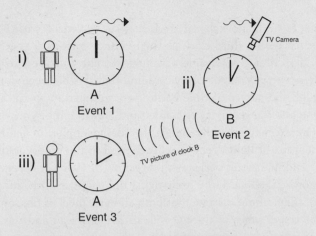

i)

ii)

A
Event 1

TV Camera

B
Event 2

iii)

TV picture of clock B

A
Event 3

Fig. 3.3.   Synchronization of distant clocks by exchange
of light (or TV) signals. (a) Observer *A* sends out a signal; his clock reads 12:00. (b) Signal arrives at *B* and
triggers a TV camera, which sends out a picture of the
clock located there. (c) Televised picture of *B*'s clock arrives at *A*; *A*'s clock reads 2:00. If the picture shows *B*'s
clock reading 1:00, *A* concludes that *B*'s clock is synchronized with his.

speed *c*. If the distance between clocks is *L*, the travel time of a light signal from one clock to the other is L/c. Hence, if a signal emitted from
clock *A* when it reads t_0 reaches clock *B* when *B* reads $t_0 + L/c$, we can
conclude that the clocks are synchronized. Alternatively, an observer,
M, could be stationed halfway between *A* and *B*. If *A* and *B* are synchronized, signals emitted from them when each reads t_0 should reach *M*
together.

A variant of this method allows clocks to be synchronized without re-
quiring any length measurement at all. Observer *A* sends out a light signal and records the reading of his clock as he does so (event 1, fig. 3.3).
When the signal reaches *B*, it triggers a television camera trained on the
clock there (event 2). The picture is transmitted back (at the speed of light)
to *A*, who records his clock reading at its arrival (event 3).

If the clocks are properly synchronized, the picture of *B*'s clock should
show it reading exactly halfway between *A*'s initial and final readings.
Suppose, for example, that *A*'s clock reads 12:00 at event 1 and 2:00 at
event 3. If the picture shows *B*'s clock reading 1:00, *A* concludes that the
clocks are indeed synchronized. If it does not, he can make the appropriate
adjustment on his clock.

It is not hard to show that if clock A is synchronized with B and B is synchronized with a third clock C, then A and C are also synchronized. All the clocks in a given frame of reference may thus be synchronized with one another. We shall always assume that this has been done.

Although the nature of time is more complicated in special relativity than in the classical relativity of Galileo and Newton, the synchronization of clocks by exchange of light signals is actually simpler. In Galilean relativity the speed of light depends on the speed of the observer relative to the ether and in general varies with direction. If Galilean observers wish to synchronize their clocks by exchanging light signals in the manner just described, they cannot assume that both signals travel at the same speed. Hence the travel times of the two signals are different and the criterion for synchronization is more complicated than the one for special relativity. (See problem 3.3.)

Synchronization by exchange of light signals works only for clocks that are at rest relative to one another. Suppose we try to apply the same procedure to two clocks in relative motion. Figure 3.4a shows the sequence of events as seen in the rest frame of A, in which B is moving to the right. In this frame the two light signals travel equal distances. Hence their travel times are equal and the criterion for synchronization is the same as before: the reading of B's clock at event E_2 should be halfway between the readings of A's at events E_1 and E_3.

A problem arises, however, when we examine the same sequence of events in the rest frame of B, in which A is moving to the left (fig. 3.4b). In that frame the return signal travels a greater distance and therefore has a longer transit time. Hence if the clocks are synchronized, the reading of B's clock at event E_2 should be less than halfway between the readings of A's clock at events E_1 and E_3.

This outcome is logically unacceptable. According to special relativity, there is no way to tell whether A or B is the moving clock; the results can depend only on their relative velocity. Yet the criteria for synchronization are different depending on whether the motion is ascribed to A or to B. We conclude that clocks in relative motion cannot be synchronized. In the next section we will learn why: the clocks keep time at different rates. Even if they could be synchronized at one instant, they would not stay synchronized.

A still more surprising result can be demonstrated. Suppose that train observers and ground observers have each synchronized their own clocks by exchanging light signals. Train observers propose to check on the synchronization of ground clocks by carrying out the following experiment.

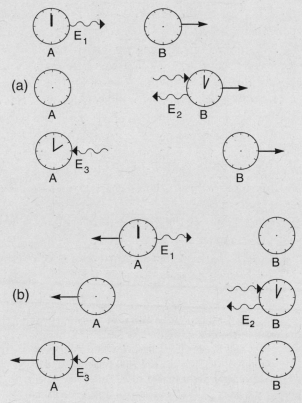

Fig. 3.4. An attempt to synchronize clocks in relative motion, using the procedure of fig. 3.3. Clock B is moving to the right relative to clock A. The sketches in (a) show the experiment in the frame in which A is at rest. The light flashes travel equal distances; hence the synchronization criterion is the same as for clocks that are stationary relative to one another. (b) The same experiment in B's rest frame, in which A is moving to the left. In this frame the two light flashes travel unequal distances; hence the reading of B at event E_2 should *not* be halfway between those of A at E_1 and E_3 if the clocks are synchronized. Since we cannot arrive at a unique criterion for synchronization, we conclude that the clocks cannot be synchronized.

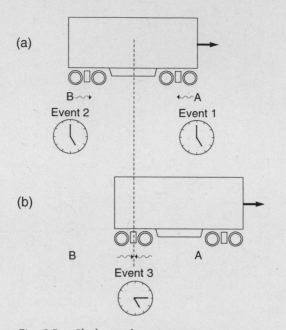

Fig. 3.5. Clock synchronization experiment as seen in the ground frame. (a) Stationary observers A and B emit light flashes when their (synchronized) clocks both read 5:00 (events 1 and 2). (b) The light flashes meet some time later (event 3) at a point halfway between A and B. The train has moved in the interim, so the light flashes meet somewhere in the rear half of the cabin.

At some prearranged train time, the train observer who sees ground clock A passing her records the reading of A. At the same (train) time, the train observer who sees ground clock B passing her records the reading of B. If the two clock readings are the same, A and B are indeed synchronized.

The results of the preceding section imply that the outcome of this test must be negative: according to train observers, ground clocks are *not* synchronized. The two events "clock A reads 5:00" and "clock B reads 5:00" are simultaneous *in the ground frame.* But if two separated events are simultaneous in one frame, they cannot be simultaneous in any other. In the train frame, therefore, when clock A reads 5:00, clock B reads some other time. The same must be true of any two ground clocks. Train observers conclude that all ground clocks are out of synchronization.

Figures 3.5 and 3.6 elucidate this all-important result. Ground observers A and B emit light pulses when their clocks read 5:00. Call the emis-

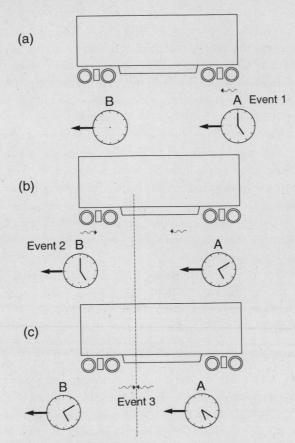

Fig. 3.6. Clock synchronization experiment as seen in the train frame. (a) Light flash leaves front platform (event 1); ground clock A, passing at that moment, reads 5:00. (b) Light flash leaves rear platform; ground clock B, passing at that moment, reads 5:00. (c) Light flashes meet at a point in the rear half of the train. Train observers infer that event 2 occurred after event 1 and therefore that clocks A and B are not synchronized. In sketch (b), clock A has moved from its position in (a) and reads a time later than 5:00.

sions of those pulses events 1 and 2 (fig. 3.5a). Just at that instant (as judged by ground observers), the front and rear ends of a passing train happen to coincide with *A* and *B*. The light pulses meet, some time later, at a point midway between *A* and *B*; call that event 3. By this time the train has moved, so the meeting occurs opposite a point in the rear half of the train (fig. 3.5b).

Figure 3.6 shows how the same three events look to train observers. At event 1 (fig. 3.6a), a light pulse leaves the front platform; ground clock *A*, which is passing just as that pulse is emitted, reads 5:00. Another pulse leaves the rear platform just as ground clock *B*, passing at that moment, reads 5:00 (event 2, fig. 3.6b). The pulses meet (event 3, fig. 3.6c) at a point closer to the rear of the train than to the front. Both sets of observers must agree on this qualitative statement even though their length measurements differ, as we shall see.

From the point of view of train observers, the pulse emitted from *B* has traversed a shorter distance than the other and must therefore have started out later. In the train frame, then, event 2 occurs after event 1. Clock *B* reads 5:00 not when *A* does but some time afterward. By the time *B* reads 5:00 (fig. 3.6b), *A* has moved some distance and reads *later* than 5:00. According to train observers, then, the two ground clocks are not synchronized. This confirms our earlier conclusion, but we have learned something more about the effect: clock *A* is running *ahead* of *B*.

If ground observers were to check on the synchronization of train clocks by a similar experiment, they would likewise find those clocks to be unsynchronized. At a given time (according to ground clocks), a train clock at the rear of the train is ahead of one at the front. This result is demanded by the first postulate: each set of observers must reach the same conclusion regarding the others' clocks. Otherwise, one frame would be singled out. We can express the general result as follows: *if a set of clocks, synchronized in the frame of reference in which they are at rest, is examined by observers in any other frame, the clocks will be found to be out of synchronization; the one that is behind in the direction of travel is ahead in its reading.*

Figure 3.7 summarizes our findings to this point. Sketch (a) is a view of ground and train clocks as seen by train observers at 10:00, train time. All train clocks read the same, whereas each moving ground clock has a different reading: *A* is ahead of *B*, which is ahead of *C*, and so on. Figure 3.7b is a similar view, showing how things look to ground observers at 8:00, ground time. Here the ground clocks are all synchronized while each train clock has a different reading.

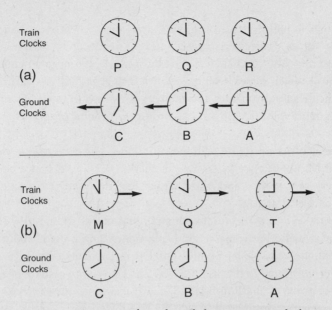

Fig. 3.7. Summary of our knowledge concerning clocks in relative motion. (a) Train and ground clocks as viewed by train observers at 10:00 (train time). All events in this picture are simultaneous in the train frame. Train clocks all read the same while ground clocks are out of synchronization. (b) View seen by ground observers at 8:00 (ground time). Here all events are simultaneous in the ground frame. Ground clocks all read the same while train clocks are out of synchronization. Notice that the spacings of the clocks are different in the two frames. The magnitudes of the differences between the readings of moving clocks are much exaggerated.

These pictures, strange though they appear, represent nothing more than the relativity of simultaneity. If each clock reading is regarded as a distinct event, all the events in figure 3.7a are simultaneous in the train frame whereas all those in figure 3.7b are simultaneous in the ground frame.

One feature of these results is worth reemphasizing. Notice that in (a), a train frame picture, ground clock *B* is opposite train clock *Q*; *B* reads 8:00 and *Q* reads 10:00. In (b), a ground frame picture, *B* is also opposite *Q*, and *the readings of both clocks are the same as in (a)*.

When two clocks are at the same place, their readings can be considered parts of a single event. All observers must agree on what those readings are. The statement "when train clock *Q* reads 10:00, ground clock *B* reads

8:00" is true in both frames. Differences arise only when one asks a question like "What does clock A read when Q reads 10 o'clock?" This is equivalent to asking, "What event involving clock A is simultaneous with the event at which Q reads 10 o'clock?" Since A and Q are in different places when that event occurs, the answers in the two frames differ. Train observers reply 9:00 (a), whereas according to ground observers the answer is 8:00 (b).

Notice also that in figure 3.7b the train clock opposite ground clock C is labeled M, not P. As I shall show in section 3.7, the distance between moving clocks is seen as contracted. Clock P in this picture is somewhere between clocks Q and M.

The differences in the readings of moving clocks in figure 3.7 were chosen arbitrarily, but some general comments can be made about what those differences should be. Because space is homogeneous, the difference must be proportional to the distance between clocks; the reading of a clock halfway between A and B in figure 3.7a, for example, must be 8:30, halfway between the readings of A and B. The difference in clock readings must depend also on V, the relative velocity of the two frames. (If V were zero, observers in both frames would agree that all clocks are synchronized.) The dependence on V will be derived in chapter 4.

The actual differences between clock readings, for reasonable separations and velocities, are much less than shown in the figure and are too small to be detected. (See problem 4.2.) Nonetheless, the nonsynchronization of moving clocks is a central element of special relativity and is essential to the further development of the theory.

Nothing has yet been said about the *rate* at which clocks in one frame run according to observers in another frame. That result ("moving clocks run slow") will come in section 3.5, after the time dilation effect has been derived. Synchronization refers solely to the question of how clocks appear at one particular instant.

3.5. Time Dilation

Another important relativistic effect is demonstrated by the thought experiment shown in figure 3.8. A light pulse emitted by a source S (event E_1) is reflected by a mirror M that is stationary relative to the source and a distance L away (event E_2). Observers in the laboratory frame, in which both light source and mirror are at rest, see the sequence of events shown in figure 3.8a; according to them, the reflected pulse returns to its starting point (event E_3).

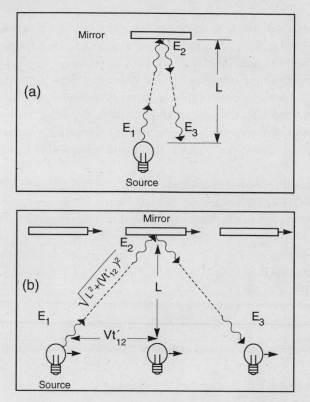

Fig. 3.8. Experiment to demonstrate time dilation. Sketch (a) shows the light path from the point of view of observers in the laboratory, for whom the apparatus is at rest. Sketch (b) shows how the same experiment looks to observers in a frame S', which is moving to the left at speed V relative to the laboratory.

Figure 3.8b shows the same sequence of events as seen by observers in a frame S' that moves from right to left at speed V relative to the laboratory. Those observers see the entire apparatus moving to the right at speed V; according to them, the light follows the zigzag path indicated in the figure. Observers in each frame measure the time interval required for the light to travel to the mirror and return.

This experiment should look familiar: it is just the transverse part of the Michelson-Morley experiment. We can repeat the analysis of section 2.3 with one important difference. In the earlier analysis, based on Galilean relativity, the speed of light was assumed to be different from c in one reference frame; here, in accord with the second postulate, we must put the speed of light equal to c along every path in both frames.

The diagrams show that the light path in frame S' (fig. 3.8b) is longer than the path in the laboratory frame (fig. 3.8a). Since the speed of light is the same in both frames, the time interval between E_1 and E_3 measured in S' must be longer than the interval between the same two events measured in the laboratory. (We assume that the distance between source and mirror is the same in both frames; that assumption will be justified in sec. 3.8.)

The quantitative relation between the travel times measured in the two frames is readily calculated. In the laboratory frame, the total path traversed by the light is $2L$; hence the time interval between E_1 and E_3 is

$$t_3 - t_1 = 2L/c \tag{3.3}$$

The analysis in frame S' follows that of section 2.3; the total travel time is given by equation (2.6):

$$t_3' - t_1' = \frac{2L}{c\sqrt{1 - \dfrac{V^2}{c^2}}} = \frac{2L}{c}\gamma \tag{3.4}$$

I have introduced the parameter γ, defined as

$$\gamma = \frac{1}{\sqrt{1 - V^2/c^2}} \tag{3.5}$$

This is standard notation in relativity theory. It is clear from the definition that γ is always equal to or greater than 1 and approaches infinity as $V \to c$; it does not differ appreciably from unity unless V is quite close to c. (See table 3.1.)

It is convenient to use the symbol Δ, followed by a variable, to denote the difference between two values of that variable. Here, let Δt denote $t_3 - t_1$ and $\Delta t'$ denote $t_3' - t_1'$. Equations (3.3) and (3.4) can be written simply as

$$\Delta t' = \gamma(\Delta t) \tag{3.6}$$

This effect is known as *time dilation*.

Equation (3.6) does not assert that the time interval between *any* two events is longer in S' than in S; such a relation would distinguish between the two frames in a manner inconsistent with the first postulate. For the

Table 3.1. Values of relativistic factor γ
as a function of V/c

V/c	γ
.0001	1.000000005
.001	1.0000005
.01	1.00005
.1	1.005
.5	1.16
.8	1.67
.9	2.29
.95	3.20
.99	7.09
.999	22.4
.9999	70.7
.99999	224

particular pair of events E_1 and E_3, the laboratory frame has a unique property: the events happen at the same place, whereas in any other frame they happen at different places. The result can be stated as follows:

> The time interval between two events is shortest when measured in the reference frame in which the events occur at the same place (if such a frame exists). In any other frame, the interval between the events is longer by the factor γ.

An interval between events that happen at the same place is called a *proper* time interval (sometimes written $\Delta\tau$); any other interval is called *improper*. Given two arbitrary events, a frame in which the events occur at the same place does not necessarily exist. If no such frame exists, the proper time interval between the events is not defined. The present analysis enables us to compare time intervals in two frames of reference only if one of the two intervals is proper. In the next chapter we shall see how to extend the result to any pair of events.

The proper time interval between two events can be read off a single clock, which is present at both events. Any improper interval necessarily represents the difference between the readings of two distinct clocks, one present at each event.

Suppose a body moves with constant speed V. In a frame moving at the same speed, the body is stationary; that frame is called the *rest frame* of the body. A clock attached to the body records the proper time interval between any two events in which the body is involved, since all such events occur at the same place in the body's rest frame.[14]

If a body travels between two points whose separation (as measured by ground observers) is L, the time for the trip measured by ground frame clocks is $T=L/V$; this is clearly an improper time interval. The proper time for the trip is measured by observers who travel with the body and is shorter than T by the factor γ:

$$\tau=T/\gamma=T\sqrt{1-V^2/c^2}$$

As the body's speed approaches c the time for the trip as measured by ground clocks approaches L/c, but the proper time approaches zero! If a spaceship travels to a distant star, the time of the trip *as measured by clocks on the spaceship* goes to zero as the ship's speed approaches c. This consequence of relativity forms the basis for much fanciful speculation about space travel. In principle, there is no limit to the distance one could travel in one's own lifetime.

EXAMPLE. A spaceship travels from earth to Alpha Centauri at speed $0.8c$. The distance between earth and Alpha Centauri is 4 light-years, measured in the earth frame. (Assume that Alpha Centauri is at rest relative to earth.)

(a) How long does the trip take according to earth clocks?

(b) How long does it take according to spaceship clocks?

SOLUTION. (a) We measure distance in light-years and time in years. In these units the numerical value of c is 1 light-year/year. In the earth frame,

$$t=L/V=4/0.8=5 \text{ yr}$$

(b) The interval between departure and arrival is a proper time interval in the spaceship frame. For this problem,

$$\gamma=1/\sqrt{1-(0.8)^2}=1.67$$

Hence $t'=t/\gamma=3$ yr.

14. If a body's speed is not constant, proper time for the body can still be defined as the time read by a clock attached to it. Although the body is not at rest in any inertial frame, at any moment it is instantaneously at rest in the inertial frame whose laboratory speed is the same as its own. Over an infinitesimal time interval, clocks in this instantaneous rest frame record the body's proper time. The proper time over a finite interval is the sum of the infinitesimal contributions thus defined.

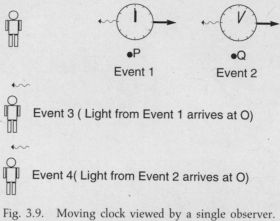

Fig. 3.9. Moving clock viewed by a single observer. When the clock is at *P*, it reads 12:00; when it is at *Q*, it reads 12:05. To observer *O*, the clock appears to be running slow. This is *not* a relativistic effect.

Equation (3.6) makes no sense if *V*, the relative velocity between the two frames in question, is equal to or greater than *c*. For $V = c$ the time dilation factor γ is infinite; for $V > c$ it is imaginary and cannot represent the ratio between two time intervals, which must be real numbers. As we shall see, the speed of light plays the role of a limiting velocity in special relativity. The results of this section, although not a proof, are already highly suggestive of that conclusion.

Moving Clocks Run Slow

Time dilation is often summarized by saying, "Moving clocks run slow." That statement is correct if properly interpreted but is also the source of much needless confusion.

There is one trivial sense in which a moving clock may be said to run slow. As already noted, an observer viewing a distant event sees it not when it occurs but later. The delay is the travel time of the light between event and observer.

Suppose a clock is moving away from an observer *O* (fig. 3.9). When the clock reads 12:00, it is at point *P* (event 1); when it reads 12:05, it is at point *Q* (event 2). Light emitted at events 1 and 2 reaches *O* at events 3 and 4, respectively. Since the light from *Q* has traveled farther than that from *P*, the interval between the time when *O* sees the clock reading 12:00 and the time when she sees it reading 12:05 is more than 5 minutes even if the moving clock is keeping time correctly. The clock therefore appears

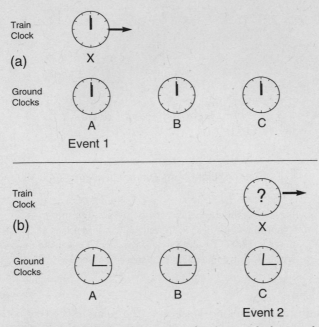

Fig. 3.10. Relativistic description of a moving clock viewed by a sequence of stationary observers. (a) Train clock X is adjacent to ground clock A (event 1); both clocks read 12:00. (b) Some time later, X is adjacent to ground clock C (event 2). The difference between the reading of C in sketch (b) and that of A in sketch (a) represents the interval between events 1 and 2 in the ground frame. With the help of fig. 3.11, one finds that X has lost time between the two events: it is running slow.

to be running slow. This argument depends only on the finite velocity of light and applies equally in Galilean relativity. Moreover, only a receding clock runs slow; a similar argument shows that an approaching clock appears to run *fast*. These results have nothing to do with time dilation.[15]

The only way to discover the "true" rate of a moving clock without introducing extraneous effects due to light travel time is to compare its reading on two occasions with those of *adjacent* stationary clocks. Two distinct stationary clocks are required.

We turn once again to the example of a train moving from left to right along a straight track; let clocks be distributed at regular intervals both on

15. They are related to the Doppler effect, discussed in chapter 4.

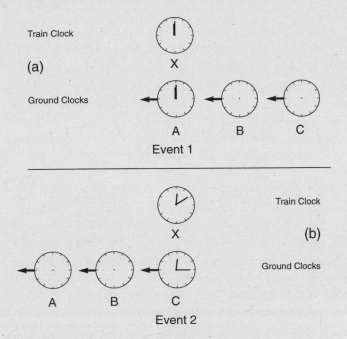

Fig. 3.11. The events of fig. 3.10 are shown as they appear in the train frame. The two events occur at the same place; hence the time interval between them is proper and is shorter than the interval between the same events measured in the ground frame. The proper time interval is read directly off a single clock, X.

the train and along the track. Each set of clocks is synchronized within its own frame.

Figure 3.10 indicates the precise measurements that ground observers carry out to determine the rate at which a particular train clock, labeled X, is running. At event 1 (fig. 3.10a), X is adjacent to ground clock A; suppose they both read 12:00 at that moment.[16] Some time later, say, at 12:15 according to ground clocks, X has moved and is now adjacent to ground clock C (event 2, fig. 3.10b).

From the point of view of train observers (fig. 3.11), all ground clocks are in motion while X is stationary. It is apparent that in the train frame, events 1 and 2 occur at the same place. Hence the interval between them,

16. The rate at which clock X is running is determined by measuring the difference between its readings at events 1 and 2. It is purely a matter of convenience and involves no loss of generality to assume that X reads the same as A at event 1.

measured by clock X, is a proper time interval and must be shorter than the ground-frame interval between the same two events. The latter quantity is 15 minutes; we conclude that X reads *less* than 12:15 at event 2. For concreteness, suppose the speed of the train is such that the time dilation factor γ (defined by eq. 3.5) has the value 1.5; the ratio of proper to improper time is then 2/3 and X reads 12:10 at event 2. Ground observers monitoring X report that it has lost 5 minutes while moving from A to C. The same must be true of any other train clock. Ground observers conclude, therefore, that *train clocks run slow.*

A logical problem arises at this point. If train clocks run slow according to ground observers, it would appear to follow that according to train observers, ground clocks run fast. Such a conclusion would run counter to the first postulate, however, since it would establish an asymmetry between the two frames. Why should train clocks run slow and ground clocks run fast rather than vice versa? Each set of observers sees the other set's clocks as moving, and they ought to reach identical conclusions concerning the behavior of each other's clocks. According to the first postulate, train observers must find that ground clocks run *slow*, not fast.

How can that be? And why can train observers not conclude from the observations of figure 3.11 that ground clocks run fast, compared to train clocks? The answer is that two *different* ground clocks appear in figure 3.11: clock A in event 1, and clock C in event 2. Since, as we have already learned, ground clocks are out of synchronization according to train observers, comparing the reading of one ground clock at one time with that of a *different* ground clock at another time yields no information about the rate at which either clock is running.

Only by examining the *same* ground clock on two occasions can train observers learn anything about the rate at which ground clocks run. Another observation, in addition to the ones shown in figure 3.11, is required. Train observers could, for example, examine the reading of clock C in figure 3.11a. That clock face has been left blank in the figure; we do not know what it reads, but from what we have learned about synchronization, we can be sure that it reads *more* than 12:00. If train observers examine the same ground clock on two occasions, they indeed find that it has lost time and conclude that ground clocks run slow, in accordance with the first postulate.

Still, a skeptic will argue, this is an absurd result. How can each of two sets of clocks be running slow with respect to the other? If the readings of a single train clock (say, X) and a single ground clock (say, A) were compared on two distinct occasions, there would be only three possible outcomes:

(i) X has lost time relative to A;
(ii) X has gained time relative to A; or
(iii) both clocks show the same elapsed time.

All observers would have to agree as to the outcome. In case (i), they would conclude that X is slow (or A is fast). In case (ii), X is fast (or A is slow). In case (iii), the clocks are going at the same rate. But *this experiment cannot be carried out.* Since X and A are moving uniformly relative to one another, they can be together at most at one instant, after which they separate forever. Hence their readings can be directly compared only once. Ground observers monitoring X must compare its readings with those of two *different* ground clocks; similarly, train observers must compare the readings of A with those of two different train clocks. The conclusions of the two sets of observers are therefore based on distinct sets of measurements. In each case, a single moving clock is compared with two different stationary clocks, and in each case the moving clock indicates the passage of a shorter time interval. Hence there is no logical contradiction.

The statement "Moving clocks run slow" means precisely that.

> If a clock is monitored by observers in a frame of reference in which the clock is moving, the clock is found to run slow by comparison with clocks that are stationary in that frame.

Astute train observers can explain why ground observers "mistakenly" think that train clocks run slow. They reason as follows: "Ground observers monitor our clock by comparing its reading on two occasions with those of their own clocks. But their comparison clocks are out of synchronization: the second is ahead of the first. That is why they erroneously conclude that our clocks run slow, whereas in fact theirs are the ones running slow." Ground observers can, of course, pursue an identical line of reasoning and reach the same conclusion. The resolution of the paradox is clearly tied to the relativity of simultaneity.

The question inevitably arises, Do moving clocks *really* run slow, or do they only appear to run slow? The question is essentially a metaphysical one: what does "really" mean in this context? If the thought experiments here described could actually be performed, they would turn out as predicted by the relativity postulates. (See the following section for a real experiment that supports the hypothesis.) The slowness of a moving clock

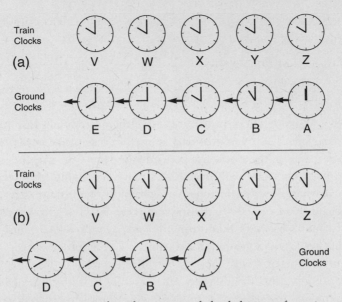

Fig. 3.12. Complete description of the behavior of moving clocks. (a) Train and ground clocks are viewed by train observers at 10:00, train time. Ground clocks are out of synchronization. (b) Clocks are again viewed by train observers at 11:00, train time. Ground clocks have moved, and each has lost 20 minutes compared to the stationary train clocks.

is as "real" as any physical effect that can be measured. The events of figure 3.10 are not an optical illusion. They could be photographed, if cameras with fast enough shutters were available; the pictures would show a clock that is clearly running slow.

The results described here concerning moving clocks complement those of section 3.4. There we examined two different moving clocks at the same time and found that they are out of synchronization; here we have examined a single moving clock at two different times and found that it is running slow. Combining these results, we obtain a complete picture of the behavior of moving clocks.

Figure 3.12a shows ground and train clocks as seen in the train frame at 10:00, train time; all ground clocks are out of synchronization. (This is merely a repetition of fig. 3.7a.) Figure 3.12b is another view seen by train observers some time later, at 11:00 train time; the ground clocks are again out of synchronization. When the two views are compared, it is apparent that each ground clock has lost 20 minutes during the intervening interval. The two sketches contain all relevant information concerning the behavior of moving clocks.

3.6. The Decay of Muons

The experiment described in section 3.5, from which we deduced the time dilation effect, is only a thought experiment. Although it demonstrates without doubt that time dilation is a logical consequence of Einstein's postulates, it cannot prove the reality of the effect because the experiment has never been carried out.

Real experiments have been performed, however, which provide striking confirmation of time dilation. In this section, I describe an experiment based on the decay of muons—unstable particles that were first discovered among the cosmic rays that continuously bombard the earth.[17]

Muons decay according to the scheme

$$\text{muon} \rightarrow \text{electron} + \text{neutrino} + \text{antineutrino}$$

The details of the decay process are irrelevant; the only feature we need be concerned with is that muon decay, like any radioactive decay, is a probabilistic process characterized by a half-life, T. Out of any group of identically prepared muons, approximately half will have decayed within a time interval T. After another interval T has passed, half the survivors will have decayed and only a quarter of the original number remain, and so on. The half-life of muons at rest is about 1.5 microseconds (1 μsec = 10^{-6} sec).

The question at issue concerns the half-life of muons in motion. According to Galilean relativity, the motion should have no effect on the probability of decay; moving muons should have the same half-life as muons at rest.

Special relativity predicts a quite different outcome. Consider a beam of muons, all moving at the same speed v. The first postulate implies that in the muons' rest frame their half-life must be 1.5 microseconds. That is, after 1.5 microseconds have elapsed *according to clocks that move with the muons*, half of them will have decayed. The 1.5 microseconds is a proper time interval.

For earth observers, the corresponding time interval is improper. The time interval during which half the muons decay, as measured by earth

17. Muons are created in the decay of another unstable particle, the pion. Cosmic ray pions are produced in collisions between energetic protons or heavy nuclei and atoms in the atmosphere.

clocks, is therefore $\gamma(v)$ times 1.5 microseconds, where $\gamma(v)$ is the time dilation factor that corresponds to the speed v. Letting T_0 denote the rest half-life and $T(v)$ the half-life for muons moving at velocity v, we conclude that according to special relativity,

$$T(v) = \gamma(v) T_0 \tag{3.7}$$

The faster the muons move, the longer they should survive according to clocks at rest in the laboratory.[18] We may regard the group of muons as a specialized clock that "ticks" once every 1.5 microseconds in its own rest frame; at each tick, half the muons decay. According to observers in the laboratory, for whom that clock is in motion, it (like any other moving clock) runs slow: it "reads" 1.5 microseconds when the true elapsed time, measured by clocks at rest in the laboratory, is longer by the factor $\gamma(v)$. This prediction is subject to direct experimental test.

The first experiment was carried out in 1940 by Bruno Rossi and D. B. Hall, who used the cosmic ray "beam" that was then beginning to be studied and was known to contain many muons moving at speeds very close to c. If the half-life of those muons were equal to T_0, the beam should advance a distance cT_0, some 450 meters, by the time half the muons had decayed. According to special relativity, with the half-life given by equation (3.7), the distance should be greater.

Rossi and Hall measured the attenuation of the cosmic ray muon beam as it proceeds down through the atmosphere; the attenuation is caused primarily by the decay of muons en route. They designed an array of Geiger counters that would register a count whenever a muon passed vertically through it and not when any other type of particle passed through. They took their equipment to several stations in Colorado, at different elevations. At each elevation, they measured the average number of counts per second. Figure 3.13 is a schematic view of the experiment, showing one trial with the detector at Echo Lake (elev. 3,200 m) and another at Denver (elev. 1,600 m).

In addition to decays, another effect depletes the muon beam as it passes through the atmosphere: some of the muons collide with oxygen or nitrogen atoms in the atmosphere and are absorbed. The experimenters corrected for this effect by placing a layer of iron above the detector at the higher elevation. Since iron is much denser than air, about 20 centimeters of iron absorbs as many muons as does all the atmosphere between the

18. This is essentially the spaceship experiment discussed on p. 72. The muons are the space travelers; in the earth frame they live much longer than they do in their own rest frame.

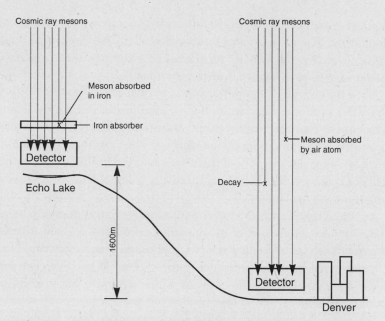

Fig. 3.13. Experiment carried out by Rossi and Hall to measure the lifetime of rapidly moving mu mesons. Identical detectors, sensitive to the passage of mu mesons, are exposed at Echo Lake (elev. 3,200 m) and at Denver (elev. 1,600 m). An iron absorber is above the detector at the higher elevation. The difference in counting rates measures the number of mesons that have decayed en route.

two elevations. Any difference in the measured counting rates at the two stations could therefore be attributed to the decay of muons in the intervening region.[19]

A muon traveling at nearly the speed of light requires 5.33 microseconds, more than three times T_0, to traverse the 1,600 m of atmosphere between the elevations of Echo Lake and Denver. According to Galilean relativity, then, practically all the muons should decay en route: the counting rate at Denver should have been reduced by a factor of about 12 compared to that at Echo Lake. Special relativity predicts that the fraction of muons surviving the trip should be considerably greater, because in the muon frame the elapsed time is less. The precise value of the enhancement factor depends on the speed of the muons.

Cosmic ray muons have a spectrum of velocities; hence the muons in Rossi and Hall's experiment did not have a unique value of $\gamma(v)$. The

19. A few muons are added to the beam by the decay of cosmic ray pions, but their number is negligibly small.

counters were designed to respond only to muons with speeds greater than $0.995c$. The value of γ for this speed is about 10. Hence according to special relativity, the half-life for the muons detected in the experiment should have been at least 10 times longer than T_0.

The result of the experiment was just as predicted by special relativity. The ratio of the counting rates at the two locations was 0.88: only about 12 percent of the muons decayed in traversing 1,600 meters of atmosphere. According to the muon clocks, the trip had lasted not three half-lives but much less than one. Because of the velocity spread in the beam, Rossi and Hall could not test the theory quantitatively; their results were, however, consistent with equation (3.7).[20]

In 1963, David Frisch and James Smith carried out an improved version of the experiment. In their setup, the array of counters was sensitive only to muons with velocities in a very narrow range. Hence they could make a definite prediction of the ratio of counting rates at two different elevations. Their results were in excellent agreement with the prediction of special relativity.

The muon lifetime experiments provide strong evidence for the reality of time dilation. No other plausible explanation has been suggested for why so many of the muons should survive a trip that lasts much longer than their rest half-life, that is, for why the half-life of muons in motion is longer than for muons at rest. Special relativity accounts for the results in a straightforward manner.[21]

In 1977, experiments carried out at the European Center for Nuclear Research (CERN) measured the lifetime of rapidly moving muons directly and confirmed the correctness of the relativistic prediction.

Experiments with real clocks have more recently been carried out. Atomic clocks were carried on jet planes circling the earth in opposite directions. Their readings at the conclusion of the trip were compared with each other and with those of clocks that had remained on the ground. The results were in conformity with the time dilation effect. (An effect due to general relativity must also be taken into account in interpreting the data.) The effect is very small because γ differs from unity by only about one

20. A decline in the counting rate of 12 percent in 5.33 μsec corresponds to a half-life of 29 μsec, about 20 times T_0. If all the muons had the same speed, the value of γ would have had to be about 20. Since the minimum value of γ accepted by the counters was 10, the result was quite reasonable.

21. According to Miller (*Albert Einstein's Special Theory of Relativity*, 266), the lifetime experiments prove only the self-consistency of special relativity, because the data analysis itself depends on relativity.

part in 10^{12}. The experiments will be described in chapter 6 because they are closely related to the twin paradox discussed there.

3.7. LENGTH CONTRACTION

One important relativistic effect remains to be derived: the change in length of an object that is moving relative to the observer. This effect, like time dilation, can be linked to the relativity of simultaneity.

To measure the length of an object, one marks the positions of its end points on some scale and subtracts one reading from the other. If the object is at rest, those measurements can be carried out at leisure. If the object is moving, however, the two position measurements must be carried out simultaneously; otherwise, the result will surely be in error.

As we have seen, simultaneity depends on the motion of the observer. If ground observers measure the position of the front and rear ends of a moving train simultaneously according to their clocks, these measurements take place at *different* times according to train clocks. According to train observers, therefore, the length measured by ground observers is incorrect.

The easiest way to determine the magnitude of the effect is to analyze a somewhat less direct, though equally legitimate, method of determining the length of an object moving at a known (uniform) speed, namely, by measuring the time required for the object to pass a stationary observer. The product of that time and the speed of the object is its length.

Figure 3.14a illustrates how ground observers can measure the length of a moving train. At event 1, the front of the train passes ground observer O; at event 2, the rear of the train passes O. After measuring the times t_1 and t_2, ground observers conclude that the length of the train is

$$L = V(t_2 - t_1) \tag{3.8}$$

where V is its speed.

Figure 3.14b shows the same events as they appear to (primed) train observers. For them the train is stationary and O is moving to the left at speed V; at t_1' O passes the front of the train and at t_2' he passes the rear. The time between those events is

$$t_2' - t_1' = L_0/V \tag{3.9}$$

where L_0 is the length of the train according to train observers, which they may measure by conventional methods since the train is at rest in their frame.

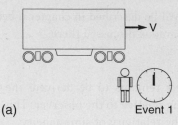

(a)

Event 1

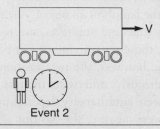

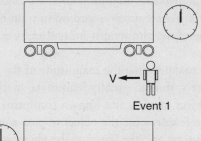

Event 2

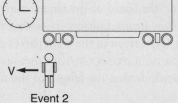

(b)

Event 1

Event 2

Fig. 3.14. Determination of the length of a moving object by measuring the time required for it to pass a stationary observer. (a) At event 1 the front of a moving train passes observer O; at event 2 the rear of the train passes O. O concludes that the length of the train is V times the difference between the two clock readings. (b) The same two events as they appear in the train frame. The time interval measured by the single ground clock in (a) is proper; that measured by the two train clocks in (b) is improper.

The time interval $t_2 - t_1$ measured by earth clocks is proper, whereas the corresponding interval $t'_2 - t'_1$ is improper. A standard time dilation argument shows that

$$t'_2 - t'_1 = \gamma(t_2 - t_1) \tag{3.10}$$

Combining equations (3.8–10), we find that

$$L = L_0/\gamma \tag{3.11}$$

In the ground frame the train is *shorter*, by the factor γ, than it is in its own rest frame. The effect is completely symmetrical: if train observers measure the length of the platform by an analogous experiment, they find it to be contracted. The length they measure is less than the value obtained by ground observers, for whom the platform is at rest.

The length of an object measured in its own rest frame is called its *proper length* and is generally designated with a subscript zero. It is the largest possible result of a length measurement; observers for whom the object is in motion measure a length contracted by the relativistic factor γ.

The contraction applies to all length measurements. In particular, a meterstick at rest in the station is less than 1 meter long when measured by train observers, and a meterstick at rest on the train is less than 1 meter long when measured by ground observers. This assertion at first appears to be self-contradictory, just like the similar statement about moving clocks running slow. If one were to measure the length of an object with a "meterstick" that is only half a meter long, one would expect to obtain twice the true length. Why then do not ground observers, with their contracted metersticks, obtain a result greater than the "true" one when they measure the length of the train, or of the train meterstick for that matter?

The resolution of this paradox relies once again on a simultaneity argument. Figure 3.15 indicates the precise sequence of measurements by which ground observers determine the length of a train meterstick. In (a) those measurements are shown as they appear in the ground frame. Ground observers scratch marks on their metersticks at the position of the right end of the train meterstick (event 1) and of the left end (event 2); those events are simultaneous in the ground frame, and the train meterstick is found to be less than 1 meter long.

Figures 3.15b and 3.15c show how the same events look to train observers. At event 1 they see a ground observer scratch a mark on his meterstick (fig. 3.15b), corresponding to the position of the right end of the

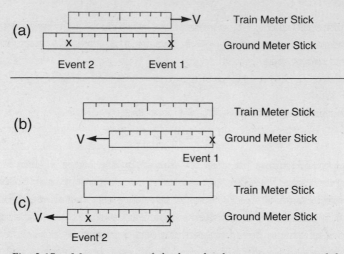

Fig. 3.15.   Measurement of the length of a moving meterstick by ground observers. In (a), ground observers mark the position of the front end of the train meterstick (event 1) and of the rear end (event 2) on their own meterstick. The two events are simultaneous in the ground frame. The distance between the marks is 0.8 m. Sketches (b) and (c) show the two events as seen by train observers. In (b), a ground observer marks the position of the front end of the train meterstick on a (short) ground meterstick. (c) *Some time later,* a different ground observer marks the position of the other end of the train meterstick. The separation between the two marks is less than the "true" length of the stick.

train meterstick at that time. The ground meterstick looks short in this picture.

Some time later, train observers see a different ground observer scratch another mark on the ground meterstick, corresponding to the position of the left end of the train meterstick (event 2, fig. 3.15c). Because the ground meterstick has moved between the two events, the distance between the two marks is less than the true length of the train meterstick. Thus train observers understand how ground observers manage to obtain too *small* a value for the length of the train meterstick, in spite of the fact that their meterstick is less than a meter long.

To measure the length of the ground meterstick, train observers must scratch a mark on their own meterstick at event 1 and another mark at the position of the other end of the ground meterstick, simultaneously with event 1. That is *not* event 2, which is simultaneous with event 1 in the ground frame. Just as in the case of moving clocks, we reconcile the·

apparently contradictory conclusions that each meterstick looks short to observers in the other frame by emphasizing that the conclusions follow from analysis of *different* pairs of events. Each pair of measurements is simultaneous in one frame but not in the other. Hence there is no logical contradiction.[22]

As the preceding discussion illustrates, length contraction is a logical consequence of the relativity of simultaneity. In fact, the contraction formula (3.11) can be deduced purely from such considerations. (See problem 3.6.)

Is a moving rod "really" contracted, or does it only appear to be contracted? A similar question was asked in connection with time dilation in section 3.5; the answer here is the same. The rod is not contracted in an absolute sense: in its own rest frame, it retains its proper length. But as far as observers in another frame are concerned, the contraction is as real as any physical effect that can be measured.

I noted in chapter 2 that before Einstein put forward his relativity theory, Lorentz and FitzGerald had proposed a contraction hypothesis to explain the outcome of the Michelson-Morley experiment. The relativistic effect is still sometimes referred to as the FitzGerald-Lorentz contraction, even though Lorentz's explanation is not correct.

In Lorentz's theory, a body is contracted when it moves relative to the ether. That contraction is absolute and is in principle measurable in any frame, including the rest frame of the body (although observers in the rest frame would find it difficult to measure because all their instruments would be similarly contracted). In Lorentz's explanation of the Michelson-Morley experiment, the absence of any fringe shift results from the cancellation of two effects: the contraction of one of the arms of the interferometer and the different speed of light along the two arms.

According to Einstein, a contraction is measurable only in a frame in which the body is moving; in the rest frame of the body its length is the proper length. The correct relativistic explanation of Michelson's result has nothing to do with contraction. The arms of the interferometer have equal lengths no matter how they are oriented, and the speed of light along the two paths is always c. Hence the travel times are always equal.

According to Lorentz, the contraction has a physical origin; it is caused by changes in the intermolecular forces in moving matter. (A discussion of this mechanism is given in sec. 4.8.) For Einstein, the contraction is a

---

22. The problem of mutual contraction is the basis for the pole and barn paradox, discussed in detail in chapter 6.

consequence of the properties of space and time. According to special relativity, the distance between two bodies is shortened when it is measured by observers for whom the bodies are in motion, even if no matter occupies the space between them. The Lorentz model cannot account for this contraction of "empty space."

As with all the relativistic effects we have discussed, length contraction is negligible unless the relative velocity between observers and the object being measured is close to the speed of light. For that reason, no experiment has directly demonstrated the effect. However, the muon lifetime experiment, described in the preceding section and cited as experimental confirmation of time dilation, can be interpreted equally well as an effect of length contraction.

Consider the events marked 1 and 2 in figure 3.12. In the ground frame, they may be described as follows:

1.   Muons pass elevation of Echo Lake at speed V;
2.   Muons pass Denver at the same speed.

Since Echo Lake and Denver are both at rest relative to the laboratory, the difference between their elevations, 1,600 meters, is a proper length. In the muon rest frame, in contrast, Echo Lake and Denver are both moving upward at speed V. The same two events in this frame may be described as follows:

1.   Echo Lake passes stationary muons at speed V;
2.   Denver passes stationary muons at speed V.

The vertical distance between Denver and Echo Lake, as measured in the muon rest frame, is an improper length. It is therefore Lorentz contracted to the value $1,600/\gamma$ m and the time interval between events 1 and 2, measured in the muon rest frame, is $1600/\gamma V$. This is the same conclusion we reached earlier on the basis of a time dilation argument. The analysis illustrates the intimate relation between time dilation and length contraction: neither effect could logically exist without the other, and either can be used to derive the other.[23]

3.8. TRANSVERSE LENGTHS

In the preceding section we compared the results of "longitudinal" length measurements. To complete the picture, we must investigate the measure-

---

23. Joseph Larmor, in his book *Aether and Matter*, pointed out that the Lorentz-FitzGerald contraction implies a change in the rate of moving clocks. Larmor's book was written in 1900, five years before Einstein published his relativity paper.

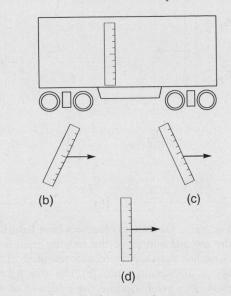

Fig. 3.16. (a) Vertical meterstick on a train.
(b) and (c) show two possible appearances the
meterstick might have when viewed by ground
observers for whom it is in motion. Since there
is no reason to prefer either of these to the
other, neither can be right. We conclude that
the moving meterstick must look vertical, as
in (d).

ment of lengths that are perpendicular to the direction of relative motion
of the observers. The outcome in this case is that there is no contraction:
observers in both frames of reference obtain the same results.

Figure 3.16 shows a vertical meterstick on a moving train. Suppose
ground observers wish to measure the length of the moving meterstick.
It is not entirely obvious that the meterstick appears vertical in the ground
frame, but we can demonstrate this with the help of a symmetry argument.

The only unique direction in the problem is defined by the motion of
the train. The meterstick is symmetric with respect to that direction: if
figure 3.16a were turned upside down, the picture of the meterstick would
be unchanged. The corresponding ground frame picture must therefore
also look the same when turned upside down (technically, when reflected
in the plane of the ground).

If the meterstick appeared inclined to the vertical to ground frame observers, as in figure 3.16b, the reflected picture would show the opposite

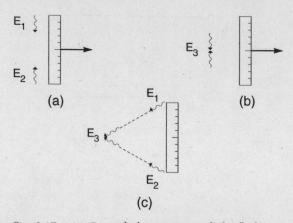

Fig. 3.17. (a) Ground observers emit light flashes as the top and bottom of the moving train meterstick cross their stationary meterstick (events 1, 2). The two events are simultaneous. (b) The two light flashes meet, at a point opposite the midpoint of the stick (event 3). (c) The same events as they look in the train frame. Train observers see the light flashes follow the indicated paths. Since the paths are of equal length, train observers agree that events 1 and 2 were simultaneous.

inclination (fig. 3.16c). But there is nothing in the problem to distinguish top from bottom, no more reason for 3.16b than for 3.16c to be the correct reflected picture. Hence neither one can be right. The only acceptable picture is the symmetric one that favors neither top nor bottom, namely, 3.16d. The moving meterstick must appear vertical.

Ground observers can measure the length of the moving meterstick by an experiment similar to that shown in figure 3.15. They agree to scratch a mark on their own vertical meterstick where the top of the train meterstick crosses it and another mark where the bottom crosses. The distance between the two scratches is the length of the moving meterstick.

Since the train meterstick is vertical, its ends cross the ground meterstick at the same time: the two scratches are made simultaneously in the ground frame. In this case, unlike that of the horizontal meterstick, train observers agree that the two scratch marks are simultaneous. To verify that, let the ground observers emit light flashes aimed directly at each other (fig. 3.17a) as they mark the passage of the top of the meterstick (event 1) and of the bottom (event 2). The two flashes meet at a point opposite the midpoint of the meterstick, which has by then moved

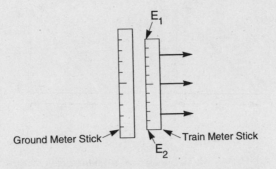

Fig. 3.18. Hypothetical result of the measurement of a vertical train meterstick. If this were the picture seen by ground observers, they would conclude that the train meterstick is contracted. But since train observers agree that events 1 and 2 are simultaneous (see fig. 3.17), they would have to conclude that the ground meterstick is stretched. Such a result would violate the first postulate and cannot occur.

along (fig. 3.17b). This is the same procedure used in figure 3.15 for longitudinal length measurements.

Observers on the train see the light flashes move along the inclined paths shown in figure 3.17c. It is apparent that the two paths are of equal length. Consequently, train observers conclude that the emissions of the two light flashes (as well as the measurements signaled by those emissions) were simultaneous events. This conclusion is in marked contrast with that of figure 3.15, in which train observers did *not* agree that the two position measurements of the ground observers were simultaneous. This accounts for the different conclusion. Figure 3.17 should be carefully compared with figures 3.5 and 3.6 and the accompanying analysis.

To train observers, the ground meterstick is in motion, but for the reasons already cited it appears vertical. Ground and train observers therefore agree that at the time of events 1 and 2 (which are simultaneous in both frames), the two metersticks are aligned with one another. Their views of the two metersticks at that moment must be identical and enable either set of observers to measure the length of the others' meterstick.

Suppose (incorrectly) that ground observers found the train meterstick to be contracted. Figure 3.18 shows how the two metersticks would appear at the time of events 1 and 2. Since this picture is valid in *both* frames, train observers would see the two scratches made by the ground observers

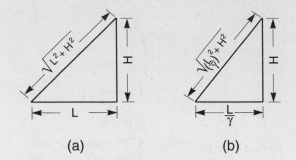

(a)   (b)

Fig. 3.19.   Measurement of a triangularly shaped object, at rest on a moving train. (a) Train observers measure the horizontal base, L, and the vertical height, H. (b) Ground observers find the base to be contracted to a length L/γ, but the height is unchanged. The hypotenuse is contracted, but by a factor less than γ.

to mark their meterstick and would conclude that the ground meterstick is longer than their own. Ground observers would conclude that the train meterstick is *shorter* than theirs. Those conclusions would violate the first postulate by creating an asymmetry between the two frames. Why should ground observers find the train meterstick to be contracted and train observers find the ground meterstick to be expanded, rather than vice versa? The only outcome that avoids this logical dilemma is that both sets of observers agree that the two metersticks are of equal length.

The all-important difference between the present situation and the one analyzed in the preceding section is the agreement over the simultaneity of the two position measurements. In measuring the length of horizontal metersticks, the two sets of observers disagree over simultaneity; they therefore employ two *different* pairs of events to measure the length of each other's meterstick. Hence it is logically acceptable that each should find the other's meterstick to be contracted. In the present case, the *same* pair of measurements serves for both sets of observers and a similar conclusion would be unacceptable.

The results of this section can be combined with those of the preceding one to yield a description of length measurements for objects with an arbitrary orientation. Consider the triangular object shown in figure 3.19, carried on our moving train. The horizontal base has length L and the vertical height is H. Both those quantities are proper lengths, since the object is at rest in the train frame. Ground observers who measure the length of the base find the contracted length L/γ. When they measure the height, they obtain the same value H as do train observers. The length of the hypotenuse, which is according to train observers, is therefore found by ground observers (fig. 3.19b) to be

$$\sqrt{(L/\gamma)^2 + H^2}$$

An obliquely oriented moving object is found to be contracted but not by the full factor γ. The shape of the object is different in the moving frame than in its rest frame.

3.9. Summary

In this chapter we have demonstrated, with the help of a few simple thought experiments, how Einstein's two postulates lead inescapably to several important and unexpected conclusions concerning time and space measurements. The principal results are as follows:

1. The concept of simultaneity, when applied to spatially separated events, is not absolute: two events that occur at the same time but at different places according to one set of observers occur at different times when measured by another set of observers moving relative to the first.

2. A set of clocks, all at rest with respect to one another, may be synchronized by exchanging light signals. When viewed by observers for whom they are in motion, the clocks are found to be out of synchronization: the one ahead in the direction of motion is found to be behind in its reading.

3. If two events occur at the same place in some frame of reference, the time interval between them measured by observers in that frame (proper time interval) is shorter than the interval between the same events measured by observers in any other frame.

Corollary: If a clock is examined by observers for whom the clock is moving, it is observed to be running slow. That conclusion is reached by comparing the reading of the moving clock on two occasions with those of (distinct) stationary clocks.

4. The length of an object is greatest when measured by observers in its rest frame (proper length). If a meterstick is measured by observers moving along the direction of the meterstick, it is found to be contracted.

5. If a meterstick is measured by observers moving perpendicularly to the direction of the meterstick, its length is found to be the same as the proper length.

PROBLEMS

3.1. Analyze the simultaneity experiment of section 3.3 from the point of view of Galilean relativity. Assume that the train is at rest in the ether frame and the ground is moving at speed V relative to the ether; figure 3.1, in which both light pulses travel at speed c, then correctly describes the experiment in the train frame. Using the Galilean velocity transformation, show that the light pulses arrive at the front and rear platforms simultaneously in the ground frame as well as in the train frame.

3.2. Refer to figure 3.3. The length of the railroad car (as measured in the ground frame) is 30 m and its speed is 100 m/sec. Find the difference between the arrival times of the two light pulses as measured in the ground frame.

3.3. For this problem, assume that Galilean relativity is valid. The objective is to synchronize clocks by a procedure similar to the one described in section 3.4. Suppose clocks A and B are both at rest in a frame S, which moves at velocity V in the x direction relative to the ether. A is located at the point $x = 0$, B at $x = L$. When A's clock reads t_0 a light pulse is emitted toward B; B's clock reads t_1 when the light pulse arrives, and a return pulse is immediately emitted toward A. A's clock reads t_2 when the return pulse arrives.
   (a) What is the value of t_1 if the clocks are synchronized?
   (b) What is the value of t_2?
   (c) Show that if the relation

$$t_1 = \frac{1}{2}\left[t_0\left(1 - \frac{V}{c}\right) + t_2\left(1 + \frac{V}{c}\right)\right]$$

is satisfied, the clocks are synchronized.

3.4. Clocks A, B, and C are all at rest relative to one another. Prove that if A is synchronized with B and B is synchronized with C by the method described in section 3.4, then A is synchronized with C. For simplicity, assume that only the x coordinates of the clocks differ.

3.5. The bright star Sirius is 8 light years from earth. (Assume that earth and Sirius are at rest relative to one another.)
   (a) How fast must a spaceship travel in order to reach Sirius in 6 years according to clocks on the spaceship? Express your answer as a fraction of the speed of light.
   (b) How long does the trip take according to earth frame clocks?
   (c) What is the distance between earth and Sirius according to observers in the spaceship frame?

3.6. For this problem, pretend that you do not know about length contraction but you do know about time dilation.
   A spaceship whose length in its own rest frame is L_0 moves at velocity V relative to earth. Let L be the length of the spaceship as measured in the earth's rest frame, S.

(a) A light pulse emitted at the rear of the spaceship (event E_1) arrives at the front (event E_2). In the spaceship frame, S', the time interval between E_1 and E_2 is $t' - t_1' = L_0/c$. Find the time interval between the same two events in frame S, in terms of L, V, and c. (See fig. 2.3a. Note that this time interval is not proper in either frame.)

(b) The light pulse is reflected and arrives at the rear of the spaceship (event E_3). Find the time interval between E_2 and E_3 in frame S.

(c) Applying a proper time argument to the interval between E_1 and E_3, show that L and L_0 are related by the length contraction formula: $L = L_0\sqrt{1 - V^2/c^2}$.

3.7. Devise a thought experiment, analogous to that of problem 3.6, by means of which the time dilation effect may be inferred from length contraction. Each effect implies the other.

3.8. A steel wire connects two trains at rest on the same track. The wire will snap if it is stretched by as much as 1 percent. The trains begin to accelerate in such a way that their velocities, as measured in the ground frame, are always equal. Eventually, the wire snaps. Explain carefully why this happens. How fast are the trains moving when the wire snaps?

3.9. Spaceships A and B, each of proper length 120 m, pass each other moving in opposite directions. According to clocks on ship A, the front end of B takes 2×10^{-6} sec to pass A, that is, to move from the front end of A to the rear end of A.

(a) What is the relative velocity of the spaceships?

(b) According to clocks on ship B, how long does the front end of A take to pass B?

(c) According to clocks on ship B, what is the interval between the time when the front end of A passes the front end of B and the time when the rear end of A passes the front end of B? Why is this answer different from the answer to (b)?

3.10. A spaceship passes earth at speed $0.6c$ (event E_1). Clocks on earth and on the spaceship read zero. Five minutes later, according to earth clocks, a light pulse is emitted in the direction of the spaceship (event E_2). At some later time the light pulse catches up to the spaceship (event E_3). Let S denote the earth frame and S' the spaceship frame.

(a) Draw sketches showing the sequence of events in each frame. Each sketch should show the position of earth, spaceship, and light pulse at a given time in the appropriate frame.

(b) Is the interval between E_1 and E_2 a proper time interval in either frame? If so, which one? What about the interval between E_2 and E_3? What about the interval between E_1 and E_3?

(c) What is the time of E_2 according to S' clocks?

(d) According to S' observers, how far away is earth when the light pulse is emitted? Express your answer in light-minutes.

(e) From your answers to (c) and (d), find the reading of the spaceship clock when the light pulse arrives.

(f) By analyzing the problem entirely in frame S, find the time of E_3 according to S frame clocks. Verify that your answers to (e) and (f) are consistent with your assertion concerning proper time in part (b).

3.11. Spaceship A passes earth at speed V when its clock (as well as the adjacent earth clock) reads zero (event E_1). When the earth clock reads T, spaceship B passes earth, moving at speed U in the same direction as the first (event E_2). Assume that $U > V$; let S be the earth frame and S' be the rest frame of ship A. Eventually, ship B catches up to A (event E_3). Note the similarity between this problem and problem 3.10.

    (a) Find the time of E_3 in frame S.
    (b) Find the time of E_2 in frame S'.
    (c) Find the time of E_3 in frame S'.
    (d) According to S' observers, how far away was earth at E_2?
    (e) From these results, find the velocity of ship B as measured by observers on A. This is the relativistic velocity transformation law, which will be derived from the Lorentz transformation in the next chapter (eq. [4.15]).

3.12. Positive pions at rest decay with a half-life of 1.8×10^{-8} sec. An accelerator produces pulses of pions; each pulse consists of 2×10^{11} pions moving at very nearly the speed of light ($\gamma = 20$). How close to the exit port of the accelerator must an experiment be located if at least 5×10^{10} pions per pulse are required?

4   The Lorentz Transformation

4.1. INTRODUCTION

The fundamental problem of relativity is the transformation of coordinates: if x, y, z, and t denote the coordinates of an event measured in some frame S, what are the coordinates x', y', z', and t' of the same event measured in another frame, S', that moves at velocity V relative to S?

As we have already noted, the classical solution to the problem, the Galilean transformation (eq. [1.1]), is inconsistent with Einstein's postulates and must be rejected. Our task here is to derive its replacement, the Lorentz transformation.

The new transformation law must satisfy several requirements:

(i)   The transformation should be *linear*. That is, x' must be expressible in the form

$$x' = ax + by + cz + dt$$

in which the coefficients a, b, c, and d depend only on V. Similar expressions hold for y', z', and t'.

The linearity requirement is imposed by the homogeneity of space. If the equations were not linear, the transformation would depend on the choice of origin, which is unacceptable in homogeneous space.

(ii)   The transformation must be *symmetrical* between the two frames: the same rules that specify the transformation from S to S' must apply also to the inverse transformation from S' to S. In other words, it should not matter which frame we choose to call S and which S'; if we interchange the labels (and change the sign of V), the equations must remain valid.

Properties (i) and (ii) are shared by the Galilean transformation.

(iii)   The velocity of light must have the same value, c, in both frames for any direction of propagation. This is just the statement of Einstein's second postulate.

(iv)   The transformation law should approach the Galilean form in the nonrelativistic limit. (That limit must be defined with some care.)

Requirements (i) through (iii) uniquely determine the form of the transformation.

4.2. THE TRANSFORMATION EQUATIONS

The literature contains many derivations of the Lorentz transformation. I present here a simple derivation that makes use of the results concerning length contraction derived in the preceding chapter.

As in the derivation of the Galilean transformation in chapter 1, we assume that the primed and unprimed axes are parallel to one another and that the relative motion between the two frames is along the x (or x') direction. When the origins pass one another, then, the two sets of axes momentarily coincide. These restrictions simplify the algebra without involving any loss of generality.

In deriving the Galilean transformation, we assumed that all clocks in both frames are set to zero at the instant when the origins O and O' coincide. Such a prescription would be inconsistent with special relativity: observers in each frame can synchronize their own clocks but will not agree that clocks in the other frame are synchronized.

Two clocks that are moving relative to one another can, however, be set to read the same when they are at the same location. We therefore stipulate that when O and O' coincide, the S and S' clocks at their common location are set to zero. In other words, the event whose S coordinates are (0, 0, 0, 0) also has the coordinates (0, 0, 0, 0) in S'. Each set of clocks can then be synchronized using the standard procedure described in section 3.4.

Figure 4.1a shows a rod of proper length L, at rest in S' and aligned along the x' axis. The left end of the rod is located at $x' = 0$ and its right end is at $x' = L$. In frame S the rod is moving to the right with velocity V and is therefore contracted to the length L/γ. At $t = 0$ its left end is at $x = 0$ and its right end is at $x = L/\gamma$ (fig. 4.1b). At some later time, t, the entire rod has moved a distance Vt (fig. 4.1c); the right end is therefore at the position

$$x = Vt + L/\gamma$$

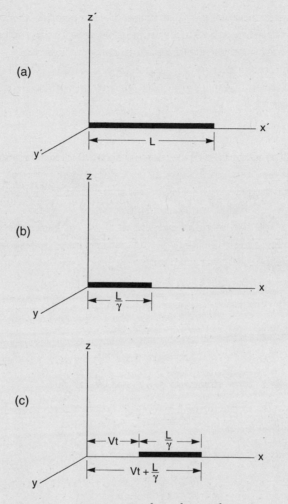

Fig. 4.1. Diagrams used to derive the Lorentz transformation. (a) A rod of proper length L is at rest in frame S'. (b) The same rod as seen at $t = 0$ in frame S; the rod is contracted to length L/γ. (c) The rod as seen at time t in frame S. The entire rod has moved a distance Vt. Hence the left end is at $x = Vt$ and the right end is at $x = Vt + L/\gamma$.

The constant position $x' = L$ in frame S' corresponds to the changing position $x = Vt + L/\gamma$ in S. Since this must be true for any value of L, we can replace L by x' in the above equation and conclude that

$$x = Vt + x'/\gamma$$

or, solving for x',

$$x' = \gamma(x - Vt) \tag{4.1a}$$

Equation (4.1a) is the first of the desired transformation relations.

The same analysis can be applied to a rod at rest in frame S, with its left end at the origin O and its right end at $x = L$. In S', this rod is moving with velocity $-V$ and is contracted by the factor γ. At time $t' = 0$, its left end was at $x' = 0$ and its right end was at $x' = L/\gamma$; hence the position of the right end at time t' is $x' = L/\gamma - Vt'$, which must correspond to $x = L$. L again being arbitrary, we may write

$$x' = \frac{x}{\gamma} - Vt'$$

or, solving for x,

$$x = \gamma(x' + Vt') \tag{4.2a}$$

Eliminating x' from equations (4.1a) and (4.2a) gives

$$t' = \gamma\left(t - \frac{V}{c^2}x\right) \tag{4.1b}$$

while eliminating x from the same equations gives

$$t = \gamma\left(t' + \frac{V}{c^2}x'\right) \tag{4.2b}$$

Equations (4.1b) and (4.2b) replace the simple relation $t' = t$ of Galilean relativity. In special relativity, the time coordinate of an event in one frame depends on its position as well as on the time in the other frame.

The remaining two equations, those involving the y and z coordinates, can be written down directly. As we saw in chapter 3, lengths perpendicular to the direction of relative motion are not contracted. Hence by considering rods oriented along the y or the z directions in either frame, we conclude that

$$y' = y \tag{4.1c}$$

and

$$z' = z \qquad (4.1\text{d})$$

just as in Galilean relativity.

Equations (4.1a–d) constitute the famous Lorentz transformation, on which all the mathematical results of special relativity are based. The inverse transformation is given by equations (4.2a–d). For future reference, the full sets of equations are exhibited together below.

Direct Transformation		Inverse Transformation	
$x' = \gamma(x - Vt)$	(4.1a)	$x = \gamma(x' + Vt')$	(4.2a)
$y' = y$	(4.1c)	$y = y'$	(4.2c)
$z' = z$	(4.1d)	$z = z'$	(4.2d)
$t' = \gamma\!\left(t - \dfrac{V}{c^2}\,x\right)$	(4.1b)	$t = \gamma\!\left(t' + \dfrac{V}{c^2}\,x'\right)$	(4.2b)

Let us check whether these results satisfy the requirements given in the introduction. The equations are indeed linear (requirement [i]) and symmetric (requirement [ii]): if all primed and unprimed quantities are interchanged and V is changed to $-V$, equations (4.1a–d) go over into (4.2a–d), and vice versa.

To verify that requirement (iii) is satisfied, consider a light pulse that leaves the origin of frame S in the positive x direction at $t = 0$. Its position at time t is $x = ct$. We transform this equation to frame S' by substituting for x and t their expressions in terms of primed quantities, given by equations (4.2a,b). The result is

$$\gamma(x' + Vt') = c\gamma(t' + \frac{V}{c^2}\,x')$$

which reduces to

$$x' = ct'$$

The light pulse travels at speed c in S', in accord with the second postulate.

Finally, we investigate whether requirement (iv) is satisfied by examining the transformation equations when V is much smaller than c. An approximation applicable in this limit is based on the fact that the square of a number much smaller than 1 is much smaller than the number itself; higher powers are smaller still. If a quantity can be expressed as a sum of powers of V/c, an approximation that should be good when V/c is very small is to keep the term proportional to V/c and neglect terms proportional to $(V/c)^2$ and all higher powers of V/c. This is called the first-order approximation.

I show in the appendix to this chapter that the relativistic parameter γ, defined by equation (3.5), can be written as the following infinite series:

$$\gamma = 1 + \frac{1}{2}\left(\frac{V}{c}\right)^2 + \frac{3}{8}\left(\frac{V}{c}\right)^4 + \ \ldots \tag{4.3}$$

The expansion (4.3) contains no term proportional to V/c; only even powers of V/c appear. To first order, then, γ is just 1 and the transformation law for x, equation (4.1a), reduces to the Galilean form, equation (1.1a). However, the time transformation, equation (4.1b), becomes

$$t' = t - \frac{V}{c^2} x \tag{4.4}$$

which is not the same as equation (1.1b): it contains the additional term $-Vx/c^2$. The presence of this term is puzzling.

The quantity of interest in most problems is not the time of a single event but the time interval between two events, say, E_1 and E_2. Writing equation (4.4) for each event and subtracting one equation from the other, we get

$$t'_2 - t'_1 = t_2 - t_1 - \frac{V}{c^2}(x_2 - x_1)$$

or

$$\Delta t' = \Delta t - \frac{V}{c^2}\Delta x = \Delta t\left(1 - \frac{V}{c}\frac{\Delta x}{c\Delta t}\right) \tag{4.5}$$

The notation

$$\Delta t = t_2 - t_1 \qquad \Delta x = x_2 - x_1 \tag{4.6}$$

was introduced in chapter 3.

Equation (4.5) reduces to the Galilean relation $\Delta t' = \Delta t$ only if the term $V\Delta x/c^2\Delta t$ on the right side can be neglected. If in addition to $V \ll c$, the condition

$$\Delta x \ll c\Delta t \tag{4.7}$$

is satisfied, the extra term is the product of two very small quantities and is indeed negligible.

If E_1 and E_2 refer to the position of a body on two occasions, $\Delta x/\Delta t$ is the body's average velocity over the time interval Δt; condition (4.7) is then just the requirement that the velocity of the body be much less than *c*, which is part of the standard definition of the nonrelativistic limit. In general, however, (4.7) is an independent requirement.

If two events occur at nearly the same time at widely separated locations, equation (4.7) is not satisfied. Any problem that involves such events must be considered intrinsically relativistic; its description is markedly different even in two frames that move slowly relative to one another.

One other feature of the Lorentz transformation becomes apparent if we treat *ct* instead of *t* as the time variable (*ct* is proportional to the time but has the dimensions of a distance). A trivial rewriting of equations (4.1a–d) gives

$$x' = \gamma \left[x - \frac{V}{c}(ct) \right] \tag{4.8a}$$

$$ct' = \gamma \left[ct - \frac{V}{c}(x) \right] \tag{4.8b}$$

Equations (4.8a,b) are completely symmetrical between the space and "time" variables: if *x* and *ct* are everywhere interchanged, the equations remain the same. This is a manifestation of the striking symmetry between space and time in special relativity. In Galilean relativity, space and time are totally distinct; in special relativity they are intimately related and it is appropriate to speak of a single entity called space-time.

4.3. SOME CONSEQUENCES OF THE TRANSFORMATION EQUATIONS

The Lorentz transformation can be used to derive, in straightforward fashion, all the relativistic effects described in chapter 3: the relativity of simultaneity, time dilation, and length contraction. It is instructive to see how these effects follow from the transformation equations; we shall also be able to extend many of the results obtained in chapter 3.

Consider two arbitrary events, E_1 and E_2. Let (x_1, y_1, z_1, t_1) and (x_1', y_1', z_1', t_1') denote the coordinates of E_1 in frames S and S', respectively, and similarly for the coordinates of E_2. We use the Δ notation (see eq. [4.6]) for the time and space intervals between the two events.

Simultaneity

The general relation between the time intervals Δt and $\Delta t'$ follows directly from equation (4.1b):

$$\Delta t' = \gamma\left(\Delta t - \frac{V}{c^2}\Delta x\right) \tag{4.9}$$

If E_1 and E_2 are simultaneous in frame S, Δt is 0 and equation (4.9) becomes

$$\Delta t' = -\frac{\gamma V}{c^2}\Delta x \tag{4.10}$$

which differs from zero unless Δx is zero as well. Equation (4.10) expresses the relativity of simultaneity: the two events in question are not simultaneous in S' unless they occur in the same place as well as at the same time in S; in that case, as already noted, they can be considered a single event.

Equation (4.10) provides a quantitative measure of the departure from simultaneity in S'. We can, for example, use it to determine how far out of synchronization two clocks appear to be in a frame in which they are moving. (See problem 4.2.)

Time Dilation

To derive the time dilation effect from equation (4.9), we simply put $\Delta x = 0$. Equation (4.9) gives $\Delta t' = \gamma(\Delta t)$. This is just the statement of time dilation. (Δt is a proper time interval since the events happen at the same place in S.) We can go further and relate the time intervals Δt and $\Delta t'$ even when neither one is proper.

EXAMPLE. E_1 and E_2 have the following coordinates in frame S:

$$E_1: x = 1 \text{ m}, \ t = 3 \times 10^{-8} \text{ sec}$$
$$E_2: x = 9 \text{ m}, \ t = 4 \times 10^{-8} \text{ sec}$$

(a) What is the time interval between the two events in a frame S' that travels at $V = 0.6c$ with respect to S?

(b) Does there exist a frame in which E_1 and E_2 are simultaneous? If so, what is its speed relative to S?

SOLUTION. From the given data, $\Delta x = 8$ m, $\Delta t = 1 \times 10^{-8}$ sec.

(a) For $V/c = 0.6$, $\gamma = 1/\sqrt{1 - .36} = 1.25$.
Using equation (4.9), we find

$$\Delta t' = 1.25\left(1 \times 10^{-8} - \frac{0.6(8)}{3 \times 10^8}\right)$$
$$= -7.5 \times 10^{-8} \text{ sec}$$

Since $\Delta t'$ is negative, the two events occur in the opposite order in S' than they do in S.

(b) If the events are simultaneous in S', $\Delta t'$ is 0. Substituting this in equation (4.9), we find

$$\frac{V}{c} = (3 \times 10^8)(1 \times 10^{-8})/8$$
$$= 0.375$$

the desired answer.

Length Contraction

A similar procedure enables us to relate the space intervals Δx and $\Delta x'$ between two arbitrary events. From equation (4.1a) we obtain

$$\Delta x' = \gamma(\Delta x - V\Delta t) \tag{4.11}$$

This general expression for $\Delta x'$ leads directly to the length contraction effect. Suppose an object is at rest in S', and let observers in S measure the position of one end (event E_1) and of the other end (event E_2) simultaneously according to S clocks. For these events, $\Delta t = 0$ and (4.11) gives

$$\Delta x = \Delta x'/\gamma$$

Since $\Delta x'$ is a proper length, this is just the Lorentz-FitzGerald contraction: observers for whom the object is in motion see it contracted by the relativistic factor γ. The result is hardly surprising, inasmuch as we used the length contraction formula to derive the transformation equations. Once again, however, we can go further and relate the two distance intervals Δx and $\Delta x'$ even when neither represents a proper length.

Returning to the previous example, we inquire:

(c) What is the spatial interval between E_1 and E_2 in frame S'?

(d) Is there a frame in which E_1 and E_2 occur at the same place? If so, what is its speed relative to S?

SOLUTION

(c) Substituting the values for Δx and Δt in equation (4.11), we find

$$\Delta x' = 1.25 \left[8 - (1.8 \times 10^8)(1 \times 10^{-8}) \right]$$
$$= 7.75 \text{ m}$$

(d) If the events are to occur at the same place in S', $\Delta x'$ must be zero. Substituting this in equation (4.9), we obtain

$$V = \frac{\Delta x}{\Delta t}$$
$$= 8 \times 10^8 \text{ m/sec}$$

Since the result exceeds the speed of light, it is not physically accept-able. We conclude that there is no inertial frame in which the two events occur at the same place. Given any two events, one can in general find *either* a frame in which they are simultaneous *or* a frame in which they occur at the same place, but not both.[1] This assertion will be proven in chapter 5.

Invariant Interval

In chapter 1, I introduced the concept of an invariant—a quantity that has the same value in all inertial frames. The time interval Δt between two events is invariant in Galilean relativity, since time does not change in a Galilean transformation. In special relativity, Δt is not invariant. There does, however, exist a combination of space and time intervals that is invariant. It is called Δs and is defined by the relation

$$(\Delta s)^2 = c^2 (\Delta t)^2 - (\Delta x)^2 \tag{4.12}$$

Since the right side of (4.12) can be positive or negative, Δs can can be either real or imaginary.

$\Delta s'$ is defined in terms of $\Delta t'$ and $\Delta x'$ by a relation analogous to (4.12). Using equations (4.9) and (4.11), we can express $(\Delta s')^2$ in terms of Δt and Δx. After some algebra, we find that

$$(\Delta s')^2 = (\Delta s)^2$$

which demonstrates that Δs is invariant as claimed.

If there exists a frame in which $\Delta x = 0$, we can calculate Δs in that frame and obtain

$$\Delta s = \pm c \Delta t \qquad (\Delta x = 0)$$

In chapter 3, we defined a proper time interval between two events as the interval measured in a frame in which $\Delta x = 0$. The invariant interval Δs is therefore just c times the proper time interval between the events, whenever the latter is defined.

In three dimensions, the definition (4.12) takes the generalized form

$$(\Delta s)^2 = c^2 (\Delta t)^2 - (\Delta x)^2 - (\Delta y)^2 - (\Delta z)^2 \tag{4.12'}$$

---

1. There is one exception: if E_1 and E_2 represent the positions of a light ray on two occasions, $\Delta x = \pm c \Delta t$ and $\Delta x' = \pm c \Delta t'$. In this case there exists neither a frame in which the events are simultaneous nor one in which they occur at the same place.

The proof of invariance proceeds just as before, since $\Delta y' = \Delta y$ and $\Delta z' = \Delta z$.

The invariant interval will prove useful in the discussion of causality in chapter 5.

4.4. THE TRANSFORMATION OF VELOCITY

I next derive the relativistic transformation law for velocity, which replaces the simple Galilean relation $v' = v - V$. Consider a body that moves with constant speed v in the x direction, as measured in frame S. Its equation of motion is

$$x = vt \tag{4.13}$$

We transform equation (4.13) to S' coordinates using (4.2a) and (4.2b). After rearranging terms, we get

$$x' = \frac{v - V}{1 - \dfrac{vV}{c^2}} t' \tag{4.14}$$

which again describes motion at constant velocity since x' is proportional to t'. The velocity in S' is

$$v' = \frac{x'}{t'} = \frac{v - V}{1 - \dfrac{vV}{c^2}} \tag{4.15}$$

Solving for v in terms of v' gives the inverse transformation

$$v = \frac{v' + V}{1 + \dfrac{v'V}{c^2}} \tag{4.16}$$

which is consistent with the symmetry requirement: if in (4.15) we interchange v and v' and reverse the sign of V, the result is (4.16).

In the nonrelativistic limit, with both v and V much smaller than c, (4.15) reduces to the Galilean form $v' = v - V$, as expected.

Although equation (4.15) was obtained by assuming motion at constant speed, the result is quite general. If the speed of a body changes with time, the instantaneous values of v and v' are related by equation (4.15). (The velocities must be calculated at different times, t and t', which are related by eq. [4.1b].)

Notice that $v' = 0$ implies $v = V$, as in Galilean relativity: an object at rest in frame S' is seen by S observers as moving with the velocity of S'.

When $v = c$, equation (4.15) gives $v' = c$, as required by the second postulate.

Limiting Velocity

An important property of special relativity is that if a body moves at a speed less than c in one frame, its speed is less than c in any other frame as well. Suppose for concreteness that v' and V are both positive. We want to prove that if $v' < c$ and $V < c$, then $v < c$. Define $\beta = v/c$, $\beta' = v'/c$, and $B = V/c$, and put

$$\beta' = 1 - \lambda \qquad B = 1 - \kappa$$

with λ and κ both positive numbers smaller than one. Equation (4.16) then reads

$$\beta = \frac{2 - \lambda - \kappa}{2 - \lambda - \kappa + \lambda\kappa} \tag{4.17}$$

In the fraction on the right side of equation (4.17), both numerator and denominator are positive numbers and the denominator is obviously greater than the numerator. Hence β must be less than one, the result we set out to demonstrate.

This result strongly supports (though it does not prove) the hypothesis that c is a limiting velocity. If a limiting velocity does exist, it must have the same value in all inertial frames; any other outcome would violate the postulate of relativity.

In Galilean relativity, no limiting velocity is possible. Suppose a body moves with speed $0.9c$ as measured in some frame S. In a frame S', which moves at speed $-0.9c$ relative to S, the Galilean speed of the body is $1.8c$. In a frame S'', which moves at $-0.9c$ relative to S', its speed is $2.7c$. It is clear that by passing through a sequence of frames, each of which moves at $-0.9c$ relative to the preceding one, we can find a Galilean frame in which the body moves with arbitrarily high speed.

The situation in special relativity is quite different. With $v = 0.9c$ and $V = -0.9c$, equation (4.15) gives

$$v' = \frac{0.9c - (-0.9c)}{1 - (-0.9)(-0.9)} = \frac{1.8}{1.81}c = 0.994c$$

instead of the Galilean result $v' = 1.8c$.

We have not yet demonstrated that c is a limiting velocity. What we have shown is that such a result is consistent with the first postulate: if all bodies move at speeds less than c in one frame, the same is true in any other frame.[2]

A similar derivation shows that if the speed of a body is greater than c in one inertial frame, it is greater than c in any other. Thus c acts as a kind of velocity barrier. I will show in chapter 7 that any material body requires an infinite amount of energy to cross this barrier. Hence a body can never attain the speed of light starting from a slower speed. But Einstein's statement that "superluminary velocities have no possibility of existence"[3] is not strictly correct.

The possibility that particles with superlight velocity might exist has been studied by Gary Feinberg, who named them tachyons. If tachyons do exist, they have very peculiar properties. They are discussed briefly in chapter 7.

Lighthouse Effect. The notion of a limiting velocity applies only to material bodies and to signals; there is no limit on the speed of purely kinematic effects. Suppose, for example, that a powerful lighthouse beacon shines on a distant cylindrical screen. If the beacon rotates at a steady rate, the spot on the screen moves at a speed that is proportional to the rotational velocity and to the distance L between lighthouse and screen. By making L great enough, we can make the spot move arbitrarily fast, even faster than c.[4] Relativity is not violated because nothing is actually moving in the direction of the spot and no information is being transmitted in that direction.

Transverse Velocities. We have thus far analyzed only velocities in the direction of relative motion between frames. What about transverse velocities? Since the y and z coordinates of any event are the same in S as in S', we might be tempted to conclude that the same is true of the y and z components of velocity, as in Galilean relativity. In special relativity, however, the time transformation brings about a change in transverse velocity.

2. It can be shown that under a Lorentz transformation, at most one speed is invariant. Hence if there is a maximum speed, it must be the speed of light.
3. "On the Electrodynamics of Moving Bodies," *Annalen der Physik* 17 (1905); p. 170 of translation in *Collected Documents*, Vol. II.
4. In recent years, powerful laser beams have succeeded in reaching the moon. If the laser is rotated at a modest angular velocity, the spot on the moon moves at a speed greater than c.

Consider a body that moves in the x-y plane with constant velocity components v_x and v_y measured in frame S. Its position is given by

$$x = v_x t \qquad y = v_y t \tag{4.18}$$

Proceeding as before, we transform these equations to S' coordinates, using equation (4.2). The x equation reproduces the result (4.15), with v and v' replaced by v_x and v'_x. That is,

$$v'_x = \frac{v_x - V}{1 - \dfrac{v_x V}{c^2}} \tag{4.19a}$$

For the y equation we get

$$y' = y = v_y t = v_y \gamma \left(t' + \frac{V}{c^2} x' \right)$$

Dividing both sides by t' gives

$$\frac{y'}{t'} = v'_y = v_y \gamma \left(1 + \frac{V}{c^2} v'_x \right)$$

Notice that v'_x appears in the expression for v'_y; this effect has no counterpart in classical relativity. Substituting for v'_x its value from equation (4.19a) we obtain after some algebra

$$v'_y = \frac{v_y}{\gamma \left(1 - \dfrac{V v_x}{c^2} \right)} \tag{4.19b}$$

A similar derivation yields the transformation law for v_z:

$$v'_z = \frac{v_z}{\gamma \left(1 - \dfrac{V v_x}{c^2} \right)} \tag{4.19c}$$

When $v_x = 0$, equations (4.19b,c) take the simple forms

$$v'_y = v_y / \gamma \qquad v'_z = v_z / \gamma$$

This result can be interpreted as a pure time dilation effect.

The inverse velocity transformation, from S' to S, is

$$v_x = \frac{v'_x + V}{1 + v'_x V / c^2} \tag{4.20a}$$

$$v_y = \frac{v'_y}{\gamma(1 + v'_x V/c^2)} \tag{4.20b}$$

$$v_z = \frac{v'_z}{\gamma(1 + v'_x V/c^2)} \tag{4.20c}$$

4.5. FIZEAU'S EXPERIMENT

A significant test of the velocity transformation is provided by Fizeau's experiment on the passage of light through moving water. Light travels more slowly in a material medium than it does in a vacuum. Its speed may be written as c/n, where the index of refraction n depends on the material. For water, n is 1.33: the speed of light in water is only three-fourths its speed in vacuum.

Suppose water flows at velocity V, and consider a light ray that propagates in the direction of the water flow. In a frame, S', that moves at the same velocity V, the water is at rest and the speed of the light ray must be c/n. Its speed in the laboratory frame is obtained by using the relativistic transformation law (4.15). The result is

$$v^+ = \frac{\frac{c}{n} + V}{1 + V/nc} \tag{4.21a}$$

A similar argument shows that the speed of a light ray moving in the opposite direction from the water is

$$v^- = \frac{\frac{c}{n} - V}{1 - V/nc} \tag{4.21b}$$

For $n = 1$ both these expressions reduce to $v = c$, as they must.

Since the speed of the water in any feasible experiment is necessarily much less than c, good approximations to equations (4.21a,b) can be obtained by making a power series expansion and retaining only terms proportional to V/c, as we did in section 4.1. The details are given in the appendix to this chapter; the results are

$$v^+ \approx \frac{c}{n} + V\left(1 - \frac{1}{n^2}\right) \tag{4.22a}$$

$$v^- \approx \frac{c}{n} - V\left(1 - \frac{1}{n^2}\right) \tag{4.22b}$$

Fizeau's experiment is shown schematically in figure 4.2. Water flows through the transparent U-shaped tube in the direction indicated in the

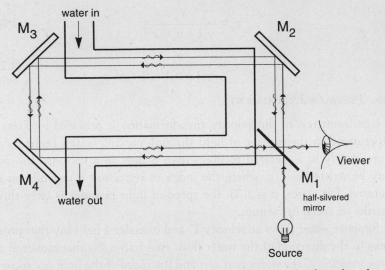

Fig. 4.2. Schematic sketch of Fizeau's experiment. For the sake of clarity, the rays that travel in opposite directions are shown as separated; in the actual setup the rays traverse the same path and create interference fringes when viewed by the observer.

figure. Light from the source S strikes the half-silvered mirror M_1 where it is partially reflected and partially transmitted. The rays follow the paths indicated in the sketch, being reflected by mirrors M_2, M_3, and M_4 before being finally rejoined. One ray travels in the same direction as the water along both long segments, while the other travels in the opposite direction to the water along both segments. The travel times for the two rays are L/v^+ and L/v^-, where L is the total length of the path within the water.

With the speeds given by equations (4.22a,b), the difference between the travel times of the two rays is

$$\Delta t = \frac{L}{\frac{c}{n} - V\left(1 - \frac{1}{n^2}\right)} - \frac{L}{\frac{c}{n} + V\left(1 - \frac{1}{n^2}\right)} \tag{4.23}$$

Since terms quadratic in V/c have already been neglected, it is appropriate to expand the fractions in equation (4.23) and retain only first-order terms, obtaining the simpler form

$$\Delta t \approx \frac{2LV}{c^2}\left(n^2 - 1\right) \tag{4.24}$$

On viewing the reunited rays, one observes a set of interference fringes. (See the discussion in sec. 2.3.) As the speed of the water is varied the fringe pattern should shift, by an amount that can be predicted from equation (4.24).

The entire setup is very reminiscent of the Michelson-Morley experiment.[5] In each case a light ray is split into two parts, which are finally reunited after having traversed different paths. There is one important quantitative difference, however: the predicted effect in Fizeau's experiment is of first order; it is proportional to V/c, not to the square of that quantity as in the Michelson-Morley experiment. This makes it possible to detect the predicted fringe shift even when V/c is only of order 10^{-8}, that is, for water velocity as low as a few meters per second.

Fizeau performed his experiment in 1851, long before Einstein had discovered special relativity. His interpretation of the experiment was based on the ether picture and Galilean relativity. In such an interpretation one can assume that the laboratory is at rest with respect to the ether, since any overall velocity relative to the ether will produce corrections only of order $(V/c)^2$.

According to the classical picture, the speed of light is c/n in the rest frame of the ether. Its speed in moving water depends on the extent to which the ether is dragged along by the moving water. (See the discussion of ether drag in sec. 2.4.) If no ether is dragged, the motion of the water has no effect on the speed of light; it is still c/n. At the opposite extreme, if the ether is dragged along at the full speed of the water, the speed of light is c/n *in the frame of the moving water;* the Galilean velocity transformation then tells us that the speed in the laboratory frame is $(c/n) \pm V$, where the plus sign applies to light moving in the direction of the water flow and the minus sign to light moving in the opposite direction.

In an intermediate case, if the ether is only partially dragged by the moving water, the speed of light in moving water can be written as $(c/n) \pm fV$, where f is the so-called ether drag coefficient, which can have any value between 0 and 1 ($f=0$ for no ether drag; $f=1$ for complete drag). Fizeau's objective in doing his experiment was precisely to measure the value of the ether drag coefficient.

The Galilean expression for the difference in travel times is

$$\Delta t = \frac{L}{\frac{c}{n} - fV} - \frac{L}{\frac{c}{n} + fV} \qquad \text{(in Galilean relativity)}$$

5. Michelson himself carried out a very accurate version of Fizeau's experiment in 1885.

which can be simplified in the same way as the relativistic expression (4.23). The result is

$$\Delta t \approx \frac{2LV}{c^2} \, n^2 f \qquad \text{(in Galilean relativity)} \qquad (4.25)$$

On comparing the prediction of special relativity, equation (4.24), with that of Galilean relativity, equation (4.25), we see that if the ether drag coefficient f has the value $1 - 1/n^2$, the two pictures give the same result (up to terms of first order in V/c). By coincidence, Fresnel had some time earlier proposed a theory of the ether that predicted just that value for f.[6] Fresnel's theory is consistent with the observations on stellar aberration, since for air ($n \approx 1$) it predicts practically no drag.

Fizeau's result, $f = 0.48$, was consistent, within the experimental error, with Fresnel's prediction $1 - 1/(1.33)^2 = 0.43$. Fizeau believed that he had confirmed Fresnel's theory. From a modern point of view, we can interpret the same result as confirming the relativistic velocity transformation. Einstein in fact cited Fizeau's experiment as an important element in support of special relativity.

By continuing the power series expansion to the next order, one obtains a more accurate expression for the predicted effect. If the coefficient of the quadratic term could be measured, it would provide a more stringent test of the theory. Unfortunately, the experiment cannot be performed with the required precision; with V/c only one part in 10^8, the second-order fringe shift is just too small to detect.

4.6. THE TRANSFORMATION OF DIRECTION AND RELATIVISTIC ABERRATION

The results of section 4.4 can be used to derive the relativistic formula for stellar aberration, which was analyzed from a Galilean point of view in section 1.5. The same derivation can be repeated using the relativistic transformation law for velocity instead of the Galilean one.

Consider again a star whose direction is perpendicular to the plane of the earth's orbit in a frame, S, in which the star is at rest and the earth is moving, as in figure 1.4a. Light from the star has the velocity components

---

6. Fresnel postulated that the density of ether within matter is proportional to the square of its dielectric constant; a given volume of water contains more ether than does an equal volume of empty space. Only the "extra" ether is dragged along by a moving medium. These assumptions lead to a drag coefficient $f = 1 - 1/n^2$.

Fresnel's theory suffers from one serious defect. The index of refraction of most substances varies with the frequency of the light (dispersion). It is hard to imagine how an ether drag coefficient could be frequency-dependent.

$$v_z = -c \qquad v_x = v_y = 0 \tag{4.26}$$

Transforming these velocity components to the earth's rest frame S', using equations (4.15) and (4.19), we obtain

$$v_z' = -c/\gamma \qquad v_x' = -V \qquad v_y' = 0 \tag{4.27}$$

where V is the earth's velocity relative to the star. The value of v_z' differs from that given by equation (1.15), which is based on the Galilean velocity transformation.

The aberration angle α, defined in figure 1.4b, is given by

$$\tan \alpha = \frac{v_x'}{v_z'} = \gamma \, \frac{V}{c} \tag{4.28}$$

The Galilean result was

$$\tan \alpha = \frac{V}{c} \tag{1.16}$$

The only difference between the two results is the factor γ in equation (4.28). In principle, this difference is subject to observational test. However, for a typical value of V, a few hundred km/sec, V/c is of order 10^{-4} and γ differs from unity by only about one part in 10^8. Since the aberration angle itself is only about $20''$ of arc, the relativistic correction would be of order $10^{-7''}$, much too small to be detected. Hence aberration measurements provide no additional confirmation of special relativity.

The same approach can be used to derive the general transformation law for the direction of any motion. We confine the discussion to two-dimensional motion.

Suppose a body moves in the x-z plane, with velocity components v_x and v_z as measured in some frame S (fig. 4.3a). The velocity components v_x' and v_z' measured in another frame S' are related to v_x and v_z by equation (4.16). Letting θ and θ' denote the angles between the body's direction and the z and z' axes, respectively, one has the trigonometric relations

$$\tan \theta = \frac{v_x}{v_z'} \qquad \tan \theta' = \frac{v_x'}{v_z'} \tag{4.29}$$

Substituting for v_x' and v_z' their values from (4.16), we find

$$\tan \theta' = \frac{\gamma(v_x - V)}{v_z}$$

Finally, using $v_x = v \sin \theta$ and $v_z = v \cos \theta$, we get

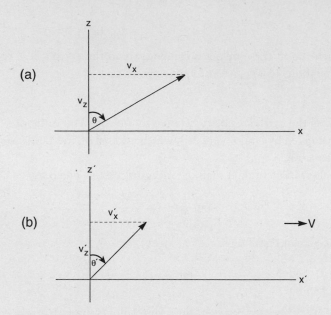

Fig. 4.3. The transformation of direction. (a) In frame S, the object in question has velocity components v_x and v_z. Its direction of motion makes an angle θ with the z axis. (b) The same motion as seen in frame S'. The velocity components v_x' and v_z' are related to v_x and v_z by eq. (4.16). The direction of motion makes an angle θ' with the z' axis; θ' is related to θ by eq. (4.30).

$$\tan \theta' = \frac{\gamma(v \sin \theta - V)}{v \cos \theta} \tag{4.30}$$

Equation (4.30) is the desired transformation law for direction. With $v = c$ it gives the general aberration formula, which applies to stars in any direction.

4.7. Acceleration in Special Relativity

The last transformation law to be examined is that for acceleration. We saw in chapter 1 that in Galilean relativity acceleration transforms in a very simple way: it is invariant ($a' = a$). Newton's second law, $F = ma$, is covariant under a Galilean transformation provided the force is invariant.

The situation in special relativity is more complicated. Consider a body that starts from rest and moves in the x direction with a constant acceleration a. Its position as a function of time is given by

$$x = \frac{1}{2} a \ t^2$$

Transforming this equation to frame S' by using equation (4.2), we obtain

$$x' + Vt' = \frac{1}{2} a \ \gamma \left(t' + \frac{V}{c^2} x' \right)^2 \qquad (4.31)$$

Equation (4.31) is *not* of the form

$$x' = (a \ \text{constant}) t' + (\text{another constant}) (t')^2 \qquad (4.32)$$

nor can it be put into that form by any algebraic manipulation. Equation (4.32) is the most general equation that describes motion with constant acceleration; hence the acceleration in S' is not constant. Motion with constant acceleration in one inertial frame appears in another inertial frame as a motion in which the acceleration changes with time.

A general expression for a' can be derived; I quote the result only for motion in one dimension:

$$a' = \frac{a}{\gamma^3 (1 - vV/c^2)^3} \qquad (4.33)$$

Notice that v as well as a appears on the right side of equation (4.33). Thus a' changes even when a is constant.

If Newton's second law is to be covariant, the quantity F/m must transform in the same complicated way as does the acceleration. We can hardly expect that to happen, and indeed it does not. The second law is not a general law of nature; it is only the low-velocity limit of a more general covariant law.

In Newtonian mechanics, a body subjected to a constant force accelerates at a constant rate and ultimately exceeds the speed of light. That cannot happen, of course, according to special relativity; the body's acceleration initially has the Newtonian value F/m but then decreases steadily and approaches zero; the speed approaches c but never reaches it. This is due in part to the relativistic increase of mass with velocity, to be discussed in chapter 7.

4.8. The Doppler Effect

An interesting problem in relativity is the theory of the Doppler effect, the change in the frequency of light detected by an observer who is moving relative to the light source. A Doppler effect occurs for any wave phenomenon. The effect for sound is familiar: the pitch of a train whistle sounds higher when the train is approaching us than when the train is at rest and lower when the train is moving away. A similar effect is detected when an observer moves toward or away from a stationary source of sound.

The Doppler effects for sound and for light differ in one important respect. Sound waves propagate in a medium. Motion of the source and of the observer are therefore distinguishable; "moving source" and "moving observer" signify motion relative to the medium. A different formula applies to each case. In the case of light, however, one cannot distinguish between motion of the source and of the observer; only their relative velocity is defined, and there can be only one Doppler formula.

Suppose a source, S, emits light of frequency f, measured in its own rest frame. The light is detected by a receiver, R, that is receding from the source at a relative velocity V. The frequency f' detected by the receiver is the number of waves per second that pass the receiver, *measured in its rest frame.*

Figure 4.4 is drawn in the source's rest frame. At event E_1 (fig. 4.4a), wave crest A is just passing the receiver; at event E_2 (fig. 4.4b), the next crest, B, passes the receiver. The distance between A and B is the wavelength $\lambda = c/f$.

Let Δt denote the interval between E_1 and E_2, measured in the source frame. During that interval the receiver has moved a distance $V\Delta t$ while each wave crest has moved a distance $c\Delta t$. From the figure we see that

$$c\Delta t = \lambda + V\Delta t \qquad (4.34)$$

Solving for Δt, we find

$$\Delta t = \frac{\lambda}{c - V} = \frac{1}{f(1 - V/c)} \qquad (4.35a)$$

If the receiver is approaching the source, a similar analysis leads to the result

$$\Delta t = \frac{\lambda}{c + V} = \frac{1}{f(1 + V/c)} \qquad (4.35b)$$

We are interested in $\Delta t'$, the interval between E_1 and E_2 measured in the receiver rest frame S'. Since E_1 and E_2 occur at the same place in S', $\Delta t'$ is a proper time interval and we can write

$$\Delta t' = \Delta t / \gamma \qquad (4.36)$$

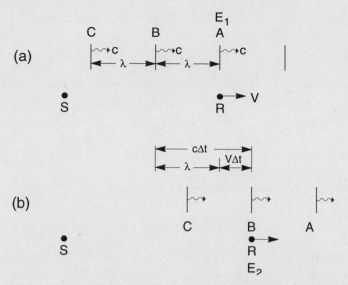

Fig. 4.4. The Doppler effect is shown in the rest frame of the source, S. The receiver, R, is moving away from the source at velocity V. In (a), wave crest A is just passing the receiver. The distance between successive wave crests is the wavelength λ; each crest moves at velocity c. In (b), the next wave crest, B, is passing the receiver. The elapsed time interval, measured in frame S, is Δt. The receiver has moved a distance $V\Delta t$, while each wave crest has moved a distance $c\Delta t$.

The desired frequency f' is the reciprocal of $\Delta t'$:

$$f' = \frac{\gamma}{\Delta t} = \gamma f \left(1 \mp V/c\right) \qquad (4.37)$$

where the minus sign refers to the case of receiver and source receding and the plus sign to receiver and source approaching.

A little algebra enables us to rewrite equation (4.37) in the forms

$$f' = f\sqrt{\frac{1 - V/c}{1 + V/c}} \qquad \begin{array}{c}\textit{Source and receiver}\\ \textit{receding}\end{array} \qquad (4.38a)$$

$$f' = f\sqrt{\frac{1 + V/c}{1 - V/c}} \qquad \begin{array}{c}\textit{Source and receiver}\\ \textit{approaching}\end{array} \qquad (4.38b)$$

Equations (4.38a,b) are the relativistic Doppler formulas. The radicals in these equations can be expanded using the binomial theorem, as discussed in the appendix to this chapter. The result is the series

$$f' = f\left[1 \mp \frac{V}{c} + \frac{3}{4}\left(\frac{V}{c}\right)^2 - \ldots\right] \qquad (4.39)$$

To first order in V/c, equation (4.39) takes the simple form

$$\frac{\Delta f}{f} = \frac{f' - f}{f} = \mp \frac{V}{c} \tag{4.40}$$

The nonrelativistic Doppler formulas, which apply to sound, are derived in any standard physics text. The result for stationary source, moving receiver is

$$f' = f\left(f \mp \frac{V}{c}\right) \tag{4.41a}$$

For stationary receiver, moving source, the result is

$$f' = \frac{f}{1 \pm V/c} \tag{4.41b}$$

The upper sign applies to source and receiver receding and the lower sign to source and receiver approaching. In lowest order, both these equations reduce to the form (4.40): when V/c is very small, the relativistic Doppler shift and both forms of the nonrelativistic shift are identical. This result is not unexpected.

EXAMPLE. A neutral pion decays by emitting two gamma rays (high-frequency light). When the pion is at rest, the gamma rays travel in opposite directions and each has a frequency f_0.

Suppose a pion moving at $0.8c$ decays and the gamma rays are emitted forward and back along the pion's original direction. What are the frequencies of the gamma rays measured in the laboratory?

SOLUTION. In the pion's rest frame, the gamma rays must both have the frequency f_0. Assume the pion was moving to the right. In the laboratory frame the frequency of gamma ray #1, moving to the left, is given by equation (4.38a) (source and receiver receding):

$$f' = f_0 \sqrt{\frac{1 - 0.8}{1 + 0.8}} = \frac{1}{3} f_0$$

The frequency of gamma ray #2, moving to the right, is given by equation (4.38b) (source and receiver approaching):

$$f' = f_0 \sqrt{\frac{1 + 0.8}{1 - 0.8}} = 3 f_0$$

The nonrelativistic formulas (4.39) and (4.41) give $0.55 f_0$ and $5 f_0$ for the two frequencies. The experiment confirms the correctness of the relativistic formulas.

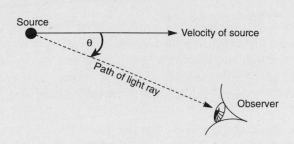

Fig. 4.5. Geometry for the Doppler effect when source and receiver are not approaching each other or receding directly. θ is the angle between the velocity of the source and the path of the light ray, as seen in the receiver's rest frame.

The formulas we have derived apply only when source and observer are moving directly toward or away from one another. The general case is shown in figure 4.5, in which θ denotes the angle between the direction of relative motion and the line that connects source and observer at the instant the light is emitted, measured in the observer's frame. A generalization of the preceding analysis shows that instead of equation (4.37), the frequency measured by the observer is given by

$$f' = \frac{f}{\gamma\left(1 - \dfrac{V}{c}\cos\theta\right)} \tag{4.42}$$

The case of head-on approach corresponds to $\theta = 0°$; in that case, (4.42) reduces to (4.38b). Similarly, the case of direct recession corresponds to $\theta = 180°$; in that case, (4.42) reduces to (4.38a).

The Doppler effect for light has been detected experimentally on countless occasions. It is not easy, however, to confirm the relativistic correction. The difficulty is the usual one: the correction is of order $(V/c)^2$, and in most experiments the relative motion between source and observer is too slow to lead to a measurable effect.

The first confirmation of the second-order Doppler effect was provided by Herbert Ives and G. R. Stilwell in 1938.[7] They observed the light emit-

7. Herbert E. Ives and G. R. Stilwell, "An Experimental Study of the Rate of a Moving Atomic Clock," *Journal of the Optical Society of America* 28 (1938):215–226. It is of interest to note that as late as 1938 Ives and Stilwell still did not

ted by a beam of hydrogen ions and compared the wavelengths of a specific spectral line emitted in the forward and backward directions. Although the Galilean and Einsteinian predictions differed by less than a tenth of an Angstrom (1 Å $= 10^{-10}$ m), the experiment was accurate enough to detect so small a difference; the results agreed with Einstein's formula.

A way to detect the relativistic Doppler effect was suggested by Einstein himself. He pointed out that when $\theta = 90°$ in figure 4.5, the source is momentarily neither approaching nor receding from the observer. Just at that instant the observed light is emitted transversely to the direction of relative motion. According to classical theory, no change in frequency should be detected in this case. The relativistic formula (4.42), in contrast, predicts $f' = f/\gamma$. This "transverse" Doppler effect is due entirely to time dilation. Although the predicted shift is small, detection of any shift whatever would confirm the relativistic effect.

The experiment is very difficult; the transverse Doppler effect is detectable only if θ is almost precisely 90°. If θ differs from 90° by even a small fraction of a degree, $\cos \theta$ in equation (4.42) differs from zero sufficiently to cause a frequency shift that totally swamps the effect of the relativistic factor γ. For a long time, the necessary precision could not be achieved.

The transverse Doppler effect was first demonstrated successfully in 1963 by Walter Kündig, who used a rotating turntable with an emitter of radiation at the center and an absorber on the rim.[8] In such an arrangement, the relative motion of source and absorber is necessarily perpendicular to the path of the radiation. The Mossbauer effect makes it possible to measure wavelengths with very high precision. The results were consistent with Einstein's formula (4.40) within 1 percent.

4.9. The Role of Lorentz and Poincaré in the Birth of Relativity

According to Max Born, an eminent physicist who witnessed its birth, "special relativity was not a one-man discovery. Einstein's work was the keystone to an arch which Lorentz, Poincaré and others had built."[9] In

---

believe in special relativity. They claimed that their results confirmed the "Larmor-Lorentz theory," based on the ether, and never even mentioned relativity.

8. W. Kundig, "Measurement of the Transverse Doppler Effect in an Accelerated System," *Physical Review* 129 (1963):2371–2375. There is also an effect due to general relativity in this experiment; see chapter 8 for discussion.

9. Max Born, *Physics in My Generation* (New York: Springer-Verlag 1969), 106.

my judgment, Born's metaphor exaggerates the contributions of Lorentz and of Henri Poincaré. Einstein did not just complete their structure; he built a much grander arch according to his own design. In this section I examine the work of Lorentz and Poincaré and comment on their influence on Einstein.

Lorentz

The reader may have wondered why the transformation that forms the basis for special relativity bears Lorentz's name rather than Einstein's.[10] The transformation equations appear in a paper published by Lorentz in 1904, a year before Einstein's relativity paper.[11] Lorentz's interpretation of those equations, however, was quite different from Einstein's.

Lorentz was considered the foremost theoretical physicist of his time. As part of a broad investigation of the electrical properties of matter, he studied electromagnetic phenomena in moving bodies. He published his results in a series of papers between 1892 and 1904.[12]

Lorentz postulated a perfectly rigid and immovable ether. He assumed that Maxwell's equations, which relate the electric and magnetic fields to the distribution of charge and current, hold in the frame of the stationary ether, S. The solutions of Maxwell's equations include waves that travel with velocity c in any direction.

10. The term "Lorentz transformation" was coined by Poincaré in 1905. Some authors call it the Lorentz-Einstein transformation.
11. The equations were in fact first published by Woldemar Voigt in 1887. Working on an elastic theory of light, Voigt noted that the wave equation retains its form under a transformation that is the same as eq. (4.1) up to a scale factor. Although Lorentz knew Voigt and corresponded with him about the Michelson-Morley experiment, Voigt's paper somehow escaped his notice. Larmor's treatise, *Aether and Matter*, published in 1900, also contained a set of equations equivalent to the Lorentz transformation.
12. Lorentz's work is most readily found in H. A. Lorentz, *Collected Papers*, 9 vols. (The Hague: Martinus Nijhoff, 1934–1938). The most important publications are "The Electromagnetic Theory of Maxwell and Its Application to Moving Bodies," *Archives Neérlandaises* 25 (1892):363 (*Collected Papers*, II:164–343); *Versuch Einer Theorie der Elektrischen und Optischen Erscheinungen in Bewegten Korpern* (Leiden: Brill, 1895) (*Collected Papers*, V:1–138); "Simplified Theory of Electrical and Optical Phenomena in Moving Systems," *Verslagen en Mededeelingen der Koninklijke Nederlandse Akademie van Wetenschappen*, English ed., 1 (1899):427 (*Collected Papers*, V:139–155); and "Electromagnetic Phenomena in a System Moving with Any Velocity Less than That of Light," *Proceedings of the Royal Academy of Amsterdam* 6 (1904):809 (*Collected Papers*, V:172–197). The 1904 paper and a short section of the *Versuch* appear in *The Principle of Relativity*, 9–34 and 1–7. Lorentz's book, *Theory of Electrons*, is a good exposition of the theory. See also Kenneth Schaffner, "The Lorentz Electron Theory of Relativity," *American Journal of Physics* 37 (1969):498–513.

Suppose a body moves with constant velocity V in the x direction, relative to the ether. Lorentz showed in his 1892 paper that if one makes a Galilean transformation to a frame S_r that moves with the body, Maxwell's equations change their form.

This result should not be at all surprising. If the ether is stationary, Maxwell's equations cannot hold in S_r. In particular, the speed of light in S_r must vary with direction.

Lorentz's aim was "to reduce, at least as far as possible, the equations for a moving system to the form of the ordinary formulae that hold for a system at rest."[13] Accordingly, he looked for a mathematical transformation to a new set of (primed) variables, in terms of which the field equations would resemble Maxwell's equations. He found the solution in steps: initially to first order in V/c, then to second order, and finally (in the 1904 paper) an exact formula valid to all orders. This was the Lorentz transformation, under which Maxwell's equations are covariant: they take the same form in terms of the primed coordinates as they do in terms of the S-frame coordinates.[14]

Lorentz never interpreted the primed variables as anything more than a mathematical construct, an "imaginary system" S' in which the body is formally at rest. The real rest system is S_r, and the real time is t in all frames. The parameter t', which Lorentz called "local time" (*Ortszeit*), is *not* the time recorded by a clock in the rest frame of the moving body.

Under such an interpretation, the covariance of the equations is of only formal interest. One can use Lorentz's transformation to generate additional solutions of Maxwell's equations, but one is not justified in interpreting those new solutions as the fields measured in the body's rest frame. Lorentz nonetheless used the S' solutions as the basis for physical arguments; this logical flaw in his theory has not been sufficiently stressed in the literature.

In the 1892 paper, Lorentz used his theory to "derive" the contraction that he and FitzGerald had earlier proposed to account for the outcome of the Michelson-Morley experiment. Consider a rod at rest on the earth and aligned in the direction of the earth's velocity V through the ether, which we take to be the x direction. Let F' denote the force on a charged particle in the rod measured in system S'; that force is purely electrostatic because everything is at rest in S'. (The thermal motion of the molecules

13. Lorentz, *Theory of Electrons*, 196.
14. Because of a technical error, Lorentz did not succeed in demonstrating the full covariance of Maxwell's equations. (An extraneous term appeared in the equation for the electric field.) The error was corrected by Poincaré in 1905.

is ignored.) The corresponding force F measured in the ether frame is partly electric and partly magnetic. Lorentz showed that the relation between F' and F is

$$F'_x = F_x \tag{4.43a}$$

$$F'_y = \gamma F_y \qquad F'_z = \gamma F_z \tag{4.43b}$$

where γ is defined in the usual way. The transverse components of force are greater in S' than in S by the factor γ, whereas the longitudinal components are the same in both frames.

Lorentz pointed out that the size and the shape of any solid body are determined by the intermolecular forces; each molecule must be in equilibrium under the forces exerted by its neighbors. He assumed that all molecular forces transform between systems S and S' in the same way as electrostatic forces do, although he conceded that "there is no reason" why that should be so.

If the rod has the same dimensions in both systems, the fact that F_y transforms differently from F_x implies that an equilibrium configuration in S' will in general *not* translate into another equilibrium configuration in S. But if the rod is shorter by a factor γ in S than in S', the transverse dimensions remaining the same, the forces in the two systems scale in the same proportion as the dimensions. For any equilibrium configuration in S', the corresponding configuration in S will then also be in equilibrium. According to Lorentz, "the displacement would naturally bring about this disposition of the molecules of its own accord, and thus effect a shortening in the direction of motion in the proportion of 1 to $\sqrt{1 - V^2/c^2}$."[15] That shortening is the FitzGerald contraction. The contraction applies to the electron itself; if it is a sphere when at rest in the ether, it is flattened into an ellipsoid when moving.

Aside from the questionable assumption concerning the transformation properties of molecular forces, Lorentz's derivation is unconvincing because it assumes that the rod's length in S' is the same as its length when at rest in the ether. Since S' is not the physical rest frame of the rod, that assumption has no justification.

In Einstein's treatment of the same problem, S' is the inertial frame in which the rod is at rest. The principle of relativity therefore demands that the rod's length in S' be its proper length; the contraction measured in frame S then follows naturally from the transformation equations.

In the *Versuch* of 1895, Lorentz proved a "theorem of corresponding

15. Lorentz, *Versuch* (1895), p. 7 of excerpt in *The Principle of Relativity*.

states," which is nothing but the covariance of Maxwell's equations to first order in V/c. Given any set of electric and magnetic fields E and B that satisfy Maxwell's equations in the ether frame S, he showed that there exists a corresponding solution in S', in which E' and B' are (to first order) the same functions of the primed variables as E and B are of the coordinates in S.

Lorentz used his theorem of corresponding states to argue that for any experiment in which both observer and light source are attached to the earth, the earth's assumed motion through the ether can have no first-order effect. He applied this argument to several experiments, including Fizeau's experiment and stellar aberration using both air- and water-filled telescopes. The 1904 paper extended the theory to all orders in V/c; in that form it applies to all terrestrial experiments.

The theorem of corresponding states has been called "the germ of modern relativistic tendencies."[16] It is indeed a relativistic theorem but only if one identifies the system S' as the actual rest frame of the body and E' and B' as the fields measured by observers in that frame. Lorentz was not prepared to take that crucial step because it would have required him to give up the privileged status of the ether as well as the concept of absolute time. As a result, his theory is an amalgam of relativistic and nonrelativistic elements that does not hold together. He insists that the Galilean frame S_r is the real rest frame but ascribes physical significance to the S' system. This position is logically inconsistent: if the S' system is only a mathematical construct, the application of the theorem to real earthbound experiments is not justified.

Even after he had grasped the significance of Einstein's theory, Lorentz was unwilling to fully abandon his beloved ether. A certain wistfulness is apparent in the following excerpt from a lecture given by Lorentz in 1913.

> According to Einstein, it has no meaning to speak of motion relative to the aether. He likewise denies the existence of absolute simultaneity. . . . The acceptance of these concepts belongs mainly to epistemology. . . . However, it depends to a large extent on the way one is accustomed to think whether one is attracted to one or another interpretation. As far as this lecturer is concerned, he finds a certain satisfaction in the older interpretations, according to which the aether possesses at least some substantiality, space and time can be sharply separated, and simultaneity without further specifi-

16. Ludwik Silberstein, *The Theory of Relativity* (London: Macmillan, 1924), 67.

cation can be spoken of. In regard to the last point, one may perhaps appeal to our ability of imagining arbitrarily large velocities. In that way, one comes very close to the concept of absolute simultaneity.[17]

These are hardly the words of a confirmed relativist.

In sum, Lorentz's theory falls short of being a theory of relativity in several important respects:

1.   His theory is firmly rooted in the ether. In his view, the Lorentz transformation does not relate two arbitrary inertial frames but is only a mathematical relation between a physical frame (that of the ether) and a nonphysical set of coordinates, the S' system. Lorentz's theory is not symmetric between the two sets of coordinates; his transformation is really not a coordinate transformation at all, in the sense that one cannot use it to change from one physical frame to another.

2.   Lorentz did not derive the transformation equations from basic principles, as did Einstein, but discovered them by trial and error.

3.   Perhaps most important was Lorentz's failure to grasp the true significance of the time transformation. Only Einstein realized that a fundamental reassessment of the nature of time is required; this is the key conceptual step in relativity.

It is instructive to look at what Einstein himself had to say about the connection between his work and that of Lorentz and Poincaré. In 1955, shortly before his death, he wrote to Carl Seelig,

> It is beyond doubt that the special theory of relativity, if we regard its development in retrospect, was ripe for discovery in 1905. Lorentz had already recognized that for the analysis of Maxwell's equations the transformation later named after him is essential, and Poincaré had deepened this knowledge. As for myself, I knew only Lorentz's important work of 1895[18] but not Lorentz's later

17. Quoted by Pais, *Subtle Is the Lord,* 166.
18. Here Einstein cites Lorentz's papers of 1892 and 1895. G. H. Keswani argues that Einstein must have suffered from a lapse of memory and had actually read Lorentz's 1904 paper when he wrote his own. G. H. Keswani, "Origin and Concept of Relativity," *British Journal for the Philosophy of Science* 15 (1965):286–306; 16 (1965):19–32. Keswani's case is not persuasive. At the very beginning of his 1905 paper, Einstein notes that "as has already been shown to the *first order* of small quantities, the same laws of electrodynamics and optics will be valid for all frames of reference for which the equations of mechanics hold good." The first-order result was just what Lorentz had obtained in 1895; in the 1904 paper, he

work, or the subsequent investigations of Poincaré. In this sense my work of 1905 was independent.

What was new about it was the realization that the significance of the Lorentz transformation transcends its connection with Maxwell's equations and is concerned with the nature of space and time in general. Also new was the understanding that "Lorentz invariance" is a general condition for every physical theory.[19]

Einstein credits Lorentz's early work with motivating him to think about the transformation properties of Maxwell's equations. It is clear, however, that Lorentz was not a major influence on him.

Lorentz was generous in acknowledging the superiority of Einstein's approach. At a conference in 1927, he explained,

> A transformation of the time was necessary. So I introduced the conception of a local time which is different for all systems of reference which are in motion relative to each other. But I never thought that this had anything to do with real time. This real time for me was still represented by the old classical notion of an absolute time, which is independent of any reference to special frames of coordinates. There existed for me only this one true time. I considered my time transformation only as a heuristic working hypothesis. So the theory of relativity is really solely Einstein's work. *And there can be no doubt that he would have conceived it even if the work of all his predecessors in the theory of this field had not been done at all.* His work is in this respect independent of the previous theories.[20] (Emphasis added.)

Poincaré

The brilliant mathematician/physicist Henri Poincaré came closer than did Lorentz to beating Einstein to the discovery of relativity. Poincaré's early writings contain many prescient ideas.

1895. Poincaré expresses skepticism about the properties of the ether. He concludes that "it is impossible to measure the absolute movement of ponderable matter, or, better, the relative movement of ponderable matter

---

extended it to all orders. If Einstein had known this result he surely would have referred to it.

19. The letter is reproduced in Born, *Physics in My Generation,* 104. I have made a few corrections in Born's translation from the German.

20. These remarks were published in the "Report on the Conference on the Michelson-Morley Experiment," *Astrophysical Journal* 68 (1928):341–402.

with respect to the ether. All that one can provide evidence for is the motion of ponderable matter with respect to ponderable matter."[21]

1898. Poincaré suggests that there are serious problems with the classical conception of time and with the simultaneity of separated events. He asserts, "We have not a direct intuition of simultaneity, nor of the equality of two intervals of time. If we think we have this intuition, it is an illusion."[22]

1899. "I regard it as very probable that optical phenomena depend only on the *relative* motion of the material bodies, luminous sources, and optical apparatus present, and that this is true not only for quantities of the order of the square or the cube of the aberration, but rigorously."[23] (Aberration here means the ratio of the earth's velocity to the speed of light.)

1900. Poincaré uses the phrase "principle of relative motion" to describe the notions expressed in the preceding quotation. In an address to an international congress in Paris, he asks the provocative question, "Does the ether really exist?"

1902. Poincaré publishes his book *Science and Hypothesis,*[24] which attracts widespread public attention. Here he first uses the phrase "principle of relativity" and elaborates on his ideas concerning time and simultaneity. Incidentally, the book contains a discussion of non-Euclidean geometry that is quite relevant to general relativity. (See chap. 8.)

1904. In an address to an international congress at St. Louis, he formulates the principle of relativity as follows:

> The laws of physical phenomena should be the same for a stationary observer as for an observer carried along in a uniform motion of translation; so that we have not and can not have any means of discerning whether or not we are carried along in such a motion.[25]

21. H. Poincaré, "A propos de la theoriè de Larmor," *L'Eclairage électrique* 5 (1895):5–14. In *Oeuvres de Henri Poincaré* (Paris: Gauthier-Villars, 1954), IX:395–413.
22. "The Measure of Time," published as chapter 2 of *The Value of Science* (Paris: E. Flammarion, 1906; English translation, New York: Science Press, 1907).
23. Published in *Electricite et optique,* 2d ed. (Paris: G. Carre et C. Naud, 1901).
24. H. Poincaré, *La Science et l'hypothese* (Paris: Flammarion, 1902); English translation first published in 1905, reprinted by Dover Books in 1952.
25. The address was published as chapters 7–9 of *The Value of Science.*

This is exactly Einstein's principle of relativity.

In this address, Poincaré discusses the synchronization of clocks by exchange of light signals, in a manner similar to Einstein's. He exhibits another insight: "Perhaps," he says, "we shall have to construct a new mechanics . . . where, inertia increasing with velocity, the velocity of light would become an impassable limit." This is just what happens according to special relativity.

After reading Lorentz's 1904 paper, Poincaré wrote two papers (published in 1905 and 1906) that analyzed the mathematical structure of the Lorentz transformation and that duplicate several results of Einstein's 1905 paper.[26] (Neither author had seen the work of the other.)

There is no denying that Poincaré articulated many of the central concepts of relativity, some of them before Einstein. His Palermo paper exhibits considerable mathematical erudition. In it Poincaré derives the relativistic transformation law for velocity and demonstrates the covariance of Maxwell's equations. He discusses the group property of the Lorentz transformation and infers the invariance of the relativistic interval Δs.

In spite of all that, Poincaré's theory is in no way a theory of relativity. In fact, as Augustin Sesmat perceptively observes, Poincaré reveals himself to be, like Lorentz, "at heart an absolutist," in the sense that he explains the apparent validity of absolute laws in moving inertial systems by compensation of effects and maintains the privileged status of the ether.[27]

Poincaré never spells out how he interprets the primed coordinates in the Lorentz transformation, but his discussion of the "contraction of electrons" reveals that his interpretation is the same as Lorentz's. He considers a spherical electron moving with constant velocity in the ether and uses the Lorentz transformation to relate the state of the "real" moving electron to that of an "ideal" electron at rest. The latter is nothing but Lorentz's S' system.

Poincaré distinguishes between two hypotheses: (a) that of Max Abraham, according to which the electron is "indeformable," that is, always remains spherical; and (b) that of Lorentz, according to which a moving electron is contracted by the factor γ along its direction of motion; the real moving electron is thus a flattened ellipsoid. In case (a), the Lorentz

26. "Sur la dynamique de l'électron," *Comptes Rendus de l'Academié des Sciences* (Paris) 140 (1905):1504–1508; "Sur la dynamique de l'électron," *Rendiconti del Circolo Matematico di Palermo* 21 (1906):129–176. In *Oeuvres de Poincaré*, IX:489–493, 494–550.
27. A. Sesmat, *Systemes de reference et mouvements (Physique relativiste)* (Paris: Hermann et Cie, 1937), 40.

transformation shows that the ideal electron is stretched out into a cigar-shaped ellipsoid, whereas in case (b), the transformation restores the electron to a spherical shape. Poincaré favors the Lorentz model because it accounts for the result of Michelson-Morley.

Despite his dedication to the principle of relativity, Poincaré fails to apply it in this problem. The principle *demands* that the moving electron be spherical in its own rest frame if it is spherical when at rest in the ether. If Poincaré had interpreted the "ideal electron" as the physical state of the electron in its rest frame, he would have had a correct relativistic description. (Abraham's model, with its rigid electron, is inadmissible in such a treatment.) But, like Lorentz, Poincaré failed to take the critical step.

Poincaré considered it necessary to account for the Lorentz contraction by some physical mechanism. He says,

> If one wants to preserve it [the theory of Lorentz] and avoid intolerable contradictions, one must assume a special force which explains both the contraction and the constancy of two of the axes. I have tried to determine this force, and found that it can be compared to a constant external pressure acting on the deformable and compressible electron, and whose work is proportional to the variations in volume of the electron.[28]

The special force, which became known as "Poincaré stress" or "Poincaré pressure," is a red herring. As Einstein showed, the contraction is inherently a kinematic effect, a direct consequence of the properties of space and time expressed through the Lorentz transformation. Whatever forces are present in matter must transform in a manner consistent with the contraction; no special force is needed.

As late as 1909, Poincaré still had not disabused himself of this fundamental misunderstanding. In a lecture at Göttingen, he asserted that the "new mechanics" is based on three hypotheses, of which the third is the longitudinal deformation of a body in translational motion.[29] (The first two were Einstein's two postulates.)

Like Lorentz, Poincaré believed in local time. In the St. Louis address of 1904, he describes Lorentz's notion of local time as a "most ingenious idea." In a review article published in 1908, Poincaré still invokes the

28. "Sur la dynamique de l'electron," *Rendiconti*, 129–176. The external pressure is presumably exerted by the ether.
29. Published in *Sechs Vorträge über ausgewählte Gegenstände ans der reinen Mathematik und Mathematischen Physik* (Leipzig: Teulener, 1910).

notion of local time to explain the principle of relativity.[30] He apparently never recognized that local time is not a relativistic concept.

Poincaré was ambivalent toward the ether. True, he often expressed doubts about the ether's properties; according to Keswani, he paid only lip service to it. Yet he continued to assign it an important role in mediating physical phenomena. In the St. Louis address, he described the interaction between electrons as being accomplished through the ether as intermediary and invoked the ether to salvage the apparent failure of Newton's law of action and reaction. In the review article cited above, he says, "The universe contains electrons, ether, and nothing else." And in 1912, the year of his death, he published a paper entitled "The Relation between Ether and Matter." Poincaré never recognized what was immediately obvious to Einstein: the principle of relativity implies that the ether can be dispensed with.

For Poincaré, the principle of relativity was an experimental fact subject to disproof. When Walter Kaufmann's early experiment seemed to support Abraham's theory, Poincaré commented, "The principle of relativity may well not have the rigorous value which has been attributed to it."[31] Einstein, however, had absolute faith in the principle. His reaction was to suspect inaccuracies in Kaufmann's experiment (as turned out to be the case; see chap. 7).

It is apparent that Poincaré was tantalizingly close to a theory of relativity. But he either did not see the all-important final step or was not bold enough to take it. Many scholars have speculated about the reasons for his failure.[32]

Poincaré's work influenced Einstein more than did that of Lorentz. During Einstein's first years in Bern he met regularly with friends in a

30. H. Poincaré, "La Dynamique de l'électron," *Revue Generale des Sciences Pure et Appliques* 19 (1908):386–402. In *Oeuvres de Poincaré*, IX:551–586.

31. H. Poincaré, "La Mécanique et l'optique," bk. 3, chap. 2 of *Science et méthode* (Paris: Flammarion, 1908), 248.

32. Pais, *Subtle Is the Lord*, 163–166; Stanley Goldberg, "Henri Poincaré and Einstein's Theory of Relativity," *American Journal of Physics* 35 (1967):934–944; Charles Scribner, Jr., "Henri Poincaré and the Principle of Relativity," *American Journal of Physics* 32 (1964):672–678; Gerald Holton, "On the Origin of the Special Theory of Relativity," in his *Thematic Origins of Scientific Thought*, 191–236, esp. 204–206. See also René Dugas, *A History of Mechanics* (Neuchatel: Griffon, 1955), 643–650; René Taton, *Reason and Chance in Scientific Discovery* (New York: Philosophical Library, 1957), 134–135; Theo Kahan, "Sur les origines de la theorie de la relativite restreinte," *Revue d'Histoire des Sciences et de Leurs Applications* 12 (1959):162; and Louis deBroglie, *Savants et découvertes* (Paris: Michel, 1951), 51.

group called the Akademie Olympia for philosophical reading and discussion. Among the works discussed was Poincaré's *Science and Hypothesis;* in a letter cited by Pais, Einstein says that the book "profoundly impressed us and kept us breathless for weeks on end."[33] This book contains many of Poincaré's early thoughts related to relativity.

Keswani points to Einstein's use of the key phrase "principle of relativity," precisely the words used by Poincaré. He concludes that Einstein took the phrase from Poincaré; the conclusion seems plausible, though not terribly significant.

More interesting is the question, Did Einstein get the idea for the principle of relativity from Poincaré? Keswani suggests that he did; I conjecture that Poincaré's work merely reinforced ideas that Einstein had entertained from an early age. More important, he pursued the principle to its logical consequences, whereas Poincaré did not.

In contrast to the relations between Einstein and Lorentz, always cordial and marked by mutual expressions of praise, those between Einstein and Poincaré were notably cool. Einstein is not cited in any of Poincaré's writings on relativity. Born tells of attending the Göttingen lectures in which Poincaré explained relativity using just the reasoning found in Einstein's paper. Yet he did not mention Einstein and gave the impression that he was recording Lorentz's work.

Einstein reciprocated in kind. In an interview published in 1920, he said, "It was found that Galilean invariance would not conform to the rapid motions in electrodynamics. This led the Dutch professor Lorentz and myself to develop the theory of special relativity."[34] Perhaps overly generous toward Lorentz, Einstein omits any mention of Poincaré; the omission must have been deliberate. Pais relates that he once asked Einstein how Poincaré's Palermo paper had affected his thinking; Einstein replied that he had never read the paper!

Although Einstein is almost universally acclaimed as the father of relativity, one inexplicable exception must be noted. The mathematician/historian Sir Edmund Whittaker, in his otherwise excellent treatise, gives Lorentz and Poincaré all the credit for the discovery of relativity and barely mentions Einstein.[35] In a chapter entitled "The Relativity Theory of Poincaré and Lorentz," he dismisses Einstein's contribution in a single

33. A. Einstein, *Lettres à Maurice Solovine* (Paris: Gauthier-Villars, 1956), VIII. Cited in Pais, *Subtle Is the Lord*, 134.
34. *New York Times*, December 3, 1920. Quoted by Pais, *Subtle Is the Lord*, 171.
35. E. T. Whittaker, *A History of the Theories of Aether and Electricity*, vol. 2 (London: T. Nelson and Sons, 1953).

sentence: "In the autumn of the same year . . . Einstein published a paper which set forth the relativity theory of Poincaré and Lorentz with some amplifications, and which attracted much attention." (!) Whittaker credits Einstein only with deriving the relativistic expressions for aberration and for the Doppler effect. That this is a gross injustice to Einstein hardly requires additional documentation. (A detailed critique is found in Gerald Holton, "On the Origins of the Special Theory of Relativity," 196–202.)

Appendix: The Low-Velocity Approximation

The approximations used in this chapter are based on the binomial theorem, which states that

$$(1+x)^n = 1 + nx + \frac{n(n-1)}{2}x^2 + \frac{n(n-1)(n-2)}{2\cdot 3}x^3 + \ldots \quad (4.A1)$$

When n is a positive integer, the series (4.A1) terminates and gives an identity. With $n=3$, for example, it gives

$$(1+x)^3 = 1 + 3x + 3x^2 + x^3 \quad (4.A2)$$

When n is any number other than a positive integer, (4.A1) is an infinite series that converges for $x<1$. If x is very small, each term in the series is much smaller than the preceding one and a small number of terms gives a good approximation to the left side. If x is small enough, the first two terms are adequate; the approximation is then just

$$(1+x)^n \approx 1 + nx \quad (4.A3)$$

The relativistic parameter γ, defined as

$$\gamma = \frac{1}{\sqrt{1-V^2/c^2}} = \left(1 - \frac{V^2}{c^2}\right)^{-\frac{1}{2}}$$

is of the form $(1+x)^n$, with $x = -V^2/c^2$ and $n = -\frac{1}{2}$. Hence we can use equation (4.A1) and obtain the series expansion

$$\gamma = 1 + \frac{1}{2}\left(\frac{V}{c}\right)^2 + \frac{3}{8}\left(\frac{V}{c}\right)^4 + \ldots \quad (4.A4)$$

The absence of a linear term in (4.A4) means that up to terms of the order V/c, γ is just 1. This is the basis for the discussion of the nonrelativistic limit in section 4.2.

In the discussion of Fizeau's experiment in section 4.5, the exact expressions for the speeds of light rays traveling in the direction of the water and in the opposite direction are given by equation (4.21a,b):

$$v^+ = \frac{\frac{c}{n}+V}{1+\frac{V}{nc}} \qquad v^- = \frac{\frac{c}{n}-V}{1-\frac{V}{nc}} \qquad (4.A5)$$

Both these expressions are of the form

$$\text{constant} \cdot \frac{1}{1+x} \qquad (4.A6)$$

with $x = \pm V/nc$. With $n = -1$, equation (4.A3) reads

$$\frac{1}{1+x} \approx 1-x$$

Hence we can write approximate forms for v^+ and v^- as

$$v^+ \approx \left(\frac{c}{n}+V\right)\left(1-\frac{V}{nc}\right) = \frac{c}{n}+V\left(1-\frac{1}{n^2}\right)-\frac{V^2}{nc}$$

$$\approx \frac{c}{n}+V\left(1-\frac{1}{n^2}\right)$$

$$v^- \approx \left(\frac{c}{n}-V\right)\left(1+\frac{V}{nc}\right) = \frac{c}{n}-V\left(1-\frac{1}{n^2}\right)-\frac{V^2}{nc} \approx \frac{c}{n}-V\left(1-\frac{1}{n^2}\right)$$

In writing the last form of these equations, we have dropped the V^2/nc terms since our expression is only accurate up to terms of order V/c. The results are equations (4.22a,b) in the text.

To simplify equation (4.23), one uses (4.A1) again. The result is equation (4.24).

PROBLEMS

4.1. The coordinates of events E_1 and E_2 in frame S are

E_1: $x=0$, $t=0$
E_2: $x=8$ light-seconds, $t=10$ sec

(a) Find the coordinates of the two events in a frame S' that moves at $0.5c$ in the $+x$ direction relative to S.

(b) In some frame S'', E_1 and E_2 occur at the same place. Find the velocity of S'' relative to S.

(c) Find the time interval between E_1 and E_2 in S'' by using the Lorentz transformation.

(d) Find the time interval between E_1 and E_2 in S'' by using the invariance of the relativistic interval $(\Delta s)^2$. The answer should of course be the same as that obtained in (c).

(e) Show that there is no physically allowed frame in which E_1 and E_2 are simultaneous.

4.2 Clocks A and B, synchronized in their own rest frame S, are observed in a frame S' in which their velocity is V. The distance between the clocks, measured in S', is L. (Clock A is ahead in the direction of motion.) Find the amount by which the clocks are out of synchronization according to S' observers. That is, find the difference between the two clock readings at a fixed time in S'. Which clock reads the later time? Find the numerical value of the difference in readings if $L = 100$ km and $V = 10$ km/sec.

4.3. A person standing at the rear of a railroad car fires a bullet toward the front of the car. The speed of the bullet, as measured in the frame of the car, is $0.6c$ and the proper length of the car is 375 m. The train is moving at $0.8c$ as measured by observers on the ground.

What do ground observers measure for
(a) the length of the railroad car,
(b) the speed of the bullet,
(c) the time required for the bullet to reach the front of the car, and
(d) the distance traveled by the bullet.

4.4. A railroad car whose proper length is 240 m is moving in the x direction at speed $0.6c$. Let S' be a reference frame fixed on the car and S a frame fixed on the ground. Let $x' = 0$ at the rear of the car.

At $t' = 0$, light pulses are emitted at the rear of the car (event E_1) and the front (event E_2). Some time later the pulses meet (event E_3). In S' it is clear that the pulses meet at the midpoint of the car at time $t_3' = 120/c$.

The following questions all refer to measurements carried out in frame S.
(a) How long is the car?
(b) Were the light pulses emitted simultaneously? If not, which one was emitted first, and how much earlier than the other?
(c) What is the spatial distance between the points at which the pulses were emitted?
(d) How much time elapsed between the emission of the first pulse and the meeting of the two pulses?
(e) Did the pulses meet at the midpoint of the car? How could this answer have been predicted?

4.5. Spaceship A passes earth at earth time $t = 0$ at speed $0.8c$ in the direction of the star Xerxes. At the same time (according to earth frame clocks) spaceship B passes Xerxes at speed $0.625c$ in the direction of earth. Assume that earth and Xerxes are at rest relative to one another and the distance between them is 10 light-years (earth frame measurement).
(a) At what rate are the spaceships approaching one another according to earth frame observers? Does this result violate the principles of relativity? Explain.
(b) Find the time when the spaceships meet according to earth frame measurements. How far is the meeting place from earth?
(c) Use the Lorentz transformation to find the time of the meeting according to clocks on ship A. (Assume the spaceship clocks read zero when it passed earth.)
(d) What is the speed of ship B according to observers on A?
(e) What is the distance between earth and Xerxes in A's frame?

(f) Consider the following argument. According to observers on A, the initial distance of B is given by the answer to (e). The time required for the ships to meet should be given by that distance divided by B's velocity, which was found in (d). This calculation, however, does not give the right answer. What is the error in the argument?

4.6. A flare is set off in Lincoln at $t=0$. A second flare is set off in Pittsburgh, 1,500 km due east, at $t = 3 \times 10^{-3}$ sec.

(a) Find the time interval between the two flares in the reference frame of a spaceship flying east at 10^5 km/sec.

(b) Find the distance between Lincoln and Pittsburgh according to observers in the spaceship frame.

(c) Find the spatial separation between the two flares according to spaceship observers. Why does this result differ from that of (b)?

(d) How fast would the spaceship be moving if the flares were simultaneous in the spaceship frame?

4.7. Prove that if a body's speed in some inertial frame is greater than c, its speed in any other inertial frame exceeds c as well. (The relative velocity between the two frames is less than c.)

4.8. A particle moving at $0.6c$ approaches a stationary particle (frame S measurement). Find the speed of a frame S', relative to S, in which the velocities of the two particles are equal and opposite. (In Galilean relativity, the answer is $0.3c$.)

4.9. A certain radioactive nucleus decays into two fragments of equal mass. When such a nucleus decays at rest, the fragments are emitted in opposite directions with equal speeds $0.4c$. A sample of this material is carried in a spaceship that moves in the positive x direction at $0.8c$ relative to the laboratory.

(a) A nucleus, at rest in the spaceship, decays. The fragments are emitted in the $\pm x$ directions. Find the speed of each fragment, measured by observers in the laboratory frame.

(b) The fragments are emitted in the $\pm y$ directions as measured in the spaceship frame. Find the speed of each fragment and its direction as measured in the laboratory.

(c) One fragment is emitted in the $+y$ direction as measured in the laboratory. Find the speed of this fragment. What is its direction in the spaceship frame?

(d) Find the laboratory direction and speed of the fragment that accompanies the one in (c).

4.10. Two events occur at the same time with a spatial separation 100 m, as measured in some frame S. In another frame S', the spatial separation of the events is 200 m. Find the difference between the times of the two events as measured in S'. (You do not have to find the relative velocity between S and S'.)

4.11. The frequency of red light is 4.5×10^{14} Hz and that of green light is 5.5×10^{14} Hz. A driver accused of going through a red light claims that the light looked green to her. How fast was the driver approaching the traffic signal? How great an error would result from using the nonrelativistic Doppler formula for this problem?

4.12. (a) Light of frequency 5×10^{14} Hz is incident perpendicularly on a mirror moving directly toward the light source at 6×10^7 m/sec. What is the frequency of the reflected light? (Hint: think of the mirror as a receiver that reemits light of the same frequency in its own rest frame.)

(b) The mirror is aligned so that the incident light makes a 45° angle with the perpendicular (as measured in the mirror frame). Find the frequency of the reflected light and its direction in the laboratory.

5   Space-Time Diagrams

5.1. WORLD LINES

This chapter introduces the space-time diagram, a geometric representation of relativity that illustrates and clarifies several important features of the theory. Consider first a problem in which events are confined to one spatial dimension. In a given frame of reference, an event is specified by two coordinates, x and t. We can therefore represent each event by a point in a two-dimensional diagram, in which distance is plotted along the vertical axis and time along the horizontal. It is convenient to use ct rather than t as the "time" variable; both scales then have the dimension of length. Such a plot is called a *space-time diagram*.

Figure 5.1 illustrates the space-time diagram; three events are shown. In the particular frame of reference to which the diagram refers, events E_1 and E_2 are simultaneous. E_2 and E_3 occur at the same location, E_3 later than E_2.

The motion of a body can be described as a succession of events, which specify its position at every instant of time. On a space-time diagram those events define a continuous line called the *world line* of the body in question.

If a body moves with constant velocity v, its position as a function of time is given by the equation

$$x = x_0 + vt = x_0 + v/c(ct) \tag{5.1}$$

where x_0 denotes the body's position at $t = 0$.

The graph of equation (5.1) is a straight line whose slope is v/c. Motion with constant velocity is therefore represented by a straight world line.

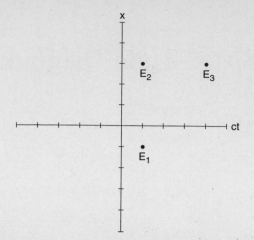

Fig. 5.1. A space-time diagram. Three events are shown. Their coordinates in this frame of reference are

for E_1: $x = -1$ m, $ct = 1$ m
for E_2: $x = 3$ m, $ct = 1$ m
for E_3: $x = 3$ m, $ct = 4$ m.

E_1 and E_2 are simultaneous, while E_2 and E_3 occur at the same location.

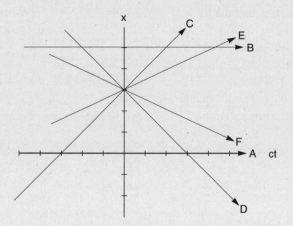

Fig. 5.2. World lines describing one-dimensional motion at constant velocity. Line A describes a body at rest at $x = 0$; line B describes a body at rest at $x = 5$ m. Lines C and D describe light rays that pass through the point $x = 3$ m at $t = 0$. The ray in C is moving to the right; that in D is moving to the left. Lines E and F describe bodies moving at constant speeds less than c; the body in E is moving faster than the body in F.

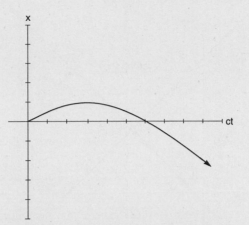

Fig. 5.3. A world line that describes one-dimensional motion with changing velocity. The body in question starts from $x = 0$ at $t = 0$, with speed $0.6c$. It proceeds to the right with diminishing speed until $ct = 3\,\mathrm{m}$ ($t = 10^{-8}\,\mathrm{sec}$), at which time it is momentarily stopped. Subsequently, the body moves to the left with increasing speed. The slope of the world line is always less than 1.

The faster a body moves, the steeper its world line.[1] A body at rest is described by a world line parallel to the time axis. A world line with negative slope describes a body moving toward lower values of x, that is, from right to left.

Since the maximum possible speed of a body is c, no world line can have a slope greater than unity; the angle between a world line and the time axis is at most 45°. A 45° world line describes the motion of a light pulse. Figure 5.2 exhibits several world lines that represent motion with constant velocity.

A general motion, in which velocity changes with time, is described by a curved world line. The slope of the line tangent to the curve at a given time measures the instantaneous velocity of the body and cannot exceed unity. An example is shown in figure 5.3. The motion of the body in question is described in the figure legend.

1. In many books, space-time diagrams are drawn with the space axis horizontal and the time axis vertical. In such a plot the slope of a world line is c/v rather than v/c and must be *greater* than unity.

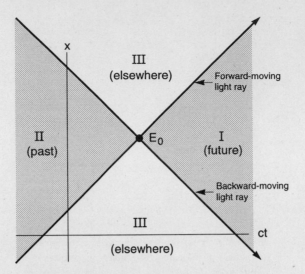

Fig. 5.4. Subdivision of the space-time plane into three invariant regions relative to some event E_0. Region I consists of events in the future of E_0; region II consists of events in the past of E_0. Region III, called the "elsewhere," consists of events that cannot be connected to E_0 by a world line of slope less than one. The boundaries between regions are the world lines of light rays that pass through E_0.

5.2. THE LIGHT CONE

The space-time diagram in figure 5.4 shows an event, E_0, and the world lines of light pulses that pass through E_0 in both directions. Those world lines divide space-time into three distinct regions. Regions I and II consist of events that can be connected to E_0 by straight lines that make angles less than 45° with the time axis; all these are possible world lines for a body that is present at E_0. Events in region I occur later than E_0, while events in region II occur earlier than E_0 (in this frame of reference). We therefore provisionally define region I as the "future" of E_0 and region II as the "past" of E_0.[2]

Region III contains the rest of space-time. A straight line between E_0 and any event in this region has a slope greater than unity and therefore cannot be the world line of a body. In fact, no curve that connects E_0 with

2. We shall soon learn that an event in region I in *any* frame occurs later than E_0 in *all* frames. The definitions of past and future are therefore absolute.

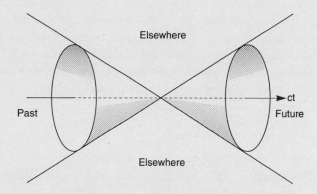

Fig. 5.5. The light cone in a world with two space dimensions. The three invariant regions are indicated.

an event in region III can be a world line, since any such curve must at some time have a slope greater than one. *All world lines that pass through E_0 are confined to regions I and II.*

If E is in region III, its space coordinate differs from that of E_0; it happens "somewhere else." I will prove in the next section that this statement is true in any frame of reference: there exists no frame in which E and E_0 occur at the same location. For this reason, region III is called the *elsewhere* for E_0. It contains events that occur both earlier and later than E_0.

These ideas are readily extended to the case of two spatial dimensions. A space-time diagram is now three-dimensional, with two space axes and a time axis, all mutually perpendicular. The world lines of light pulses still make 45° angles with the time axis. The set of all such lines that pass through a particular event, E_0, now defines a pair of cones with their axes along the time direction and with opening angles of 45°; they are called *light cones.* One opens in the direction of increasing time and is called the future light cone; the other is called the past light cone (fig. 5.5).

The interiors of the light cones constitute the generalization of regions I and II of figure 5.4. Any world line of a material body that passes through E_0 must lie inside the light cones; the world line of a light pulse lies on the surface of the light cones. The exterior of the light cones is the elsewhere, the generalization of region III.

The final generalization, to three spatial dimensions, poses a problem. A space-time diagram would have to have three space axes and a time axis, all mutually perpendicular. No such diagram can be constructed in our three-dimensional space. We can, however, conceptualize it and even describe it analytically. The light cones become hypercones, three-dimensional "surfaces" in a four-dimensional space; the world lines of light pulses are on the surfaces of the hypercones, and those of material bodies are confined to the interiors of the hypercones. We shall have no occasion to employ such esoteric concepts.

5.3. RELATIVITY AND CAUSALITY

A space-time diagram refers to a single frame of reference. To describe events in two frames, we must construct a diagram for each frame. To each point (event) in one diagram there corresponds a unique point in the other; the correspondence is given by the Lorentz transformation. In fact, the Lorentz transformation can be viewed as a *mapping* from one space-time diagram to another. Every world line in one diagram is mapped into a unique world line in the other. A straight world line, which describes motion at constant velocity, is mapped into another straight line; the two velocities are related by the velocity transformation, equation (4.15).

Figure 5.6 exhibits space-time diagrams for two frames, S and S', whose relative velocity is $0.6c$. The world lines of light rays that pass through event E_0, which define the light cones for E_0, have been constructed in each diagram. The same three events, E_1, E_2, and E_3, are plotted in each diagram.

Notice that E_1, which is in region I in the S diagram, is also in region I in the S' diagram. Likewise, E_2 is in region II in both diagrams and E_3 is in region III in both.

These results are not accidental. We shall see that the subdivision of space-time into three regions is invariant. Events in the future light cone of E_0 (region I) in one frame are in the future light cone in any other frame; events in the past light cone (region II) in one frame are in the past light cone in any other frame; and events in the elsewhere (region III) are in the elsewhere in all frames. This result is an important part of the logical structure of relativity.

The surfaces of the light cones contain the world lines of all light rays that pass through E_0. The invariance of the speed of light ensures that those surfaces are mapped into one another by the transformation. Moreover, since any speed less than c in one frame corresponds to a speed less than c in the other, the interiors of the light cones must be mapped into one another, and similarly for the exteriors. Region III in S is therefore mapped into region III in S', and vice versa.

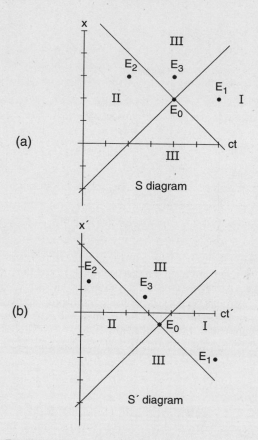

Fig. 5.6. Space-time diagrams for (a) frame S and (b) frame S'. The relative velocity between S and S' is $0.6c$. The three invariant regions for event E_0 ($x = 2$ m, $ct = 4$ m) are shown in each diagram. Three events, E_1, E_2, and E_3, are shown in each diagram. It is apparent that each event is in the same region in both diagrams.

To complete the proof of invariance, we have only to show that the sense of past and future inside the light cones is preserved, that is, the future light cone of E_0 in one diagram is mapped into the future light cone in the other and the past light cone is mapped into the past.

Figure 5.7 exhibits the space and time intervals between E_0 and events in each of the three regions in frame S. For E_1, within the future light cone of E_0, Δx and Δt satisfy the inequalities

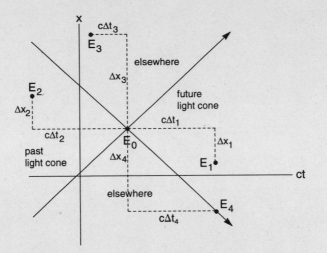

Fig. 5.7. Plot showing the space interval Δx and the time interval $c\Delta t$ between event E_0 and events in each of the three invariant regions. E_1 is inside the future light cone of E_0, E_2 is inside the past light cone, and E_3 is outside both light cones. E_4 is on the future light cone. The relations between the intervals in each case are given in the text.

$$\Delta t_1 > 0 \qquad \textit{event in}$$
$$c|\Delta t_1| > \Delta x_1 \qquad \textit{future light cone} \qquad (5.2a)$$

E_2, within the past light cone of E_0, is characterized by

$$\Delta t_2 < 0 \qquad \textit{event in}$$
$$c|\Delta t_2| > \Delta x_2 \qquad \textit{past light cone} \qquad (5.2b)$$

whereas for E_3, outside the light cones,

$$c|\Delta t_3| < |\Delta x_3| \qquad \textit{event in elsewhere} \qquad (5.3)$$

For E_4, on the light cone itself,

$$c|\Delta t_4| = |\Delta x_4| \qquad \textit{event on light cone} \qquad (5.4)$$

To prove that the sense of past and future within the light cones is invariant, we must show that $\Delta t'$ and Δt have the same sign whenever E is inside either light cone of E_0. We make use of equation (4.3), which gives the general relation between Δt and $\Delta t'$:

$$\Delta t' = \gamma \left(\Delta t - \frac{V}{c^2} \Delta x \right) \tag{5.5}$$

where V, as usual, is the velocity of S' relative to S.

It is clear from equation (5.5) that the signs of $\Delta t'$ and Δt can in general differ. When E is inside either light cone of E_0, however, the magnitude of Δt is greater than that of $\Delta x/c$. (See fig. 5.7 and eqs. [5.2a,b].) Since V/c is always less than one, it follows that

$$|\Delta t| > \left| \frac{V}{c^2} \Delta x \right| \tag{5.6}$$

and the second term on the right side of equation (5.5) is smaller in magnitude than the first. Hence the subtraction cannot lead to a change of sign: $\Delta t'$ must have the same sign as Δt.

We have shown that if E is within the future light cone of E_0 in one frame, it occurs after E_0 in *all* frames, while if E is within the past light cone in one frame, it occurs before E_0 in all frames. This is the property we set out to demonstrate; it justifies the use of the terms "absolute future" and "absolute past" to characterize the two regions.

When E is in region III, the magnitude of Δx is greater than that of $c\Delta t$. In that case, the second term on the right side of equation (5.5) can be larger in magnitude than the first and Δt and $\Delta t'$ can have opposite signs. This is the reversal of time ordering that was cited in chapter 1 as one of the unexpected implications of special relativity. We have proven that such a reversal can occur only if the events in question are outside each other's light cones. No single observer can be present at both events.

These results have direct implications for the important problem of causality. If E_0 is the cause of E, it must precede E by a time interval Δt sufficient to enable a signal emitted at E_0 to reach E. Since the maximum speed of any signal is c, Δt must exceed $\Delta x/c$, the time required for light to travel the distance Δx from the location of E_0 to that of E. This is just the condition that defines the future light cone: if E is an effect of E_0, it must be within (or on the surface of) the future light cone of E_0. Likewise, if E is the cause of E_0, it must be within (or on the surface of) the past light cone of E_0.

If these relations were not invariant, observers in some frame could see an effect before its cause. We have demonstrated that such a disaster cannot occur: *whenever two events are causally related, their time order is absolute.* Observers in any frame see the cause before the effect. The order

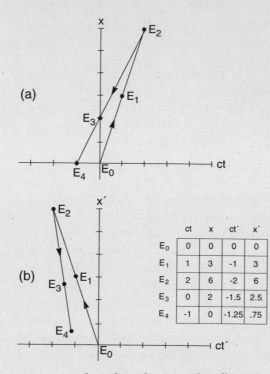

	ct	x	ct′	x′
E_0	0	0	0	0
E_1	1	3	-1	3
E_2	2	6	-2	6
E_3	0	2	-1.5	2.5
E_4	-1	0	-1.25	.75

Fig. 5.8. A hypothetical motion that illustrates the difficulties that would arise if signals could travel faster than light. The world line $E_0 - E_1 - E_2$ describes a signal emitted at the origin with speed $3c$ as measured in frame S (a). In frame S' the same signal propagates backward in time at speed $3c$; its world line is shown in (b). At event E_2 a return signal is emitted at speed $7c$ as measured in S'. That signal propagates backward in time in S and returns to its starting point (event E_4) before the first signal was emitted. The x and ct coordinates of each event in both frames are tabulated; they are related by a Lorentz transformation. The relative velocity between S and S' is $0.6c$.

of two events can depend on the observers' motion only if the events are causally independent (i.e., neither one is the cause of the other). Relativity is therefore a fully causal theory.

The causal character of relativity depends directly on the premise that no signal can propagate faster than the speed of light. If "superluminal" signals (with speed greater than c) existed, E_0 could be the cause of an event E outside its light cones. Observers in some frame would detect E before E_0, in violation of causality.

Einstein commented as follows in a paper published in 1907: "Even though, in my opinion, this result does not contain a contradiction from a purely logical point of view, it conflicts so absolutely with the character of all our experience that the impossibility of the assumption $v > c$ is sufficiently proved by this result."[3]

An even more bizarre scenario can be constructed with superluminal signals. Figure 5.8a shows the world line of a hypothetical signal that leaves the origin at $t = 0$ (event E_0) at speed $3c$, measured in frame S. The signal reaches the point $x = 6$ m at $ct = 2$ m (event E_2).

Sketch (b) shows the world line of the same signal in a frame S' that moves at velocity $0.6c$ relative to S. According to equation (4.15), the velocity of the signal in S' is $-3c$. As the figure shows, in S' the time of E_2 is less than that of E_0: S' observers see the signal arrive *before* it departs. The S' world line seems to be propagating backward in time.

A return signal is emitted at speed $7c$, measured in S'. The return signal propagates forward in time in S' but backward in time in S (at speed $-2c$). The signal arrives at the original point of departure (event E_4) *before* the first signal was emitted! By making use of such a combination of superluminal signals it would be possible to communicate with one's own past.[4] Logical disasters of this kind are avoided only if no signal can exceed the speed of light.

Finally, we can justify the use of the term "elsewhere" to characterize region III. We start from the general relation between Δx and $\Delta x'$, equation (4.7):

$$\Delta x' = \gamma(\Delta x - V\Delta t) \tag{5.7}$$

Suppose we want to find a frame in which the two events in question occur at the same location. Setting $\Delta x'$ in equation (5.7) equal to zero, we obtain for V/c the following expression:

3. A. Einstein, "On the Inertia of Energy Required by the Relativity Principle," *Annalen der Physik* 23 (1907):371–384. (*Collected Papers*, doc. 45.)
4. The problem is accurately described by the following limerick, attributed to Arthur Buller:

> There was a young lady named Bright
> Whose speed was much faster than light
> She set out one day
> In a relative way
> And returned on the previous night.

$$\frac{V}{c} = \frac{\Delta x}{c\Delta t} \tag{5.8}$$

When E is in region I or II, the value of V/c determined by equation (5.8) is less than one. (See eq. [5.2].) When E is in region III, however, the magnitude of Δx is greater than that of $c\Delta t$. In this case the relative velocity V determined by equation (5.8) is greater than c. Hence there exists no frame in which $\Delta x'$ is zero. Two events outside each other's light cones take place at different locations in any frame; each is indeed "elsewhere" of the other.

A similar argument shows that the condition $\Delta t' = 0$ can always be satisfied if E is in region III but not if it is in region I or II.

To summarize, if two events are inside each other's light cones, there always exists a frame in which they occur at different times at the same location but never one in which they occur simultaneously at different locations. Conversely, if two events are outside each other's light cones, there always exists a frame in which they are simultaneous but never one in which they occur at different times at the same location.

If E is on one of the light cones of E_0, Δx and Δt satisfy equation (5.3). In this case there exists neither a frame in which the events are simultaneous nor one in which they occur at the same place.

5.4. TIMELIKE AND SPACELIKE INTERVALS

In chapter 4, we defined the relativistic interval

$$(\Delta s)^2 = c^2(\Delta t)^2 - (\Delta x)^2 \tag{5.9}$$

and showed that it is invariant: if $(\Delta s')^2$ is defined in terms of $\Delta t'$ and $\Delta x'$ by a relation similar to (5.9), then

$$(\Delta s')^2 = (\Delta s)^2$$

for any two inertial frames S and S'.

The inequalities (5.2) through (5.4) that define the three regions of the space-time diagram can be expressed succinctly in terms of Δs. The condition $|c\Delta t| > |\Delta x|$ becomes simply

$$(\Delta s)^2 > 0$$

whereas the opposite condition $|c\Delta t| < |\Delta x|$ becomes

$$(\Delta s)^2 < 0$$

Hence the three regions of space-time are defined by the invariant conditions:

region I (future light cone): $(\Delta s)^2 > 0$, $\Delta t > 0$ (5.10a)

region II (past light cone): $(\Delta s)^2 > 0$, $\Delta t < 0$ (5.10b)

region III (elsewhere): $(\Delta s)^2 < 0$ (5.10c)

On the surface of the light cones

$$\Delta s = 0 \qquad\qquad (5.10d)$$

These relations can be used to distinguish in an invariant way among three types of intervals. Suppose first that $(\Delta s)^2$ is positive; the events in question are within each other's light cones. In this case there exists a frame in which $\Delta x = 0$ and

$$\Delta s = \pm c\Delta t$$

Such an interval is called *timelike*. If two events are separated by a time-like interval, there is no frame in which they are simultaneous: if $\Delta t = 0$, $(\Delta s)^2$ cannot be positive.[5] Δs for a timelike interval is just the proper time interval between the events, as defined in chapter 3.

When $(\Delta s)^2$ is negative (the events are outside each others' light cones), there exists a frame in which $\Delta t = 0$ and

$$\Delta s = \pm i |\Delta x|$$

Such an interval is called *spacelike*. There is no frame in which the events occur at the same place: if $\Delta x = 0$, $(\Delta s)^2$ cannot be negative. For events separated by a spacelike interval, proper time is not defined.

Finally, if $\Delta s = 0$, the two events can be connected by the world line of a light pulse in any frame. Such an interval is called *lightlike*.

5.5. The Loedel Diagram[†]

The space-time diagrams constructed up to this point refer to a specific frame of reference. If a problem involves two frames, a separate diagram must be constructed for each, as in figure 5.6. Although there is a one-to-one correspondence between points in the two diagrams, that correspondence is not apparent from the plots. To determine which point in the S'

---

5. In some books, $(\Delta s)^2$ is defined with the opposite sign:

$$(\Delta s)^2 = (\Delta x)^2 - c^2 (\Delta t)^2$$

With this definition, $(\Delta s)^2$ is positive for spacelike intervals and negative for time-like intervals.

†Sections 5.5 through 5.7 and the appendixes can be omitted with no loss of continuity.

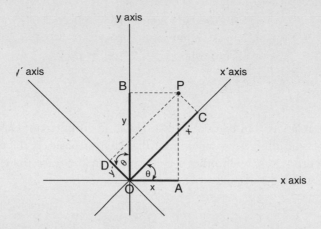

Fig. 5.9.   Two rectangular coordinate systems are plotted with a common origin. The x' and y' axes make an angle θ with the x and y axes. The coordinates of a point P in both systems are indicated.

diagram corresponds to a given point in the S diagram, one has to use the Lorentz transformation.

Can anything better be done? Is it possible to include both frames of reference in a single diagram and thereby display the Lorentz transformation geometrically? The motivation for such a plot is provided by analogy with ordinary spatial diagrams.

Figure 5.9 shows two rectangular coordinate systems whose axes are labeled (x, y) and (x', y'). The two sets of axes have a common origin, O; one set is rotated relative to the other by the angle θ. The coordinates of an arbitrary point, P, in either system can be found from this single diagram, as shown by the dotted lines. The two sets of coordinates are related by the equations

$$x' = x \cos \theta + y \sin \theta \qquad (5.11a)$$
$$y' = -x \sin \theta + y \cos \theta \qquad (5.11b)$$

Equation (5.11) can be considered a "transformation," mathematically similar to the Lorentz transformation, equation (4.1). In each case, the primed coordinates are expressed as linear combinations of the unprimed ones.

A naive idea would be to try to treat space-time diagrams in precisely the same way, that is, to plot the x and ct axes of frame S and those of S'

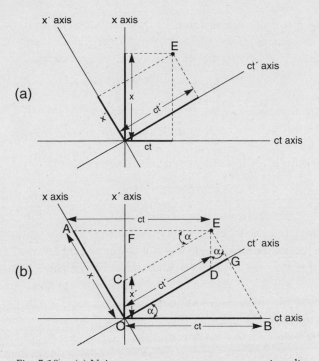

Fig. 5.10. (a) Naive attempt to construct a space-time diagram for two frames of reference, by analogy with fig. 5.9. The primed and unprimed coordinates of an event *E* are indicated. These coordinates are not related by a Lorentz transformation; hence the construction is useless. (b) The Loedel diagram. Notice the difference between this plot and the one in (a): in (b), the *ct* axis is perpendicular to the *x'* axis, and the *x* axis is perpendicular to the *ct'* axis. The coordinates of event *E* in both frames are constructed according to the prescription described in the text.

with a common origin and at some angle to each other, as in figure 5.10a. This approach fails because the two sets of coordinates thereby assigned to a given event are not related by a Lorentz transformation.

It is not hard to discern the underlying reason for the failure. Under a rotation of axes, described by equation (5.11), the quantity $x^2 + y^2$ is invariant (it is the square of the distance between O and P), whereas the corresponding quantity $(ct)^2 + x^2$ is not invariant under a Lorentz transformation.

An ingenious alternative approach was discovered by Enrique Loedel.[6] Loedel's construction is based on the realization that although $(ct)^2 + x^2$ is not invariant in relativity, the quantity $(ct)^2 - x^2$ *is* invariant. The relation

$$(ct)^2 - x^2 = (ct')^2 - x'^2 \tag{5.12}$$

which expresses that invariance, can be trivially rewritten as

$$x^2 + (ct')^2 = x'^2 + (ct)^2 \tag{5.13}$$

Equation (5.13) is formally analogous to the equation

$$x^2 + y^2 = x'^2 + y'^2 \tag{5.14}$$

which expresses the invariance of the distance to point P in figure 5.9. This analogy led Loedel to construct the diagram shown in figure 5.10b, in which the x axis is perpendicular not to the ct axis but to the ct' axis; the x' and ct axes are likewise perpendicular. (The physical significance of the angle α between the ct and ct' axes will become apparent presently.)

Diagrams in which the coordinate axes are not perpendicular are unfamiliar to most readers. Accustoming oneself to such a construction requires some effort, but the reward is that one can see the effects of relativity displayed geometrically. Although the Loedel diagram does not provide any new results, it complements the standard algebraic presentation of the theory. Many students find it helpful.

The mathematical basis for the Loedel diagram is developed in appendix 5A. Here I explain how the diagram is constructed and illustrate some of its applications.

Our first task is to specify how the coordinates of a given event E in the diagram are defined. Loedel's prescription is the following. (See fig. 5.10b.) Through point E construct lines parallel to each of the coordinate axes; two parallelograms, $OAEB$ and $OCED$, are thereby defined, one for each frame. The coordinates of E in each frame are by definition the lengths of the sides of those parallelograms:

$$x = OA \qquad ct = OB$$

6. E. Loedel, "Aberracion y Relatividad," *Anales de la Sociedad Científica Argentina* 145 (1948):3–13. A somewhat similar diagram has been devised by Robert W. Brehme: "A Geometric Representation of Galilean and Lorentz Transformations," *American Journal of Physics* 30 (1962):489–496. A space-time diagram for two frames of reference was first proposed by Hermann Minkowski in 1908. The Minkowski diagram is awkward to apply because different scales must be used to measure distances in the two frames.

and

$$x' = OC \qquad ct' = OD$$

It is shown in appendix 5A that the two sets of coordinates thus defined are correctly related by a Lorentz transformation. The angle α between the x and x' axes (and between the t and t' axes) is determined by the relative velocity V between S and S':

$$\sin \alpha = V/c \tag{5.15a}$$

$$\cos \alpha = \sqrt{1 - V^2/c^2} = 1/\gamma \tag{5.15b}$$

When $V = 0$, $\alpha = 0$ and there is no transformation; the primed and unprimed axes coincide. As V approaches c, α approaches 90°.

World Lines

In the Loedel diagram in figure 5.11, line AA is parallel to the ct axis. According to Loedel's prescription, the x coordinate of every point on that line is the same. A world line parallel to the time axis therefore describes a body at rest in frame S, just as for a regular space-time diagram. The

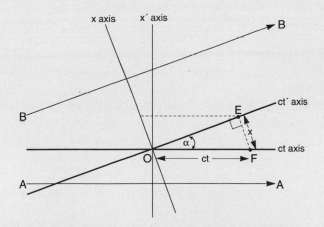

Fig. 5.11. World lines of a body at rest in frame S (line AA) and in frame S' (line BB). The coordinates in S of E, an event on the ct' axis, are constructed. As shown in the text, they satisfy the relation $x = Vt$, which describes a body moving with velocity V.

time axis itself is the world line of a body at rest at $x = 0$. Line *BB*, which is parallel to the t' axis, is similarly the world line of a body at rest in frame *S'*.

As an exercise, let us find the *S* equation of the world line $x' = 0$, which describes a body at rest at the origin of *S'*. The dotted lines in figure 5.11 determine the *S* coordinates of point *E* on the t' axis. Its *x* coordinate is *EF* and its *ct* coordinate, *OG*, is equal to *EF*. Since *OEF* is a right angle,

$$\sin \alpha = EF/OF = x/ct$$

Using equation (5.15a), we obtain

$$x = Vt$$

which is the expected result. One can similarly verify that a world line along the *t* axis, which describes a body at rest in frame *S*, has in *S'* the equation $x' = -Vt'$, which describes a body moving with velocity $-V$.

One limitation of the Loedel diagram is that only two frames can be included in a single plot. The angle between axes is fixed by their relative velocity, according to equation (5.15). We cannot add a third frame by constructing another pair of axes, as one can do with the rotated axes of figure 5.10a. To compare measurements in *S* with those in some third frame *S″*, we must construct an entirely new diagram.

5.6. Applications of the Loedel Diagram

The Light Cones

Figure 5.12 displays the light cones and the three invariant regions of space-time in the Loedel diagram. The figure looks quite busy but merits careful study.

World line *POA* describes a light pulse that passes through the origin at $t = 0$ in the positive *x* direction. *E* is an event on that world line. According to Loedel's prescription, the *S* coordinates of *E*, *x* and *ct*, are the sides of parallelogram *ODEF*. Since the *S* equation of the world line is $x = ct$, all four sides of that parallelogram are equal and *POA* bisects the angle between the positive *x* and the positive *t* axes. The world line makes an angle $45° + \frac{1}{2}\alpha$ with each axis.

Parallelogram *OMEN* similarly defines the coordinates of *E* in frame *S'*. It too has equal sides and corresponds to the equation of motion $x' = ct'$. *POA* also bisects the angle between the positive x' and the positive t' axes.

World line *QOB* describes a light pulse that passes through the origin

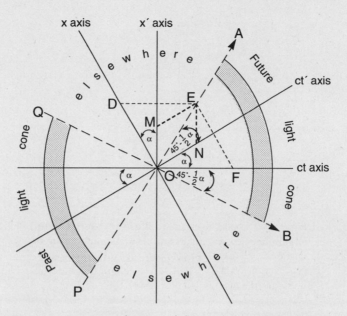

Fig. 5.12. Invariant division of the Loedel diagram. *POA* and *QOB* are the world lines of light rays that propagate in the forward and backward directions. The region *AOB* is the future light cone for an event at *O*, while *QOP* is the past light cone. The remainder is the elsewhere. Notice the angles made by the world lines of the light rays with the two times axes.

at $t=0$ in the negative x direction. Its S equation is $x=-ct$. *QOB* bisects the angle between positive t and negative x axes and also the angle between positive t' and negative t'. Notice that the world lines of the two light rays that pass through the same event in opposite directions are perpendicular to one another in the Loedel diagram.

The past and future light cones of an event at the origin are indicated by shading in the figure; the exterior of the light cones is the "elsewhere." The invariance of the three regions is quite apparent in the Loedel diagram.

Simultaneity

The Loedel diagram in figure 5.13 exhibits the relativity of simultaneity. Events E_1 and E_2 are simultaneous in frame S. We see directly from the diagram that the two events are not simultaneous in S'. (In this particular case, $t_2' < t_1'$.)

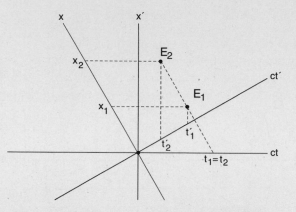

Fig. 5.13. Loedel diagram to illustrate the relativity of simultaneity. Events E_1 and E_2 are simultaneous in frame S $(t_1 = t_2)$ but happen at different places $(x_2 > x_1)$. In frame S' the two events are not simultaneous: as the diagram shows, $t_2' < t_1'$.

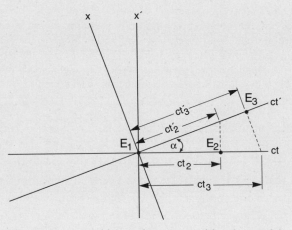

Fig. 5.14. Time dilation demonstrated by the Loedel diagram. Events E_1 and E_2 occur at the same place in frame S; the time interval between them is longer in S' than in S. Events E_1 and E_3 occur at the same place in S'; the time interval between them is longer in S than in S'.

Time Dilation

Figure 5.14 is a Loedel diagram that demonstrates time dilation. Event E_1 takes place at the origin in both frames ($x_1 = x_1' = 0$, $t_1 = t_1' = 0$). Event E_2 takes place at the same location as E_1 in frame S. Hence $t_2 - t_1$ is a proper time interval whereas $t_2' - t_1'$ is improper. The figure shows that

$$\frac{ct_2}{ct_2'} = \cos\ \alpha = \frac{1}{\gamma}$$

Hence

$$t_2' = \gamma t_2$$

as required by a time dilation argument.

Also shown is event E_3, which occurs at the same place as E_1 in S'. In this case the time interval $t_3' - t_2'$ is proper whereas $t_3 - t_1$ is improper. The geometry shows that $t_3 = \gamma t_3'$, again consistent with a time dilation argument.

Length Contraction

Finally, figure 5.15 demonstrates length contraction. A rod of proper length L_0 is at rest in frame S, with its left end at $x = 0$ and its right end at $x = L_0$. AA and BB are the world lines of the two ends of the rod. AA is along the t axis, while BB is parallel to that axis.

The intersections of AA and BB with a line $t = $ constant show how the rod appears to S observers at that time. The shaded bands indicate the appearance of the rod in frame S at $t = 0$ and at two later times.

To see how the rod looks to S' observers, we have to examine the intersections of the world lines at each of its ends with lines of constant t'. Such lines are parallel to the x' axis. The dotted bands show the appearance of the rod at $t' = 0$ and at a later t'. It is apparent that the rod looks contracted in S'; its length is $L_0 \cos\ \alpha = L_0/\gamma$, in agreement with what we learned about length contraction in chapter 3.

Also shown in the figure are the world lines CC and DD that describe a rod at rest in S'. Observers in S see this rod as contracted, as the figure clearly shows. The Loedel diagram makes the symmetry between the two frames quite explicit; this is one of its major attractive features.

In the following chapter, we shall employ the Loedel diagram to help clarify some of the paradoxes of relativity.

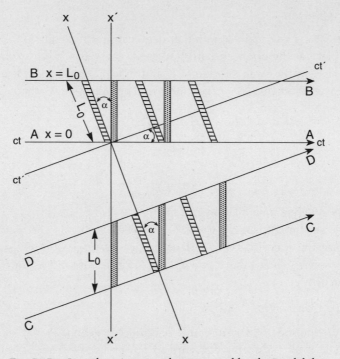

Fig. 5.15. Length contraction demonstrated by the Loedel diagram. World lines AA and BB represent the two ends of a rod at rest in frame S; lines CC and DD are those of a rod at rest in S'. The shaded bands are snapshots taken in S; each one shows the appearance of the rod at a specific time in S. The dotted bands are similar snapshots taken in S'. It is apparent that rod AB looks contracted in S', whereas rod CD looks contracted in S.

5.7. THE DOPPLER EFFECT REVISITED

The Loedel diagram provides a very nice geometric representation of the relativistic Doppler effect, which was treated analytically in section 4.8. With a little trigonometry, the formulas for the Doppler effect can be derived.

Figure 5.16 contains the Loedel diagram for the Doppler problem. Frame S is the rest frame of the source and S' that of the receiver, with relative velocity V between them. (Notice I do not specify which of the two is moving.) The source is located at $x = 0$; its world line is along the t axis. The receiver is at $x' = 0$; its world line is along the t' axis. For $t < 0$, the source and receiver approach one another. At $t = t' = 0$, the world lines

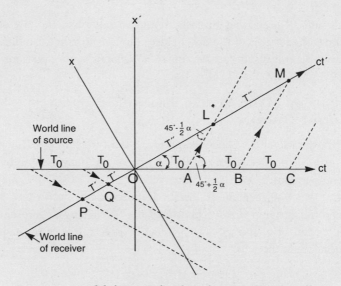

Fig. 5.16. Loedel diagram for the relativistic Doppler effect. The source, whose world line is the *ct* axis, emits light flashes every *T* seconds (source time). The receiver, whose world line is the *ct'* axis, at first approaches the source; it passes the source at $t=t'=0$ and thereafter recedes. During the approaching phase, the light rays arrive every *T'* seconds (receiver time), where $T' < T$. During the receding phase, the rays arrive every *T''* seconds, where $T'' > T$. Formulas for *T'* and *T''* are derived in the text.

cross as the source and receiver pass one another. Thereafter they recede from one another.

The source emits light of frequency f and period $T_0 = 1/f$. Once every T_0 seconds (as measured in S) it emits a wave crest that propagates in both directions. The world lines of the wave crests are shown as dotted lines in the figure. The arrivals of the wave crests at the receiver are marked with solid circles.

Let T' be the interval between arrival of successive crests at the receiver during the approach phase and let T'' be the corresponding interval during the recession phase, *as measured by the receiver's clock*. The reciprocals of T' and T'' are the frequencies measured by the receiver during each phase.

The diagram clearly shows that T' is less than T_0 while T'' is greater than T_0, as expected. Finding the values of T' and T'' requires some trigonometry; we have to solve the triangles OAL and OPQ of the figure. Since one side and all the angles of each triangle are known, the law of sines determines the lengths of the other sides. The calculation is carried out in appendix 5B. The results are of course consistent with equations (4.40) and (4.41), obtained analytically.

APPENDIX 5A: MATHEMATICAL BASIS FOR THE LOEDEL DIAGRAM

To justify the use of the Loedel diagram we have to show that the primed and unprimed coordinates of point P in figure 5.10b are related by a Lorentz transformation.

Notice that line segment AD is the hypotenuse of both right triangle OAD and right triangle EAD. The legs of the first triangle represent x and ct', while the legs of the second represent x' and ct. Hence we have from the Pythagorean theorem

$$x^2 + (ct')^2 = x'^2 + (ct)^2 \tag{5.A1}$$

As noted in the text, equation (5.A1) gave Loedel the idea for his construction.

To obtain the Lorentz transformation and determine the significance of the angle α between axes, we proceed as follows. From right triangle OAF one has

$$x \cos \alpha = OF = OC + CF = OC + CE \sin \alpha$$
$$= x' + ct' \sin \alpha \tag{5.A2}$$

If we put

$$\sin \alpha = V/c \tag{5.A3}$$

$$\cos \alpha = \sqrt{1 - V^2/c^2} = 1/\gamma \tag{5.A4}$$

equation (5.A2) becomes

$$x = \gamma(x' + Vt') \tag{5.A5}$$

which is just equation (4.2a) of the Lorentz transformation.

Similarly, from right triangle OGB one has

$$ct \cos \alpha = OD + DG = OD + DE \sin \alpha$$
$$= ct' + x' \sin \alpha \tag{5.A6}$$

Using (5.A3) and (5.A4), this becomes

$$t = \gamma\left(t' + \frac{V}{c^2} x'\right) \tag{5.A7}$$

which is the other part of the Lorentz transformation. Thus, with α defined by equation (5.A3), the Loedel diagram is consistent with the Lorentz transformation.

APPENDIX 5B: THE DOPPLER FORMULA

In this appendix we derive the relativistic Doppler formulas from the Loedel diagram of figure 5.16.

For the case of source and receiver receding we must find the length of *OL*, which represents the time interval T'' between arrival of successive wave crests at the receiver. In triangle *OAL*, side *OA* is T_0. Angle *AOL* is α. The exterior angle *BAL* is the angle between the world line of light ray *AL* and the t axis. That angle is $45° + \frac{1}{2}\, \alpha$. Angle *OAL*, its supplement, is therefore $135° - \frac{1}{2}\, \alpha$. Since the sum of all three angles of the triangle is 180°, angle *OLA* must be $45° - \frac{1}{2}\, \alpha$.

The ratio between T'' and T_0 is given by the law of sines:

$$\frac{T''}{T_0} = \frac{OL}{OA} = \frac{\sin\,(\sphericalangle OAL)}{\sin\,(\sphericalangle OLA)} = \frac{\sin \frac{1}{2}(270° - \alpha)}{\sin \frac{1}{2}(90° - \alpha)} \tag{5.B1}$$

This expression can be simplified by using the trigonometric identity

$$\sin \frac{\theta}{2} = \sqrt{\frac{1 - \cos\,\theta}{2}}$$

Using this formula in both the numerator and the denominator of equation (5.B1), we obtain

$$\frac{T''}{T_0} = \sqrt{\frac{1 - \cos(270° - \alpha)}{1 - \cos(90° - \alpha)}} = \sqrt{\frac{1 + \sin\,\alpha}{1 - \sin\alpha}}$$

Finally, substituting for $\sin\,\alpha$ its value V/c, we get

$$\frac{T''}{T_0} = \sqrt{\frac{1 + V/c}{1 - V/c}} \tag{5.B2}$$

Since $T_0 = 1/f$ and $T'' = 1/F''$, (5.B2) becomes

$$f'' = f \sqrt{\frac{1 - V/c}{1 + V/c}} \tag{5.B3}$$

which is the Doppler formula for the case of source and receiver receding, equation (4.41a).

The formula for the case of source and receiver approaching is obtained by solving for side *OQ* in triangle *OPQ*. The derivation proceeds exactly as in the other case, and the result reproduces equation (4.41b).

PROBLEMS

5.1. Event E_0 has the following coordinates in some frame S:

$$x = 2 \text{ m} \qquad ct = 4 \text{ m}$$

Locate E_0 on a space-time diagram for frame S. Draw the lines that define the three invariant regions of space-time for E_0. On the diagram, locate each of the following events:

$$E_1: x = 0, \ ct = 1 \text{ m}$$
$$E_2: x = 4 \text{ m}, \ ct = 2 \text{ m}$$
$$E_3: x = 4 \text{ m}, \ ct = 5 \text{ m}$$

Specify whether each event is on the light cone of E_0, in the past or the future of E_0, or in the elsewhere. If E is within one of the light cones, find the speed (relative to S) of a frame in which E occurs at the same place as E_0. If E is in the elsewhere of E_0, find the speed of a frame in which it is simultaneous with E_0.

5.2. Refer to the three events E_1, E_2, and E_3 of problem 5.1. For each pair of events, (1,2), (2,3), and (1,3), find the value of $(\Delta s)^2$ and characterize the interval between the events as spacelike, timelike, or lightlike. Which pairs of events could be causally related?

5.3. Consider problem 3.8, in which a light pulse catches up to a spaceship.

(a) Construct the space-time diagram for the earth frame, S. Use the light-minute as the unit of distance. Draw the world lines of the earth, the spaceship, and the light pulse. Find graphically the time when the light pulse catches up to the spaceship and the distance at which this happens according to earth frame observers.

(b) Construct the space-time diagram for S', the spaceship frame. Draw the world lines of the earth and the spaceship. Using the result of problem 3.8, locate the point on the earth's world line at which the light pulse is emitted and draw the world line of the light pulse. Find the time of E_3 in S'. Check your graphical results for (a) and (b) with those obtained analytically in problem 3.8.

(c) Construct the Loedel diagram for the problem. Draw the three relevant world lines and find the time of E_3 in each frame. Compare with the results of (a) and (b).

5.4. Work problem 3.9 using space-time diagrams. Follow the procedure outlined in problem 5.3. Use the following numerical values: $V = 0.6c$, $U = 0.8c$, $T = 4$ min.

5.5. Frame S' moves at velocity $V = 0.5c$ relative to frame S.

(a) Construct the Loedel diagram for frames S and S'. Label the x, ct, x', and ct' axes.

(b) Event E_0 has the following coordinates in S: $x = 2$ m, $ct = 3$ m. Locate E_0 on the Loedel diagram and find graphically its coordinates in S'.

(c) Find the coordinates of E_0 in S' using the Lorentz transformation and compare with the results of (b).

(d) Draw the world lines of light pulses that pass through E_0 in both directions.

Find graphically the time in S' at which a light pulse emitted at E_0 reaches the point $x' = 0$.

5.6. A spaceship is approaching earth at speed $0.8c$. At $t = 0$ the distance of the spaceship is 10 light-minutes (earth frame measurements). At this time a radar signal is emitted from earth in the direction of the spaceship. When the signal reaches the spaceship, a return signal is immediately emitted toward earth.

Construct the Loedel diagram for this problem, with the earth at $x = 0$ and the spaceship at $x' = 0$. Draw the world lines of earth, spaceship, and both radar signals. Find the time when the return signal reaches earth in both the earth frame and the spaceship frame.

6 Paradoxes of Relativity

6.1. INTRODUCTION

A paradox is an apparent inconsistency or contradiction in a theory or in
any logical system. If the paradox cannot be satisfactorily resolved, the
theory in question fails and must be abandoned or at least modified.

In this chapter we analyze several paradoxes based on special relativity
which have attracted attention over the years. All of them involve length
contraction or time dilation. One, the twin paradox, has generated a great
deal of controversy. Careful study of the paradoxes and of their resolution
helps clarify many subtleties in the theory. The space-time diagram, introduced in chapter 5, proves useful in analyzing the paradoxes and identifying their resolution.

6.2. THE "POLE AND BARN" PARADOX

Figure 6.1 is a sketch of the pole and barn problem, the prototype for a
number of paradoxes based on the length contraction effect. A barn has
proper length L. A pole, also of proper length L, is carried by a runner
who moves at nearly the speed of light. In the rest frame of the barn,
therefore, the pole is observed as contracted to length L/γ and should fit
with ease inside the much longer barn.

In the rest frame of the runner, however, the barn is moving and appears contracted. The pole cannot fit in so narrow a barn; both ends must
protrude for a definite interval while the pole is passing through the barn.
By analyzing the situation in two different frames of reference, therefore,
we arrive at apparently contradictory conclusions.

The space-time diagram exhibits clearly the sequence of events seen in
each frame. Figure 6.2 is a space-time diagram for frame S, the rest frame
of the barn. The world lines of the front door (F) and of the rear door (R)

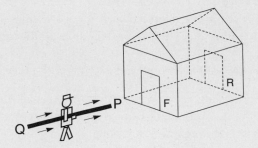

Fig. 6.1. The pole and barn paradox. Does the pole fit inside the barn, or doesn't it?

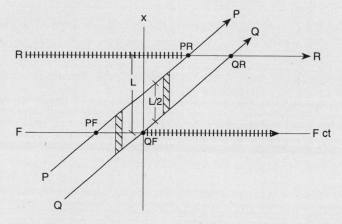

Fig. 6.2. Space-time diagram for the pole and barn problem constructed in frame S, the rest frame of the barn. Shown are the world lines of F and R, the front and rear doors of the barn, and of Q and P, the two ends of the pole. Events PF, QF, PR, and QR, which denote the passage of the ends of the pole through each door, are indicated with solid circles. Cross-hatching on the world line of a door indicates that the door is closed. During the interval between QF and PR, both doors are shut and the pole is entirely inside the barn.

of the barn are shown; they describe bodies at rest at $x = 0$ and $x = L$, respectively. Also shown are the world lines of P and Q, the two ends of the pole; they describe a body of length L/γ moving from left to right at speed V.[1]

The four events of interest in the problem are those at which P and Q pass through each door; call those events PF, PR, QF, and QR. On the

1. For the diagram and the calculations, we take V/c to be $\sqrt{3}/2$, which makes $\gamma = 2$.

diagram, *PF* is found at the intersection of the world lines of *P* and *F* and similarly for the other three events. Event *QF* is arbitrarily assigned the time $t = 0$.

The diagram shows that the order of events[2] in the barn frame is:

1. *PF* (*P*, the leading end of the pole, passes through the front door)
2. *QF* (*Q*, the trailing end of the pole, passes through the front door)
3. *PR* (*P* passes through the rear door)
4. *QR* (*Q* passes through the rear door)

This order is consistent with the barn observers' assertion that the pole is shorter than the barn. During the interval between event *QF*, when the trailing end *Q* enters through the front door, and event *PR*, when the forward end *P* leaves through the rear, the pole is entirely within the barn.

Figure 6.3 is the space-time diagram for *S'*, the rest frame of the runner. The same four world lines are shown. In this frame the positions of *Q* and of *P* are constant ($x' = 0$ and $x' = L$) while the world lines of *F* and *R* describe a body of length L/γ (e.g., the roof beam of the barn) moving from right to left at speed *V*.

According to figure 6.3 the order of events[3] in *S'* is:

1. *PF*
2. *PR*
3. *QF*
4. *QR*

The all-important difference between the two sets of observations is that events *PR* and *QF* occur in the opposite order.[4] According to *S'* observers, the forward end of the pole exits through the rear door (event *PR*) *before* the trailing end enters through the front door (event *QF*). During the interval between those events, therefore, the pole protrudes from both doors. *S'* observers maintain that, contrary to the assertion

---

2. The times of the four events can be found analytically by writing the equations of motion for *P*, *Q*, *F*, and *R* and solving the appropriate pairs of equations simultaneously. The result is $t(PF) = -L/\gamma V$, $t(QF) = 0$, $t(PR) = (L/V)(1 - 1/\gamma)$, $t(QR) = L/V$.

3. The times of the four events in *S'* are $t'(PF) = -L/V$, $t'(PR) = -L(1 - 1/\gamma)/V$, $t'(QF) = 0$, $t'(QR) = L/\gamma V$. The times and positions of each event in the two frames are of course related by a Lorentz transformation.

4. Those two events must be separated by a spacelike interval; the invariant $(\Delta s)^2$ can be calculated in either frame and is negative.

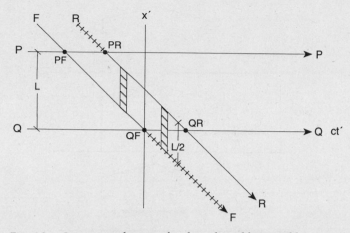

Fig. 6.3. Space-time diagram for the pole and barn problem constructed in S', the rest frame of the runner. See the legend to fig. 6.2.

of the barn observers, the pole is never entirely inside the barn. This contradiction constitutes the paradox.

The disagreement can be put into even sharper focus by embellishing the story a little. Suppose the rear door of the barn is initially shut and the front door is initially open. (Cross-hatching on the world line of a door indicates that the door is shut.) Immediately after event QF (i.e., just after the trailing end of the pole has passed through the front door), a barn observer stationed at that door shuts it. Just before event PR, when P is about to collide with the closed rear door, another barn observer opens that door, permitting the pole to pass through.

According to barn frame observers, during the interval between QF and PR *both doors are shut.* Figure 6.4 contains a sequence of pictures that summarizes the entire history of the problem as seen in the barn frame. These pictures "prove" that the pole is shorter than the barn; note particularly figure 6.4c, in which the pole is entirely inside the barn with both doors shut. What better proof could one ask for?

Observers in the runner's frame remain unconvinced. According to them, event PR occurred before QF; hence the rear door was opened *before* the front door was shut. At no time in the space-time diagram of figure 6.3 are both doors shut. On the contrary, during the interval between events PR and QF both doors are *open.* Figure 6.5 contains a sequence of pictures analogous to those of figure 6.4 but showing the story as it appears in S'. In figure 6.5c, both ends of the pole are seen protruding

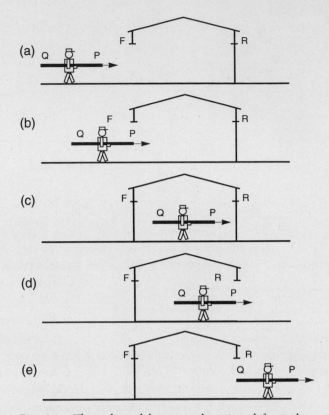

Fig. 6.4. The pole and barn paradox viewed from the barn frame. (a) Runner approaching the barn; the front door is open, the rear door is closed. (b) Leading end of the pole enters the barn (event *PF*). (c) Trailing end of the pole has just entered the barn; the front door is now closed. The pole is entirely within the barn, with both doors closed. (d) Leading end of the pole is about to exit through the rear door, which has just been opened. (e) Trailing end of the pole leaving the barn (event *QR*).

from the barn, with both doors open; this picture "proves" that the pole is longer, in direct contradiction to the assertion of the barn observers.

Which observers are right? The key to the resolution of the paradox is the realization that the question, Is the pole ever entirely inside the barn? is, in essence, a question about the simultaneity of separated events. It is logically equivalent to asking whether there exists a pair of events, E_1 and E_2, such that

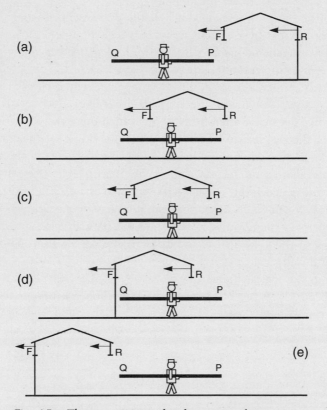

Fig. 6.5. The story as seen by the runner (frame S'). (a) Barn approaching the stationary pole; the front door is open, the rear door is shut. (b) Rear of barn just reaching end P of the pole (event PR). Rear door has just opened. (c) Pole sticks out of both ends of the barn; both doors are open. (d) Front of the barn has just passed end Q of the pole (event FQ). Front door is now shut, rear door is open. (e) Barn has completely passed the pole (event QR has just occurred).

(i)   at event E_1 end P of the pole is inside the barn;

(ii) at event E_2 end Q is inside the barn; and

(iii) E_1 and E_2 are simultaneous.

If the answer to the question is affirmative, the pole is shorter than the barn. If not, the pole is longer.

In a Galilean world, in which time and simultaneity are absolute, the question posed above must have a definite answer, independent of any frame of reference. A situation such as the one described would be logically unacceptable. In special relativity, however, simultaneity is not absolute; hence we should not be surprised to find that observers in the two frames answer the question differently.

The two sets of observers do agree on one important aspect of the story: no part of the pole comes in contact with either door. Such an occurrence would leave detectable aftereffects, about which there can be no disagreement. If a theory predicted that observers in one frame see the pole smash into a closed door while those in the other frame see it pass through the same door, that would indeed constitute an unacceptable contradiction. As we have seen, however, special relativity makes no such embarrassing prediction. *S'* observers agree that the pole passes successfully through the barn, even though it is longer than the barn. It manages to do so, according to them, because of the delay in closing the front door. But it does get through.

Within the logical framework of relativity, then, the results are not contradictory. Two sets of observers disagree about the time order of certain separated events, but such disagreements are nothing new; they are

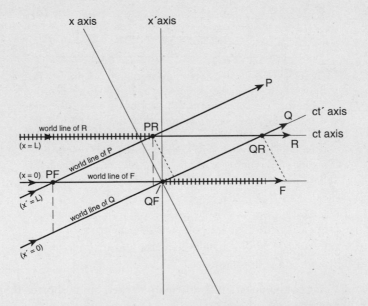

Fig. 6.6.   Loedel diagram for the pole and barn problem. This figure contains all the information found in both fig. 6.2 and fig. 6.3. The reversal of the time order of events *QF* and *PR* in the two frames is apparent.

unavoidable consequences of Einstein's postulates, as we saw in chapter 3. On all questions that involve physically verifiable effects, the observers agree. Thus the paradox is resolved.

Is the pole ever "really" completely inside the barn? Tempting though that question may be, it is simply not a legitimate question in Einsteinian relativity. The descriptions of the two sets of observers are equally real and equally valid, each within their own frame of reference. Since no preferred frame exists, there is no objective basis for ascribing any more reality to one description than to the other.

For the benefit of readers who have had the courage to follow the derivation of the Loedel diagram in chapter 5, figure 6.6 presents the Loedel diagram for the pole and barn problem. This figure contains all the information found in both figure 6.2 and figure 6.3. All the salient features of the story, including the reversal in time order of events *QF* and *PR*, are apparent in the diagram.

6.3. WHAT IF THE POLE STOPS?

Persistent barn observers decide to make one final attempt to demonstrate that the pole is indeed shorter than the barn: they will bring it to rest while it is inside. This simple-sounding proposal requires careful analysis. In fact, any change in the motion of an extended body is a complicated process from the point of view of special relativity. In describing such a process it is essential to specify the timing of the changes in velocity of each part of the body.

In a Galilean world, changes in velocity present no difficulty. All parts of a rigid body have the same velocity at any given time. As the velocity changes, observers in any frame of reference see all parts slow down or speed up in step. When a moving pole is brought to rest, barn frame observers see the two ends begin to slow down together and stop at the same time. Observers for whom the pole was initially at rest see both ends begin to move simultaneously and reach their final velocity together.

In special relativity, the story is not so simple. Spatially separated events can be simultaneous in only one frame of reference. If both ends of the pole begin to slow down at the same time according to barn frame clocks, then according to *S'* clocks, the two ends begin to move at *different* times. But if one end is moving while the other end remains at rest for a finite period of time, *the length of the pole must change.* Conversely, if the two ends begin to move simultaneously according to *S'* clocks, barn frame observers see them begin to slow down at different times. In that case, the length of the pole as measured in the barn frame changes.

These results should not be surprising, since we know that the measured length of any object depends on its velocity in the frame in which the measurement is being carried out. When that velocity changes, the measured length should likewise change.

At first glance, the length contraction formula appears to provide a complete description of the effect: as the velocity of a body measured in some frame changes from an initial value v_i to a final value v_f, the length of the body measured in that frame should change from L/γ_i to L/γ_f, where L is the body's proper length and γ_i and γ_f are the relativistic factors that correspond to the initial and final speeds.

In the present problem, as the pole comes to rest relative to the barn, barn observers should see it stretch from its contracted length $L/2$ to its proper length L; thus their attempt to prove that the pole is truly shorter than the barn fails. Meanwhile, S' observers should see the pole shrink as it picks up speed relative to them, finally reaching the contracted length $L/2$ when its speed is V. Thus S' observers also change their minds about the relative length of the pole and the barn.

All of the preceding is true, however, only if the proper length of the pole remains unchanged as its speed changes. Although that seems a reasonable supposition, we cannot assume that it is automatically satisfied. The simple statement "The velocity of the pole changes from v_i to v_f" is not a sufficiently precise description of a change in the pole's motion. To determine what happens to the proper length, we have to specify precisely how the changes in velocity of the various parts of the pole are correlated. This is another of the unexpected subtleties of special relativity.

To illustrate the phenomenon, we shall analyze three different ways by which the pole could hypothetically be brought to rest relative to the barn. Each leads to a different conclusion as to what happens to the proper length. Although all three stopping procedures are idealized, there is no reason in principle why they could not be carried out.

Figure 6.7 illustrates the first stopping method. Identical thin but powerful clamps are fixed in the barn and spaced along the path of the pole. At a prearranged time *according to barn frame clocks*, all the clamps snap shut and each one grips the portion of the pole immediately adjacent to it, bringing it quickly to rest. If the clamps are sufficiently strong, the pole travels only a very short distance before stopping (fig. 6.7a).[5]

5. Although special relativity sets a strict upper limit on the speed any body may attain, it imposes no such restriction on acceleration. There is no reason, in principle, why the pole could not stop in an arbitrarily short time.

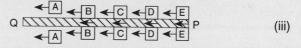

Fig. 6.7. One way of bringing the pole to rest. (a) The process as seen in the barn frame S. In (i), the pole is moving and all clamps are open. In (ii), all clamps have snapped shut simultaneously; the pole has stopped with no change in length. (b) The same process as seen in frame S'. (i) Pole is initially at rest, open clamps are moving from right to left. (ii) Clamp E has just snapped shut, setting end P in motion. Other clamps are still open. (iii) Clamp B has just snapped shut. Everything to the right of clamp B is in motion, while everything to the left of B is still at rest. The pole is considerably shrunken. (iv) The last clamp, A, has just snapped shut. The pole is now all in motion, having shrunk to length L/γ^2.

This stopping scheme has been designed so that all parts of the pole slow down in step, as seen by barn frame observers. The length of the pole *in the barn frame* therefore does not change; it remains L/γ. But since the pole is finally at rest in the barn frame, L/γ is by definition its new proper length. In the process of coming to rest, the pole's proper length has been reduced by a factor of γ.

Let us analyze the same scenario from the point of view of S' observers, for whom the pole is initially at rest and the clamps are moving (fig. 6.7b, sketch [i]). If our analysis is self-consistent, those observers must see the pole contract as it speeds up, in such a way that the final measured length is just $1/\gamma$ times the new proper length, or L/γ^2. We shall verify that this is indeed what S' observers see.

As each moving clamp snaps shut, it sets the corresponding portion of the pole into motion. The clamps do *not*, however, engage simultaneously. As a result, each portion of the pole begins moving at a different time. Specifically, clamp E snaps shut first and the right end of the pole begins to move to the left while the remainder of the pole is still at rest (fig. 6.7b, sketch [ii]). Thus the pole begins to shrink. The other clamps then engage in sequence; after clamp A, at the extreme left, has engaged, the entire pole is in motion.

To determine how far the right end of the pole moved while the left end remained at rest, we need to know the interval between the times when clamps E and A engage. That time interval can be calculated from the results of chapter 4. According to equation (4.10), the time interval $\Delta t'$ measured in frame S' between two events that occur simultaneously in S with a spatial separation Δx is given by

$$\Delta t' = -\frac{V}{c^2}\gamma(\Delta x) \tag{6.1}$$

For the events in question here, Δx has the value $-L/\gamma$. Hence the interval (measured in S') between the times when the two ends of the pole begin moving is

$$\Delta t' = LV/c^2 \tag{6.2}$$

From this information it is straightforward to calculate the amount by which the pole shrank. End P traveled at velocity V for a time interval $\Delta t'$ given by equation (6.2), during which the other end remained at rest. The distance traversed by P during that interval (and hence the amount by which the pole shrank)[6] is

$$\text{change in length} = V\Delta t' = LV^2/c^2 \tag{6.3}$$

---

6. In this calculation we have ignored the distance traveled by each end of the pole while its speed is changing. As noted earlier, special relativity imposes no restriction on acceleration; hence the change in velocity can be assumed to take place instantaneously and the distance traveled while the velocity changes can be made arbitrarily small.

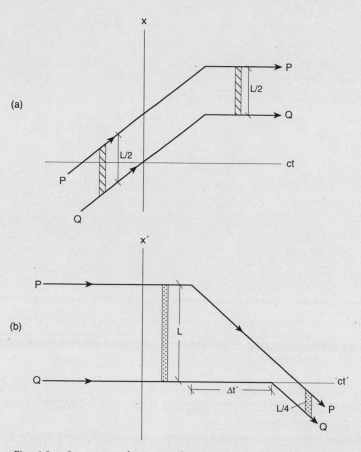

Fig. 6.8. Space-time diagrams showing the world lines of the two ends of the pole during the stopping procedure of fig. 6.7. (a) In frame S, both ends come to rest simultaneously; the length of the pole remains $L/2$. (b) In frame S', end P begins to move first; after a time interval $\Delta t'$, end Q begins to move. As a result, the length of the pole diminishes from L to $L/4$.

The final length of the pole, after all of it is in motion, must be

$$L_{\text{new}} = L - LV^2/c^2 = L/\gamma^2$$

This is the value we inferred earlier from the analysis in the barn frame.

Figure 6.8 contains the space-time diagrams in both frames for the stopping process we have analyzed, with $\gamma = 2$. The world lines of P and Q are shown in each diagram. In frame S (diagram [a]), the two ends come to rest simultaneously and the length of the pole remains unchanged. In

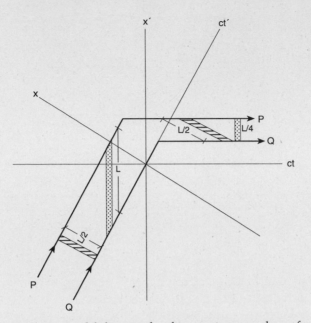

Fig. 6.9.   Loedel diagram for the stopping procedure of
figs. 6.7 and 6.8. This diagram contains the information
found in both parts of fig. 6.8.

S' (diagram [b]), Q begins moving some time after P does; the shrinking
of the pole is apparent.

The Loedel diagram for the process is shown in figure 6.9, which contains all the information found in both diagrams shown in figure 6.8. The
change in length of the pole in frame S' is apparent.

Figure 6.10 illustrates a different stopping procedure, designed so that
all changes in velocity take place simultaneously in S' instead of in S. An
array of S' observers is stationed alongside the pole, and each one is instructed to give the adjacent portion of the pole a sharp backward impulse
at some prearranged time. All the impulses are simultaneous *according to
S' clocks* and are just strong enough to impart to each part of the pole a
backward velocity V. Since the impulses are synchronized, all parts of the
pole speed up in step; hence its length, as measured in S', remains L (fig.
6.10a). This, however, is no longer a proper length, inasmuch as the pole
is now in motion in S'. Since L is a contracted length, the pole's new
proper length must be γL.

Figure 6.10b shows this process from the point of view of barn frame
observers. They see each part of the pole receive an impulse that brings it

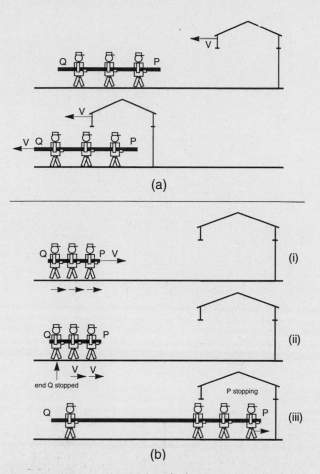

(a)

(b)

Fig. 6.10. A second procedure for bringing the pole to rest. In this scheme all changes in velocity take place synchronously in the runner's frame, S'. (a) The process as seen in S'. In the top sketch, pole and S' observers are at rest and the barn is approaching. The length of the pole is L, its proper length; the barn is contracted. In the second sketch, S' observers have just pushed back on the pole, imparting to it a backward velocity, V. The impulses were delivered simultaneously; hence the length of the pole remains L. (b) The same process as seen in the barn frame. (i) Pole and S' observers are all moving at speed V, the barn is at rest. The length of the pole is L/γ. (ii) An S' observer has just pushed on end Q, bringing it to rest; the remainder of the pole is still moving. (iii) An S' observer has just pushed on end P, bringing it to rest. During the interval between (ii) and (iii), the pole has been stretching. In (iii), its length is γL.

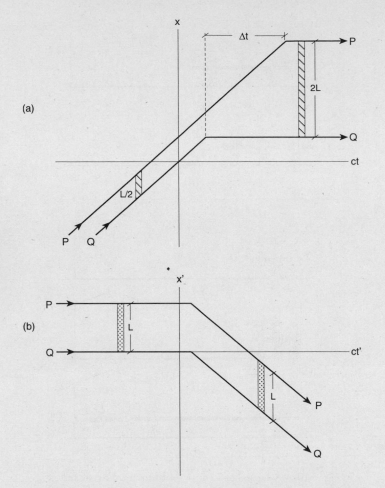

Fig. 6.11. Space-time diagrams for the stopping procedure of fig. 6.10. Again the world lines of both ends of the pole are constructed in each frame. (b) In S', both ends begin to move simultaneously; the length of the pole remains L. (a) In S, end Q stops first. After a time interval Δt, end P stops. As a result, the length of the pole increases from $L/2$ to $2L$.

to rest, but the impulses occur at different times. End Q stops first, followed in turn by adjoining sections until finally end P stops. The interval between the stopping of the two ends is given by an expression similar to (6.1). While the pole is coming to rest, therefore, S observers see it stretch. It is not hard to verify that the final length in frame S is γL, the new proper length, in accord with the conclusion obtained above. (See problem 6.2.)

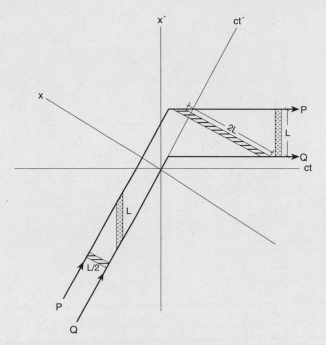

Fig. 6.12. Loedel diagram for the stopping procedure of figs. 6.10 and 6.11. This diagram contains all the information found in both parts of fig. 6.11.

Figure 6.11 exhibits the space-time diagrams for the second stopping procedure. The stretching of the pole in frame S is apparent. Figure 6.12 contains the Loedel diagram for the process. The reader should compare these figures with figures 6.8 and 6.9.

In each of the stopping procedures analyzed thus far, the proper length of the pole changes as it comes to rest relative to the barn. Real physical changes must have taken place in the structure of the pole. Such an outcome surely cannot be a necessary consequence of the change in the pole's velocity. It must be possible to change the speed of a body without altering its proper length; indeed, one would expect that to be the normal outcome.

A hypothetical slowing-down procedure that preserves proper length is the following. Closely spaced retro-rockets are attached to the pole. The rockets are programmed to fire a sequence of identical tiny bursts of propellant; each burst provides a small impulse to the portion of the pole to which the rocket is attached. Each set of bursts takes place simultaneously *according to clocks that move with the pole.*

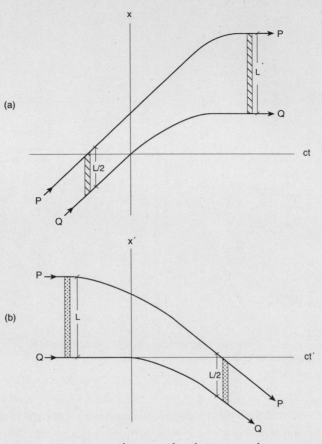

Fig. 6.13. Space-time diagrams for the retro-rocket stopping procedure, which preserves the proper length of the pole. The changes in velocity of the two ends of the pole take place at different times in each frame. As a result, the pole stretches from length $L/2$ to length L in S (a) and shrinks from length L to $L/2$ in S' (b). The proper length remains L.

Let us follow in detail what happens in this scenario. The first set of bursts takes place simultaneously in S', the pole's initial rest frame. As a result, the pole acquires an infinitesimal velocity, δ, in S', with no change in length (in that frame) since all portions changed their velocity together. The pole is now at rest in a new frame we call S_1, which moves at velocity δ relative to S'. Its proper length is unchanged.

A second set of bursts takes place simultaneously in S_1; as a result of these bursts, the pole acquires a velocity δ in S_1, again with no change in

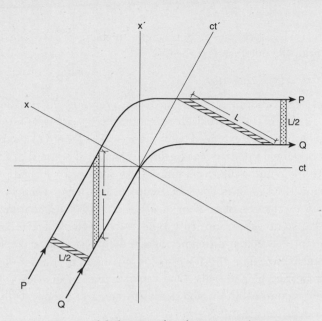

Fig. 6.14. Loedel diagram for the retro-rocket stopping procedure. This diagram contains all the information found in both parts of fig. 6.13.

length in that frame. It is now at rest in a frame S_2 that moves at velocity δ relative to S_1. The process continues until the pole is finally at rest in frame S, the barn frame. The proper length remains unchanged throughout the entire procedure.

The distinguishing feature of this scenario is that the pole passes through a succession of frames S', S_1, S_2, . . . , S, in each of which it is instantaneously at rest and has the same length (its proper length). The rest frame of the pole coincides in turn with S', S_1, S_2, . . . , and so on; all changes in velocity take place simultaneously in this frame. The rest frame is, however, not an inertial frame; it is undergoing acceleration.

Special relativity is not equipped to describe observations in noninertial frames. By using a succession of inertial frames, with which the pole's rest frame coincides in turn, we finesse that problem.

Figure 6.13 displays the space-time diagrams for the retro-rocket stopping scheme in frame S (diagram [a]) and frame S' (diagram [b]). Figure 6.14 is the corresponding Loedel diagram. As the diagrams show, the increments in velocity of the two ends do not take place simultaneously in either frame. In frame S, the pole stretches as it slows down; in S' it shrinks as it speeds up. The proper length remains constant. Any stopping procedure that leaves the proper length unchanged must be logically equivalent to the one just described.

Rigid Bodies in Relativity

The principal qualitative conclusion of the preceding discussion is that changes of length are unavoidable, according to special relativity, whenever the velocity of an extended body changes. The length of the body can remain unchanged in one frame (or one sequence of frames) at the most.

This result requires us to reexamine the concept of a rigid body. By definition, a rigid body is one whose dimensions never change. Even in Newtonian mechanics, that is an idealization. Any real material is deformable to some extent; no truly rigid body exists. There is, however, an important conceptual difference between the classical and relativistic viewpoints. Newtonian physics imposes no inherent limitation on the elastic properties of materials. One can imagine substances with arbitrarily low compressibilities and postulate an ideal "rigid" body as the limiting case of zero compressibility.

In special relativity, however, the concept of a rigid body is unacceptable even as an idealized limiting case. For even if the dimensions of an object were to remain forever fixed in some particular frame, they would change by arbitrarily large amounts in frames that move rapidly relative to that one. Such changes in length have nothing to do with the elastic properties of the material. They are consequences of the properties of space-time.

In each of the stopping procedures we have described, the change in speed of every part of the pole was specified independently, through the action of a clamp or retro-rocket attached directly to the segment in question. Any other procedure introduces ambiguities from the point of view of special relativity.

Suppose, for example, I simply push on one end of a block. This is, after all, the most common way to change the motion of an object. How should the subsequent motion of the block be described?

If the block were rigid, all portions would begin moving at once and the block would maintain a constant length. But how does the far end of the block "know" that it is supposed to move? According to Newton's second law, changes in motion are brought about only by forces. By pushing on one end, I exerted a force only on the atoms there. The atoms in the rest of the block cannot move until the force is somehow transmitted to them.

We can visualize the process in terms of a simple model of matter, in which the interatomic forces are represented by springs. Pushing on one end of the pole causes the atoms there to move a little and thus compresses the first row of springs. The compressed springs push against the second row of atoms, setting them into motion and compressing the next row of springs. The process continues until the impulse finally reaches the far end of the pole. But all this takes time. How long it takes depends on the detailed structure of the block—in our model, on the masses of the atoms and the stiffness of the springs—and is not easily determined.

We can be certain of one thing, however: the transmission of the force down the block constitutes a signal (actually an elastic wave). This signal, like any other, can travel no faster than the speed of light. If the length of the block is L, the far end cannot possibly begin to move until at least a time interval L/c after the near end was pushed.[7] During that interval, the block is being compressed.

If the block were rigid, a force exerted on one end would cause the other end to change its velocity immediately. The signal that transmits the force would have to propagate at infinite speed. That, of course, is forbidden by special relativity.

We conclude from this analysis that the concept of a rigid body is inconsistent with the postulates of relativity. An even more dramatic example of relativistic nonrigidity is given in the next section.

6.4. OTHER LENGTH PARADOXES

In this section we analyze briefly two other paradoxes based on the contraction effect. Each one presents some points of interest.

Paradox of the Fast Walker[8]

A man walks very fast over a rectangular grid, of the type used in some bridge roadways. The rest length of the walker's foot is equal to the spacing between grid elements. In the rest frame of the grid, his Lorentz contraction makes him narrower than the grid spacing; observers in that frame expect him to fall in. In the rest frame of the walker, in contrast,

7. Until that time, the far end is outside the light cone of the event that initiated the motion. Real elastic waves travel much more slowly than light; hence the actual time delay before the far end begins to move would be considerably longer than the ideal lower limit L/c.
8. This paradox was invented by Wolfgang Rindler: "Length Contraction Paradox," *American Journal of Physics* 29 (1961):365–366.

the grid spacing is contracted and he should pass over the grid without any difficulty. The two predictions are contradictory.

Although this paradox bears some similarity to that of the pole and barn, the two differ in one important respect. In the pole and barn problem, we concluded, the analyses in both frames are correct and there is in fact no paradox. Such a conclusion is not acceptable in the present problem. The two predictions are definitely incompatible: the question of whether or not the walker falls into the gridwork is subject to direct observation and must have an unambiguous answer. Either he falls in or he does not. If special relativity is self-consistent, one of the analyses must be wrong. Which is the right one, and why?

For ease of discussion we replace the walker with a rectangular block that moves along a tabletop containing a rectangular hole. Both the block and the hole have proper length L. V is the velocity of the block in frame S, the rest frame of the table, and γ is the corresponding Lorentz factor. Figure 6.15 shows the story as it appears in frame S, where the block's length is only L/γ.

One additional modification simplifies the analysis without altering the essential features of the paradox. In the problem as originally stated, the block would topple over the edge of the hole and would then rotate as it falls, in a rather complicated motion. To avoid this complication, we assume that the block is suspended from vertical threads that move along at the same speed as the block (sketches [a] and [b]). As soon as the block is entirely over the hole (sketch [c]), all the threads are cut simultaneously (in frame S) and the block begins to fall, remaining horizontal as it does so. E_1 and E_2 denote the events at which the threads that support the two ends of the block, Q and P, are cut.

Sketch (d) shows the block falling, and entirely within the hole. A short while later the forward end of the block, Q, collides with the far side of the hole (event E_3, sketch [e]).

The resolution of the paradox becomes apparent when we translate the pictures of figure 6.15 to frame S', in which the block was originally at rest and the (narrow) hole moves from right to left at speed V (fig. 6.16). Once again simultaneity plays a critical role.

The key to the argument is that in S', the threads are *not* all cut simultaneously. The first one to be cut is the one that supports end Q (event E_1, sketch [b]). At the time of E_1, part of the block protrudes into the hole but is still suspended. The remainder still rests on the tabletop.

Immediately after the thread that supports end Q has been cut, that end begins to fall while the rest of the block remains stationary. We are

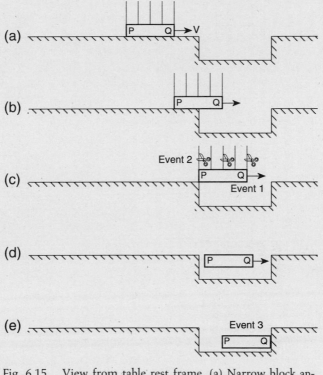

Fig. 6.15. View from table rest frame. (a) Narrow block approaching, suspended from strings. (b) Block entirely over the edge; strings prevent it from toppling. (c) When the block is entirely over the hole, the strings are cut simultaneously (events 1 and 2). (d) Block is inside the hole and falling. (e) Q, the leading end of the block, is about to smash into the side of the hole (event 3).

led to a surprising conclusion: as seen in S', the block cannot maintain its rectangular shape but must become deformed.

Sketch (c) shows the situation a short time later. The thread at point X is just being cut. The part of the block to the left of X is still stationary, while everything to the right of X is falling. End Q, which was the first to begin falling, has fallen the greatest distance. Detailed analysis shows that the shape of the curved portion is a parabola.

This result is a striking manifestation of the fact that rigid bodies are inconsistent with relativity. Not only can the length of a body change when it is in motion but its shape is liable to change as well when viewed in different frames.

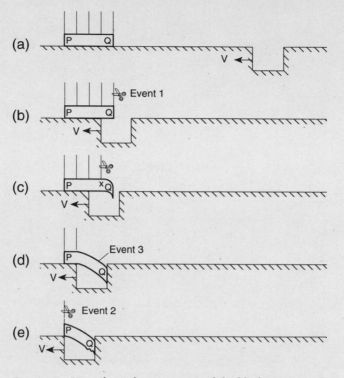

Fig. 6.16. View from the rest frame of the block. (a) Narrow hole is approaching stationary block. (b) String at Q is being cut (event 1); Q begins to fall. (c) String at X is being cut. Everything to the right of X is falling, while everything to the left of X is still stationary. (d) Far side of the hole is about to smash into end Q (event 3). Notice that the hole has not yet reached end P. (e) String at P is being cut (event 2) just as the hole reaches it. The right side of the block is mangled as a result of its earlier collision with the hole.

Sketch (d) shows the situation seen by S' observers at the time of E_3, when end Q is about to collide with the far side of the hole. The moving hole has still not reached the left end of the block, and event E_2 has not yet occurred. (In frame S, E_2 occurs before E_3.) By the time the last thread is cut (sketch [e]), the right-hand portion of the block has already been mangled.

The pictures in figures 6.15 and 6.16 are consistent with one another and provide the resolution of the paradox. The walker does indeed fall into the grid!

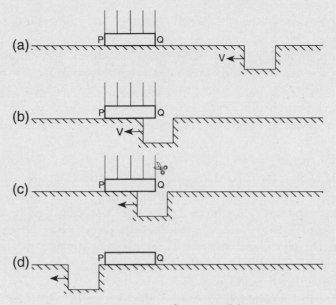

Fig. 6.17. Incorrect sequence of pictures of the experiment as viewed from the original rest frame of the block. The narrow hole passes under the block, which remains stationary. The error in this scenario is explained in the text.

Our explanation is still incomplete in one respect. We obtained figure 6.16 by translating everything from frame *S* (fig. 6.1.5), in which the block is always horizontal. But suppose we had analyzed the problem from the beginning in the block's original rest frame, *S'*. How would we have been led to conclude that the sketches in figure 6.16 give the correct description of the story? And what is wrong with the argument that predicts that the moving narrow hole should simply pass under the stationary block, as in figure 6.17?

To answer these questions, let us picture the block as being composed of many small vertical segments, each attached to one thread. Before any threads have been cut (fig. 6.17b), every segment is in equilibrium. For each segment that is over the hole, the force of gravity is balanced by the upward pull of a thread.

Figure 6.17c shows the thread that holds up the end segment *Q* being cut. That segment is no longer in equilibrium because there is nothing to balance the downward pull of gravity. Therefore figure 6.17c cannot be right. The end segment must begin to fall, initially with the acceleration of gravity. The rest of the block is still suspended and momentarily remains horizontal.

If no additional threads were cut, the fall of the end segment would be quickly arrested by upward forces, called shear forces, exerted by the adjacent segment. The block would resemble a cantilevered beam set into a wall. Each part of such a beam is held up by shear forces exerted by the adjoining parts; the beam does not remain strictly horizontal but "droops" a little.

In the present problem, however, the second segment itself begins to fall shortly after the first. Moreover, the information that the first thread has been cut and the end segment is falling is not communicated instantaneously to the rest of the block. The news can propagate no faster than the speed of light and cannot reach the adjacent segment before a time interval D/c has passed, where D is the distance between the two segments. Until then the shear forces cannot act. For at least a time interval D/c, therefore, the end segment is in free fall.

Before the interval D/c has passed, the next thread has been cut and the next segment of the block has begun to fall. This must be true because the interval between the cutting of the two threads is *spacelike*. (Remember the two events are simultaneous in frame S.) The process continues until all the threads have been cut and the entire block is falling freely.

Under these circumstances, the shear forces never get a chance to act. Each segment of the block is in free fall and never "knows" that the adjacent segments are at different heights. The shape of the block in frame S' is determined solely by how long each segment has been falling and has nothing to do with the elastic properties of the material of which it is composed. A rubber block and a steel block would have precisely the same shape.

How Does the Plate Pass through the Hole?

Our final length paradox is due to R. Shaw.[9] A thin plate of proper length L moves in the x direction at a relativistic speed V, as seen in some frame S. A tabletop with a hole of proper width L is parallel to the plate and moves in the y direction at speed u. (See fig. 6.18.) The motions are arranged so that the plate and the hole in the tabletop arrive at the origin at the same time. Since the plate is contracted to length L/γ, it slips easily through the hole in the tabletop (fig. 6.18a).

9. R. Shaw, "Length Contraction Paradox," *American Journal of Physics* 30 (1961):72.

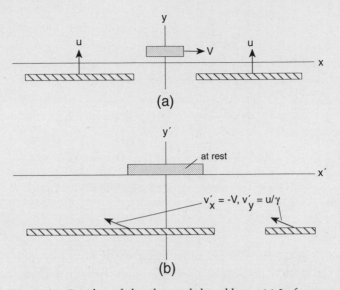

Fig. 6.18. Paradox of the plate and the tabletop. (a) In frame
S, the plate moves in the x direction at velocity V while the
tabletop moves in the y direction at velocity u. The plate is
Lorentz-contracted, hence narrow; the hole in the tabletop is
wide. The plate passes easily through the hole. (b) View in
frame S', in which the plate is at rest and wide. The hole is
moving and narrow. The velocity components of the tabletop
are indicated in the figure. How does the plate get through?
The answer is found in fig. 6.19.

Now consider the situation as seen in frame S', in which the plate is at
rest. In that frame the tabletop has both x and y components of velocity;
these are given by (see eq. 4.19)

$$v'_x = -V \qquad v'_y = u/\gamma \tag{6.4}$$

In the plate's frame the hole is contracted; its width is only L/γ. The
plate, however, has its proper length L. Figure 6.18b purports to show the
situation in S'. How can the plate pass through such a narrow hole?

The resolution of this paradox is that figure 6.18b is in fact incorrect.
If the tabletop is parallel to the x axis and is moving in the y direction in
frame S, then in S' *it is no longer parallel to the x axis*. One can use the
Lorentz transformation to show that the tabletop makes an angle θ with
the x axis, where θ is given by the equation

$$\tan \theta = \frac{u}{c}\gamma \tag{6.5}$$

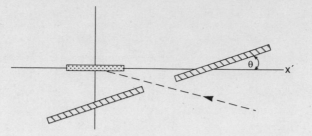

Fig. 6.19. Correct picture of the plate and tabletop as seen in frame S'. The tabletop is inclined to the x axis; this makes it possible for the long plate to slip though the narrow hole.

The actual picture in S' is given not by figure 6.18b but by figure 6.19; S' observers agree that the plate passes through the hole without difficulty.

Alternatively, we could work in the frame in which the tabletop is at rest. In that frame the plate moves in a diagonal direction and is inclined at the angle θ. The conclusion is the same.

6.5. The Paradox of the Twins

The twin paradox (also known as the clock paradox) is the most famous and most vigorously debated of the paradoxes of relativity. Arthur and Barbara are twins. Arthur stays at home while Barbara sets out on a space-ship that travels at a uniform fast speed V relative to earth. After some time the ship quickly turns around and returns to earth at the same speed V. The twins get together and examine each other's appearance.

We analyze the problem first in the frame of reference of Arthur, the stay-at-home twin. While Barbara is on the outward leg of her journey, her clock (like all moving clocks) runs slow. During the return trip, Barbara's clock likewise runs slow. Hence when Barbara arrives home her clock should read less than Arthur's, which has been at rest through the entire episode. Assuming that biological clocks behave like all others (i.e., that the aging process is governed by the time elapsed on a clock that follows the individual in question), we conclude that Barbara has aged less than Arthur: she should look younger than her twin when they are reunited.

From Barbara's point of view, however, it is Arthur who has been moving all the time and whose clock has therefore been consistently slow. Hence an identical argument leads to the conclusion that Arthur's clock ought to read less than Barbara's and Arthur ought to look younger when the twins are reunited. This contradiction constitutes the paradox.

A third argument leads to still another conclusion. Since there is no such thing as absolute motion, so this argument goes, all results must be symmetrical between the two twins. Hence they should be the same age when they are reunited. Thus there are three possible outcomes, of which only one can be right.

When we first investigated time dilation in chapter 3, we analyzed in detail the logical problem associated with the fact that two sets of observers each think the others' clocks are running slow. That result is self-consistent, we concluded, because the judgments of the two sets of observers are based on different sets of measurements; in each case, the reading of a single "moving" clock is compared on two occasions with the reading of an adjacent "stationary" clock. The same two clocks cannot be directly compared more than once if they are moving uniformly relative to one another.

That argument does not apply to the present problem because here the *same* two clocks are together at the beginning and at the end of Barbara's trip. The comparison of their readings can be made directly; there is no need to exchange information with distant observers.

Under these circumstances, the question of which clock has the greater reading must have a definite answer, about which all observers must agree. When the twins are reunited, either Arthur looks younger or Barbara looks younger or both look the same age. These are the only logical possibilities. Thus we are confronted by a true paradox.

The astute reader will object that our analysis has ignored several important parts of the problem: the startup period during which the spaceship gets up to full speed, the turnaround at the midpoint of the trip, and the slowdown at the end. Perhaps the times recorded by the twins' clocks during those intervals differ in just such a way as to make the total times equal.

This speculation cannot be right, for even if the times recorded by the twins' clocks during the three acceleration periods were to differ substantially, the difference would be a fixed amount independent of the duration of the trip. By contrast, the difference in elapsed times attributable to time dilation during the constant-speed legs of Barbara's trip is proportional to the duration of those legs; if the trip were made twice as long, the difference in aging due to time dilation would be doubled. An effect that is independent of the duration of the trip cannot cancel another that is proportional to the duration. If the trip is long enough, in fact, the time elapsed during the startup, turnaround, and slowdown periods is an arbitrarily small fraction of the total travel time.

The turnaround nonetheless holds the key to the resolution of the paradox because it establishes an asymmetry in the problem. While the spaceship is reversing its direction, Barbara is accelerating and is fully conscious of that fact. The engines on the spaceship must be fired. During the turnaround period, Newton's laws do not hold in the spaceship; fictitious inertial forces appear to act. If Barbara looks through a telescope, she sees the positions of all the stars shift because of the change in stellar aberration.

Although Barbara sees Arthur receding during the first portion of her trip and approaching her during the second portion, she infers from all these observations that her velocity (and not Arthur's) has changed. Thus the problem is not symmetric after all, and we can dismiss the argument that the reunited twins ought to be the same age because of symmetry considerations.

Special relativity is restricted to the description of measurements made in inertial frames. In the present problem Arthur's rest frame, S, is inertial but Barbara shifts from one inertial frame, S', to another, S'', at the turnaround.[10] The interval between the start and the turnaround is a proper time interval in S', while the interval between the turnaround and Barbara's arrival home is proper in S''.

Let T be the duration of the complete trip as measured in frame S; each leg lasts $T/2$. A standard time dilation argument shows that the duration of the outward leg is $T/2\gamma$ according to clocks in S', while that of the return leg is $T/2\gamma$ according to clocks in S''. For Barbara, who changes from frame S' to S'' at the turnaround, the total elapsed time must be T/γ. The traveling twin is indeed younger on her return. This conclusion was already stated, in somewhat different form, by Einstein in his 1905 paper; it is nearly universally accepted today. Some experimental evidence to support it will be presented below.

It is important to distinguish what the twins can actually see from what they learn by exchanging information with other observers in their own frame of reference. Suppose each twin trains a powerful telescope on the other throughout the duration of the trip. What do they see?

10. Strictly speaking, Barbara changes frames three times: first as the spaceship is getting up to speed at the beginning of the trip, then at the turnaround, and again as the spaceship slows down at the end. The problems associated with the first and last changes can be circumvented by making the initial age comparison just after the ship has reached full speed and the final one just before deceleration begins. The problem of the turnaround, however, cannot be evaded in such fashion.

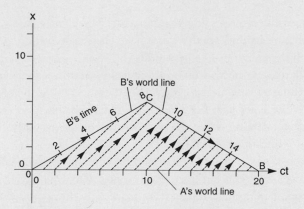

Fig. 6.20. Space-time diagram for the twin paradox, constructed in Arthur's rest frame. *OB* is Arthur's world line and *OCB* is Barbara's world line. Barbara's time is indicated along her world line. The dashed lines are the world lines of light signals emitted by Arthur once a year. Between *O* and *C*, those signals reach Barbara at the rate of one every two years; between *C* and *B*, they reach her at the rate of two a year.

For concreteness, let $T = 20$ years and let the speed of the spaceship be $0.6c$; γ is then 1.25 and all the numerical results come out to be simple numbers. Each leg of Barbara's journey lasts 10 years according to Arthur's clock and $10/\gamma = 8$ years according to Barbara's.

Figure 6.20 displays the space-time diagram for the problem, constructed in Arthur's rest frame. He is located at $x = 0$; his world line is along the time axis. Barbara's world line is *OCB*; it starts out at $t = 0$ and rejoins Arthur's world line at $t = 20$ years.

The S coordinates of event C, the turnaround, are $t = 10$ years, $x = 6$ light-years; its coordinates in S' are $t' = 8$ years, $x' = 0$. Armed with that knowledge, we divide each half of Barbara's world line into 8 equal parts and mark her time along the world line at intervals of one year.

Also shown in the diagram, as dashed lines, are the world lines of light signals emitted from earth at intervals of 1 year. Each signal carries a picture of Arthur at the time the signal is emitted. (More prosaically, each signal simply carries the reading of Arthur's clock at the time it is emitted.)

The diagram shows that during the outward leg of Barbara's trip, a signal arrives every 2 years according to her clock. She therefore "sees"

Arthur age at the rate of half a year per year. The picture that arrives just as she is turning around, when 8 years have elapsed according to her clock, shows Arthur as having aged by 4 years.

While Barbara is on her homeward journey, Arthur's signals reach her at the much faster rate of two per year. Altogether, then, she sees Arthur age at half a year per year for 8 years and at 2 years per year for another 8 years. His total aging is $4 + 16 = 20$ years, consistent with the conclusion that she finds him older than she is at her return.

Barbara realizes, of course, that each picture shows Arthur's appearance not when it arrives but at an earlier time, when the signal was emitted. The rate of aging that she observes represents the combination of two effects: (a) relativistic time dilation and (b) the changing travel times of the signals. During the outward leg, both factors slow down Arthur's apparent rate of aging, while during the return leg, the decreasing light travel time speeds up his apparent rate of aging. Time dilation slows it down during both legs.

The apparent rates of aging can be interpreted in terms of the Doppler effect, which was discussed in section 4.8. The "source," Arthur, emits signals at a frequency $f_0 = 1/\text{yr}$. With $V/c = 0.6$, equation (4.40a) shows that the frequency at which the signals are detected during the outward leg is $f' = \sqrt{(1 - 0.6)/(1 + 0.6)} = 0.5/\text{yr}$. For the return trip, equation (4.40b) gives $f' = \sqrt{(1 + 0.6)/(1 - 0.6)} = 2/\text{yr}$. These are precisely the frequencies we have inferred from the space-time diagram.

Arthur's observations can be similarly analyzed. To avoid cluttering the figure, we have redrawn the space-time diagram for the problem in figure 6.21, this time showing the world lines of light signals emitted by Barbara. At first Arthur receives signals at the rate of 0.5 per year, and at the end he receives them at the rate of 2 per year, just as Barbara does.

There is, however, an important difference between their observations. As figure 6.20 shows, Barbara begins to receive signals at the fast rate as soon as her ship turns around; hence she sees Arthur age at the fast and slow rates for equal periods. Arthur, however, continues to receive signals at the slow rate until the signal emitted by Barbara at the turnaround reaches him (fig. 6.21). That signal arrives at $t = 16$ years. Hence Arthur sees Barbara age at the slow rate, $0.5/\text{yr}$, for 16 years, and at the fast rate, $2/\text{yr}$, for only 4 years. He sees her age by a total of $8 + 8 = 16$ years, in accord with the earlier analysis. The asymmetry in the twins' observations is apparent in the two figures.

One puzzling aspect of the story remains to be cleared up. According to Barbara, Arthur's clock runs slow throughout her trip. During the out-

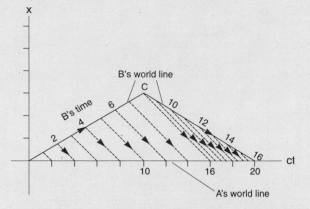

Fig. 6.21. The space-time diagram of fig. 6.20 is repeated to show the world lines of light signals emitted by Barbara once a year according to her clock. Those signals reach Arthur at the slow rate (0.5/yr) for the first 16 years. Only for the last four years do signals arrive at the fast rate (2/yr). Figs. 6.20 and 6.21 demonstrate the asymmetry in the problem: Arthur receives only 16 of Barbara's signals while she receives 20 of his.

ward leg, which lasts 8 years according to her clock, only 6.4 years $(8/\gamma)$ elapse on his clock. During the return leg, another 6.4 years elapse on Arthur's clock, for a total of 12.8. Yet his clock reads 20 years when she arrives home; 7.2 years of Arthur's life seem to be unaccounted for.

We learned in chapter 3 that moving clocks not only run slow but they are also out of synchronization. The one that is ahead in its direction of motion is behind in its reading. When Barbara reaches the turnaround, event C, her clock reads 8 years. Observers in every frame agree that an S clock at the same location reads 10 years. But the answer to the question, What does Arthur's clock read when Barbara turns around? is frame-dependent.

In frame S the answer is, of course, 10 years. (S clocks are synchronized in their own frame.) According to S' observers, S clocks are out of synchronization. Because Arthur's clock is ahead of the one at C in the direction of motion, it is behind in its reading. (It in fact reads 6.4 years.)

When Barbara's ship turns around, she changes from frame S' to frame S''. Observers in S'' agree with those in S' that S clocks are out of synchronization, but they disagree as to which S clock is ahead. (The two sets of observers see frame S moving in opposite directions.) According to S'

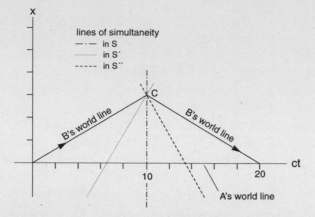

Fig. 6.22. The space-time diagram of fig. 6.20 has been redrawn once more. Shown here are the lines of simultaneity for each of the three frames of reference in the problem: frame S (Arthur's rest frame), S' (Barbara's rest frame during the outward leg of her trip), and S" (Barbara's rest frame during the return leg). At the turnaround point C, Barbara's definition of simultaneity changes abruptly. Her answer to the question, What does Arthur's clock read *now*? changes abruptly from 6.4 years to 13.6 years.

observers, Arthur's clock is 3.6 years *behind* the S clock at C; it reads only 6.4 years when that clock reads 10 years. According to S" observers, Arthur's clock is 3.6 years *ahead* of the one at C; it reads 13.6 years when the one at C reads 10 years.

The question, What does Arthur's clock read when Barbara is turning around? means precisely, What does Arthur's clock read at an event that takes place at his location simultaneously with the turnaround? When Barbara changes frames she acquires a new set of fellow-observers with whom to consult concerning distant events; we should not be surprised that her answer to the question changes abruptly (from 6.4 years to 13.6 years).

The preceding is a purely formal statement; nothing strange happens either to Barbara's clock or to Arthur's when she changes frames. When looking through her telescope, she does *not* see Arthur age abruptly by 7.2 years as she turns around. Barbara's direct observations were analyzed earlier; they include nothing startling at the turnaround. We have, however, accounted for the "lost" 7.2 years of Arthur's life. With these remarks, the paradox is fully resolved.

In figure 6.22 we have redrawn once more the space-time diagram for the problem and constructed lines of simultaneity through event C for each of the three frames. The one for frame S is shown in long dashes; it intersects Arthur's world line at $t = 10$ years. The one for frame S' is shown as dotted; it intersects Arthur's world line at $t = 6.4$ years. Finally, the line of simultaneity for frame S'' has been drawn with short dashes. It intersects Arthur's world line at $t = 13.6$ years.

During the 1950s and 1960s, a lively controversy raged over the twin paradox. Herbert Dingle published a paper in which he claimed that Einstein had made a "regrettable error."[11] Dingle argued that the twin paradox could not be resolved and that therefore special relativity is logically inconsistent. Dingle's argument was challenged by W. H. McCrea, Frank Crawford, and others, and a long series of rebuttals and counterrebuttals followed. In the opinion of practically everyone except Dingle, the controversy was settled in favor of Einstein.[12]

An experiment to test the prediction of asymmetric aging was proposed by J. C. Hafele in 1970[13] and carried out by him in collaboration with Richard Keating in 1971.[14] Hafele pointed out that if the earth were not rotating, a jet plane that traveled around the world would play the part of the traveling twin; a clock carried on the plane should, on its return, have lost time compared to one that had remained fixed at the starting point.[15]

11. Herbert Dingle, "Relativity and Space Travel," *Nature* 177 (1956):782–784; followed by a reply by W. H. McCrea and a rebuttal by Dingle.
12. "Crackpot" papers claiming to disprove Einstein's theory are circulated to this day; they are generally not published. Dingle was a serious scientist and not a crackpot; nonetheless, he was wrong. For details on the controversy, see the papers cited in note 11; also H. Dingle, "Relativity and Space Travel," *Nature* 178 (1956):680–681; W. H. McCrea, "Relativity and Space Travel," *Nature* 178 (1956): 681–682; H. Dingle, "The 'Clock' Paradox of Relativity," *Nature* 179 (1957):865– 866, 1242–1243; Frank Crawford, "The 'Clock Paradox' of Relativity," *Nature* 179 (1957):1071–1072; J. H. Fremlin, "Relativity and Space Travel," *Nature* 180 (1957): 499–500, with a reply by Dingle; Edwin McMillan, "The 'Clock' Paradox and Space Travel," *Science* 126 (1957):381–384; H. Dingle, "The Case against Special Relativity," *Nature* 216 (1967):119–122; W. H. McCrea, "Why the Special Theory of Relativity Is Right," *Nature* 216 (1967):122–124, and other references cited therein.
13. J. C. Hafele, "Relativistic Behaviour of Moving Terrestrial Clocks," *Nature* 227 (1970):270–271; "Relativistic Time for Terrestrial Circumnavigation," *American Journal of Physics* 40 (1971):81–85.
14. J. C. Hafele and Richard Keating, "Around-the-World Atomic Clocks: Predicted Relativistic Time Gains," *Science* 177 (1971):166–167; "Around-the-World Atomic Clocks: Observed Relativistic Time Gains," *Science* 177 (1971):168–170.
15. The change in velocity of the traveling clock takes place continuously instead of just at the midpoint of the trip, as in the standard version of the twin paradox; that, however, does not affect the argument.

The rotation of the earth complicates matters; even the "stay-at-home" twin is not in an inertial frame. The desired test can still be accomplished, however, by sending two planes around the world in opposite directions and comparing their clock readings at the end of the trip with that of a clock that remained fixed on earth. All three clocks are accelerated and should run slow compared to a hypothetical clock in an inertial frame that does not take part in the earth's rotation. The amount of time lost by each clock depends on its speed in the inertial frame. For the stay-at-home clock, the speed is just the earth's rotational speed (about 0.5 km/sec at the equator); for the clock that travels from west to east the plane's ground speed must be added to the earth's rotational speed, whereas for the clock that travels from east to west the plane's ground speed must be subtracted from the earth's rotational speed. Special relativity therefore makes a definite prediction for the relative readings of the three clocks at the conclusion of the experiment.[16]

Because both the earth's speed and the speeds of jet planes are small fractions of the speed of light, the magnitude of the predicted effect is very small. The predicted differences in clock readings after one trip around the earth are only about 10^{-7} sec. However, atomic clocks are sufficiently accurate (and stable) to permit the detection of so small a time difference. When the experiment was carried out, the results were in excellent agreement with the predictions based on relativity.

PROBLEMS

6.1. In the pole and barn problem, let the speed of the runner as measured in the barn frame be $0.866c$ ($\gamma = 2$). Let the origin of frame S be the front door and the origin of frame S' be end Q of the pole. The proper length of both pole and barn is 5 m. At event QF, $t = t' = 0$.

(a) Write the equation that describes the position of end P in barn frame coordinates as a function of time. Find the time of event PR in the barn frame.

(b) Find the time of PR in the runner frame. Verify that PR occurs before QF in the runner's frame and after QF in the barn frame, as needed to resolve the paradox.

(c) Is the interval between PR and QF spacelike, timelike, or lightlike? Explain.

6.2. In the pole and barn paradox, the pole is brought to rest by the method illustrated in figures 6.10 and 6.11. In the runner's frame, the length of the pole

16. Another effect must be taken into account, namely, the influence of the earth's gravity on the rate at which a clock keeps time. This effect, called gravitational time dilation, is discussed in chapter 8. It causes both plane clocks to run fast relative to the stay-at-home clock; its magnitude is comparable to that of special relativity.

does not change. Analyze the motion of the two ends of the pole in the barn frame. Which end of the pole stops first? How long does the other end continue to move after the first end has stopped? Verify that after the pole has come to rest in the barn frame, its proper length is 2L.

6.3. Refer to the paradox of figures 6.15 and 6.16. Let the proper length of the block and the hole be 0.1 m, and suppose the velocity of the block in the table frame is 0.6c. Events 1 and 2, the cutting of the two end strings, occur at $t = 0$ in the table frame.

(a) Find the interval between events 1 and 2 in S', the block's original rest frame. Where is end P at the time of event 1 in S'?

(b) Where is end P at the time of event 3 in S'?

(c) How far does end Q fall before it strikes the far end of the hole at event 3?

6.4. The Apollo X spacecraft took about 1 day to reach the moon. How much younger than his twin would an Apollo astronaut have been at the end of the mission? Assume uniform velocity. (The earth-moon distance is about 400,000 km.)

7 Relativistic Mechanics

7.1. The Equivalence of Mass and Energy

We conclude our study of special relativity by examining its implications for energy, momentum, and mass. It is in this area that most of the applications of the theory are found. Several fundamental ideas of classical mechanics must be modified.

The famous relation between mass and energy, which for many people epitomizes relativity, appeared first in a short paper published by Einstein in September 1905,[1] just three months after his first paper on relativity. Einstein's argument was based on the transformation properties of electromagnetic radiation. In the first paper, he had shown that if an electromagnetic plane wave of energy ϵ travels at an angle ϕ with the x axis in some frame of reference S, its energy in frame S' is

$$\epsilon' = \frac{\epsilon\left(1 - \dfrac{V}{c}\cos\,\phi\right)}{\sqrt{1 - V^2/c^2}} = \gamma\epsilon\left(1 - \frac{V}{c}\cos\,\phi\right) \tag{7.1}$$

where V, as usual, is the velocity of S' relative to S, assumed to be in the x direction. We will see in section 7.7 that equation (7.1) can be derived also from the photon model of light, together with the formula for the relativistic Doppler effect.

Einstein analyzed the following thought experiment. A body at rest in some inertial frame S simultaneously emits a light wave of energy $L/2$ at an angle ϕ with the x axis and another wave of equal energy in the oppo-

---

1. A. Einstein, "Does the Inertia of a Body Depend upon Its Energy Content?" *Annalen der Physik* 18 (1905):639–641. English translations appear in *The Principle of Relativity* and in *Collected Papers* (doc. 24).

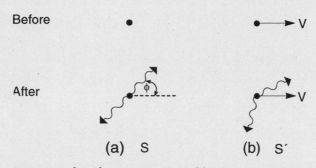

Fig. 7.1. Thought experiment used by Einstein to derive the equivalence between mass and energy. (a) Body at rest emits light waves of equal energy in opposite directions; the body remains at rest. (b) The same experiment is viewed in frame S', in which the body has velocity V both before and after the emission. The light waves do not travel in opposite directions in this frame. The energies of the light waves in the two frames are related by eq. (7.1).

site direction (fig. 7.1a). Since the emissions are symmetrical, the body remains at rest. Let E_0 and E_1 denote its energy before and after the emission; conservation of energy gives

$$E_0 = E_1 + L \tag{7.2}$$

Figure 7.1b shows the same process in a frame S' that moves to the left at speed V relative to S. In S', the body's velocity is V both before and after the emission. Applying equation (7.1) to each of the light waves, we find that their total energy in S' is γL. (The terms proportional to $\cos \phi$ cancel.) Conservation of energy in S' therefore gives

$$E_0' = E_1' + \gamma L \tag{7.3}$$

where E_0' and E_1' denote the body's initial and final energies in S'.

Taking the difference between equations (7.2) and (7.3) gives

$$E_0' - E_0 = E_1' - E_1 + (\gamma - 1)L \tag{7.4}$$

According to Einstein, the difference $E' - E$ is the body's kinetic energy in S'.[2] That is to say,

2. This is not entirely obvious. Kinetic energy is the energy attributable to a body's motion. It is the difference between the body's energies when moving and when at rest, *measured in the same frame*. In the present problem, the energies E' and E are measured in different frames. The principle of relativity demands, however, that the energy of a body at rest (in a given internal state) be the same in any frame. Einstein's conclusion follows.

$$E'_0 - E_0 = K_0 \tag{7.5a}$$
$$E'_1 - E_1 = K_1 \tag{7.5b}$$

where K_0 and K_1 denote the kinetic energies before and after the emission, measured in S'.[3] Substituting these expressions in (7.4), we obtain

$$K_0 - K_1 = (\gamma - 1)L \tag{7.6}$$

In Newtonian mechanics, kinetic energy is defined as

$$K = \tfrac{1}{2}\, mV^2 \tag{7.7}$$

We cannot assume that relation (7.7) is valid in special relativity, and indeed it is not. (The relativistic definition of kinetic energy is given by eq. [7.12] below.) Equation (7.7) must, however, apply in the low-velocity limit. We therefore assume that V/c is very small and expand γ using the binominal theorem as we have done on several prior occasions (see the appendix to chap. 4):

$$\gamma \approx 1 + \tfrac{1}{2}(V/c)^2 \tag{7.8}$$

With K given by (7.7) and γ by (7.8), the factor $\tfrac{1}{2}V^2$ appears on both sides of equation (7.6) and cancels. We thus obtain

$$m_0 - m_1 = L/c^2 \tag{7.9}$$

where m_0 and m_1 are the masses of the body before and after the emission. *As a result of the emission, the body's mass has decreased.* The decrease, $m_0 - m_1$, is $1/c^2$ times the energy radiated, measured in the body's rest frame. A similar effect must occur even when the emission results in a change in the body's velocity.

Equation (7.9) contradicts one of the basic tenets of classical physics—that mass is a conserved quantity. (It can be neither created nor destroyed.) Einstein has demonstrated with a simple example that in special relativity, mass need not be conserved. This result has far-reaching implications.

Einstein comments boldly that "since obviously here it is inessential that the energy withdrawn from the body happens to turn into energy of radiation rather than into some other kind of energy, we are led to the more general conclusion: the mass of a body is a measure of its energy content; if the energy changes by L, the mass changes in the same sense

3. Because these quantities are measured in S', they strictly ought to be labeled K'_0 and K'_1. But since the body's kinetic energy in frame S is always zero, no confusion can result from the use of the simpler notation.

by L/c^2."[4] This is the first statement of the relation between mass and energy, the most famous result of relativity. Several other derivations were subsequently published, both by Einstein and by others.

The mass-energy relation applies to any kind of energy. When a spring is compressed or stretched from its equilibrium configuration, it acquires elastic energy; as a result, its mass increases. Similarly, the mass of a capacitor increases when the capacitor is charged (electrostatic energy). Whenever a body is heated or cooled (thermal energy), its mass changes. In all these cases, however, the effect is too small to measure.

It would be wrong to conclude that because Einstein assumed V to be small, his result is in any way approximate. All the quantities in equation (7.9) are measured in the body's rest frame. The relation between them cannot depend on the velocity of the auxiliary frame S' in which the kinetic energy is expressed. We are free to assign to S' any velocity we please. Hence the result is rigorously true.

In his first paper, Einstein had derived the relativistic expression for kinetic energy, equation (7.12) of section 7.2. If one employs that expression for K in equation (7.6), the desired result follows directly for any value of V. Einstein did not use that argument because his derivation of the kinetic energy formula applied only to a pointlike structureless particle. The kinetic energy of a massive body might depend on its internal state as well as on its velocity. Einstein remarked in passing that the difference $K_0 - K_1$ depends on velocity exactly like the kinetic energy of an electron, but he pointedly did not assume that each K has that form.

Einstein was at first uncertain as to the significance of his result. He wrote to his friend Conrad Habicht, "The line of thought is amusing and fascinating, but I cannot know whether the dear Lord doesn't laugh about this and has played a trick on me."[5]

Because of the factor c^2 in the denominator of equation (7.9), the mass loss associated with the emission of a moderate amount of energy is very small. Einstein noted, however, that radioactive decay is accompanied by the release of "enormous" amounts of energy. "Is the reduction of mass

4. *Collected Papers*, doc. 24, 174. In 1952, Herbert Ives published a paper claiming that Einstein's derivation was fallacious because he had implicitly assumed the very result he was trying to prove. H. E. Ives, "Derivation of the Mass-Energy Relation" *Journal of the Optical Society of America* 42 (1952):540–543. In fact, however, Ives's criticism is fallacious; Einstein's argument is perfectly sound. See John Stachel and Roberto Torretti, "Einstein's First Derivation of Mass-Energy Equivalence," *American Journal of Physics* 50 (1982):760–763, for a thorough discussion.
5. Quoted in Pais, *Subtle Is the Lord*, 148.

in such a process not large enough to be detectable?" he inquired in his 1907 review.[6]

Einstein suggested that the mass-energy relation could be tested by comparing the atomic weight of a radioactive atom with the sum of the atomic weights of the end products of the decay. The difference should be equal to $1/c^2$ times the energy released in the disintegration of one mole of the radioactive material. The latter can be calculated from the measured rate of energy release and the radioactive lifetime.[7] Einstein estimated that for radium the fractional loss in mass would be .00012. He concluded that the predicted effect could be verified if the atomic weights were known to an accuracy of five decimal places. "This, of course, is impossible," he said. "However," he added, "it is possible that radioactive processes will be detected in which a significantly higher percentage of the original mass will be converted into energy than in the case of radium." This remark proved to be prescient, as we shall soon see.

Rest Energy

The energy content of a body, measured in a frame in which the body is at rest, is called the rest energy and is labeled E_0. It of course depends on the body's internal state—its temperature, physical and chemical structure, and so on.

Einstein has shown that the rest energy contributes an amount E_0/c^2 to a body's mass. Presumably, a body would have some mass even if it gave up all its internal energy; that quantity might be called "intrinsic" mass and labeled μ. We can write the total mass m as

$$m = \mu + E_0/c^2 \tag{7.10}$$

Einstein commented as follows in the 1907 review:

6. *Collected Papers*, doc. 47, 287.
7. Einstein's idea is sound, but it must be applied to the masses of the particular isotopes involved in the decay and not to the atomic weights, which are averages over isotopic masses. Einstein did not know about isotopes, whose existence was discovered only in 1913. Isotopic masses can be measured with high precision by mass spectroscopy.

The isotope of radium isolated by Marie and Pierre Curie was ^{226}Ra, which decays by alpha emission to ^{222}Rn with a half-life of 1,620 years. The radon is itself radioactive; several other alpha and beta decays follow in rapid succession until finally the stable isotope ^{206}Pb is reached. Five alpha particles and four betas are emitted during the chain of decays. When one adds the masses of all the decay products and compares the sum to the mass of ^{226}Ra, the fractional mass loss is .00014, in good agreement with Einstein's estimate.

This result [eq. (7.10)] is of extraordinary theoretical importance because the inertial mass and the energy of a physical system appear in it as things of the same kind. With respect to inertia, a mass μ is equivalent to an energy content of magnitude μc^2. Since we can arbitrarily assign the zero-point of E_0, we are not even able to distinguish between a system's "actual" mass and its "apparent" mass without arbitrariness. It seems far more natural to consider *any* inertial mass as a reserve of energy.[8] (Emphasis added)

Einstein is saying that since the zero-point of energy is arbitrary, we can include the term μc^2 in the rest energy and write simply

$$E_0 = mc^2 \tag{7.11}$$

The values of E_0 defined by equations (7.11) and (7.10) differ by a constant, μc^2. When a body changes its internal state, the relevant quantity is the change in energy. Adding the same constant to the energy of each state can have no measurable consequences. Unless the actual value of the rest energy can be measured, we cannot discriminate between equations (7.10) and (7.11). Writing E_0 in the form (7.11) appears to be just a matter of convenience.

There is, however, one potential way to obtain an absolute measure of rest energy. If, as Einstein suggests, all inertial mass is truly a reserve of energy, the intrinsic mass μ must be zero. This implies that all the mass of a system could disappear as the result of some process, its entire rest energy being converted to other forms. The energy release in that process, which can be measured, would determine the value of E_0.

Disappearance of matter is routinely observed today in elementary-particle physics. For example, the neutral pion, discussed in chapter 2, decays into two gamma rays (high-frequency radiation). π^0 decay constitutes a realization of Einstein's thought experiment; after the decay the pion no longer exists and no other massive particle has been created. The change in mass in the process is therefore the entire pion mass: m_1 in equation (7.9) is zero.

According to equation (7.11), when a π^0 decays at rest the total energy of the gamma rays should be $m_\pi c^2$, where m_π is the mass of the pion.[9] Measurements confirm the prediction. Einstein must have been gratified

8. *Collected Papers*, doc. 47, 286.
9. If the pion decays in flight, as generally happens, the gamma ray energy includes the pion's kinetic energy as well.

to learn about π^0 decay; in a sense, he had anticipated such a possibility already in 1907.

One way to describe what happens in radioactive decay is to say that some mass is converted to energy. A better statement is that mass and energy are equivalent: all mass carries energy and all energy contributes to mass. When a π^0 decays, one form of energy (the rest energy of the pion) is converted into another form (radiant energy).

According to equation (7.11), the rest energy associated with even a small amount of mass is enormous: for $m = 1$ gram, the rest energy is nearly 10^{14} joules, comparable to the output of a typical power plant in one day. The possibility of converting that huge energy into other forms has obvious implications of a practical nature. Although all of a neutral pion's rest energy is converted into radiant energy when it decays, on a macroscopic scale only a very small fraction of mass energy can be converted into useful energy. That small fraction forms the basis for nuclear power generation, as we shall see in section 7.4.

7.2. KINETIC ENERGY AND TOTAL ENERGY

Relativistic Kinetic Energy

In his first paper on relativity, Einstein derived an expression for the kinetic energy of a charged point particle by calculating the work done by electric forces in accelerating the particle from rest to velocity v. The result[10] was

$$K = \left(\frac{1}{\sqrt{1 - v^2/c^2}} - 1 \right) mc^2 = (\gamma - 1)\, mc^2 \qquad (7.12)$$

In the nonrelativistic limit, with γ given by the approximate form (7.8), equation (7.12) reduces to the Newtonian form $\frac{1}{2} mv^2$, as it must.

Einstein argued that formula (7.12) should apply also to a "ponderable material point," since such a body can be considered to have an infinitesimal electric charge. (The kinetic energy does not depend on the particle's charge.) In fact, the result (7.12) turns out to apply to *any* material body.

The validity of equation (7.12) is confirmed by a vast body of experimental evidence from nuclear and elementary-particle physics. In an ex-

10. The quantity $1/\sqrt{(1 - v^2/c^2)}$, which appears in eq. (7.12), depends on the body's velocity in the same way the parameter γ, defined in chapter 3, depends on the relative velocity V between frames. Following standard usage, we employ the same symbol for both. The reader should bear in mind, however, that the two quantities are quite different. At times I shall use the notation $\gamma(v)$ to identify the velocity to which a particular γ refers.

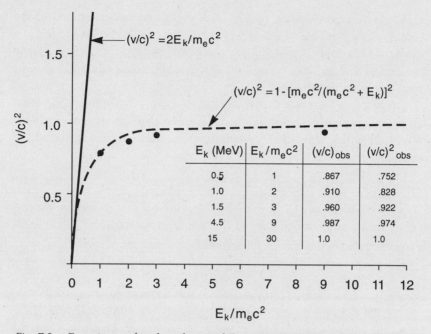

Fig. 7.2. Experiment that directly tested Einstein's formula for kinetic energy. Plotted is v^2 as a function of kinetic energy for electrons. The solid line is the Newtonian relation $v^2 = 2E_k/m$; the dashed line is Einstein's formula, eq. (7.12). The data points clearly confirm the validity of Einstein's equation.

periment designed specifically for this purpose,[11] electrons accelerated to nearly the speed of light in a linear accelerator were brought to rest in a container of water. The increase in temperature of the water measured how much kinetic energy the electrons had lost. The energy loss per electron is plotted in figure 7.2 as a function of v^2.

If kinetic energy were given by the Newtonian expression $\frac{1}{2} mv^2$, the data points would lie on a straight line; Einstein's formula, equation (7.12), coincides with that line for small v but departs from it more and more as v increases. The data clearly confirm Einstein's formula.

According to equation (7.12), the kinetic energy of a body approaches infinity when its speed approaches c. This important prediction is confirmed by the experimental data. Notice that in figure 7.2 the electrons' velocity increases very little between the last two data points even though their kinetic energy is more than doubled.

11. William Bertozzi, "Speed and Kinetic Energy of Relativistic Electrons," *American Journal of Physics* 32 (1964):551–555.

This result provides us, finally, with a physical understanding of the role of c as a limiting speed, at least for a material body. To accelerate a material body to the speed of light, one would have to transfer an infinite amount of energy to it. Since that is impossible, the body's speed must always remain less than c.

For any speed greater than c, the kinetic energy according to equation (7.12) is imaginary. This is another indication that speeds greater than c are incompatible with special relativity. (See, however, the discussion of tachyons in sec. 7.8.)

A General Expression for Energy

A body's total energy is the sum of its rest energy and its kinetic energy. Since we have identified mc^2 as the rest energy, equation (7.12) leads directly to an expression for total energy:[12]

$$E = \gamma mc^2 \tag{7.13}$$

When $v = 0$, equation (7.13) reduces to the rest energy.

For the thought experiment of the preceding section, equation (7.13) gives the energies in S' as $E_0' = \gamma m_0 c^2$, $E_1' = \gamma m_1 c^2$. Equation (7.3), the conservation of energy in frame S', is then consistent with the result (7.9). Equation (7.13) plays a central role in relativistic mechanics.

Units

All the applications of relativistic dynamics are microscopic; they deal with elementary particles such as electrons and protons and with atomic nuclei. Ordinary Mks units of mass and energy, kilograms and joules, are not well suited to the description of processes that involve such particles; all the numbers are very small. The mass of a proton, for example, is 1.7×10^{-27} kg.

It is standard practice to measure energy in nuclear physics in units of million electron volts, written MeV.[13] The energy acquired by a particle of electronic charge when it passes through a potential difference of

12. Eq. (7.13) appears first in a paper by Max Planck, "The Principle of Relativity and the Fundamental Equations of Mechanics," *Verhandlungen der Deutschen Physikalischen Gesellschaft* 4 (1906):136–141, the first paper on relativity published by anyone other than Einstein. Einstein's first use of the equation was in "On the Inertia of Energy Required by the Relativity Principle," *Annalen der Physik* 23 (1907):371–384; also *Collected Papers*, doc. 45.
13. In high-energy physics, the unit GeV (for giga-electron volt) is also used. 1 GeV = 1,000 MeV.

one million volts is 1 MeV. The conversion factor between MeV and joules is

$$1 \text{ MeV} = 1.602 \times 10^{-13} \text{ J} \tag{7.14}$$

Dividing by c^2, we obtain a unit of mass:

$$1 \text{ MeV}/c^2 = 1.783 \times 10^{-30} \text{ kg} \tag{7.15}$$

A more commonly used unit of mass is the *atomic mass unit*, written as u. By definition, an atomic mass unit is one-twelfth the mass of the neutral atom ^{12}C; the conversion factor between atomic mass units and kilograms is

$$1 \text{ u} = 1.660565 \times 10^{-27} \text{ kg} \tag{7.16}$$

Another useful relation is

$$(1 \text{ u})c^2 = 931.5 \text{ MeV} \tag{7.17}$$

The mass of a proton is 1.007276 u;[14] its rest energy is (1.007276)(931.5) = 938.28 MeV.

Relativistic Mass

Einstein's expressions for rest energy (7.11) and for total energy (7.13) may be rephrased by postulating that when a body is moving with velocity v its mass increases by the factor $\gamma(v)$. If we call m the *rest mass* and define a velocity-dependent quantity

$$M(v) = \gamma(v) \, m \tag{7.18}$$

as (relativistic) mass, we can write the energy[15] simply as

$$E = Mc^2 \tag{7.19}$$

Equations (7.18) and (7.19) contain no new information; they are merely an alternative way of expressing equation (7.13). The idea of a velocity-dependent mass does have a certain appeal, however. Mass is associated with inertia, which measures a body's resistance to a change in its motion. The electrons' inertia in the experiment shown in figure 7.2

14. The reason that the proton mass was not selected as the unit is that with the ^{12}C scale, the masses of most isotopes are close to whole numbers.
15. We shall see in section 7.3 that relativistic momentum can be written in an analogous manner as $p = Mv$. The kinetic energy, however, cannot be written as $\frac{1}{2} Mv^2$.

clearly increases as their speed increases; when the speed is close to c, a large increment of energy brings about almost no change in speed.

The increase in mass implied by equation (7.18) is $M - m = (\gamma - 1)m$, which is just $1/c^2$ times the kinetic energy, equation (7.12). Thus by adopting the concept of velocity-dependent mass, we can include kinetic energy in the general statement that *any* energy E contributes an amount E/c^2 to a body's mass. This is another attractive feature of that interpretation.

Whether or not to speak of velocity-dependent mass is largely a matter of taste. Although it is currently unfashionable to do so, Einstein did and we shall as well. Whenever there is a possibility of confusion, we shall refer explicitly to rest mass or relativistic mass.

The possibility of a velocity-dependent mass had been discussed long before Einstein. Theories proposed by Lorentz and by Abraham had predicted that mass increases with velocity.[16] An experiment performed by Walter Kaufmann in 1901 gave qualitative support to the idea.

Kaufmann measured the deflection of beta rays (high-speed electrons emitted in beta decay) by magnetic fields. According to standard theory, the magnetic deflection depends on the ratio of a particle's charge to its mass. Although Kaufmann's experiment was not very accurate, it suggested that the electron's mass does increase with velocity.

In 1906, Kaufmann published a paper containing new results, which he claimed disproved Einstein's theory; Kaufmann's data were in better agreement with Abraham's theory, which predicted a dependence of mass on velocity different from (7.18).

Einstein ignored Kaufmann's criticism because he was convinced that Kaufmann's experimental work was wrong. In 1908, a much more accurate experiment performed by Alfred Bucherer confirmed Einstein's suspicion. Not only did the dependence of beta ray mass on velocity measured by Bucherer agree with Einstein's relation (7.18) but the value of the rest mass agreed closely with that measured by J. J. Thomson for low-velocity electrons (cathode rays). Bucherer had thus demonstrated that beta rays and cathode rays are the same particles. His results were influential in gaining acceptance for Einstein's theory.[17]

16. In these theories the electron has a "longitudinal" mass and a "transverse" mass, which differ. Einstein's first paper also employed two masses. Planck showed in 1906 that if force is defined as the rate of change of momentum instead of as mass times acceleration, the need for two masses disappears.
17. Many years later, C. T. Zahn and A. H. Spees called attention to problems with the velocity filter employed by Bucherer and concluded that Bucherer's re-

7.3. RELATIVISTIC MOMENTUM

We saw in chapter 1 that conservation of momentum is covariant under a Galilean transformation. If momentum is conserved in one inertial frame it is conserved in any other, subject to the condition that mass is conserved.

By repeating the analysis of section 1.7, we can easily demonstrate that the conservation law is *not* covariant in special relativity if momentum is defined in the same way as in classical mechanics—as the product of (rest) mass and velocity.

Equation (1.22) describes conservation of momentum for a general two-body collision in frame S:

$$m_A v_A + m_B v_B = m_C v_C + m_D v_D \qquad (7.20)$$

Transforming all the velocities in this equation to frame S' using the relativistic velocity transformation, equation (4.15), we obtain

$$m_A \frac{v_A' + V}{1 + v_A' V/c^2} + m_B \frac{v_B' + V}{1 + v_B' V/c^2} = m_C \frac{v_C' + V}{1 + v_C' V/c^2} + m_D \frac{v_D' + V}{1 + v_D' V/c^2} \qquad (7.21)$$

If the conservation law were covariant, a relation identical to (7.20) would hold in S', that is,

$$m_A v_A' + m_B v_B' = m_C v_C' + m_D v_D' \qquad (7.22)$$

Since equation (7.21) does not reduce to the form (7.22), the conservation law is not covariant.

The simple collision shown in figure 7.3a illustrates the result. The colliding bodies, A and B, have equal masses. As seen in frame S, they move in opposite directions with the same speed, v. After the collision both bodies are at rest. Since the total initial and total final momenta are both zero, momentum is conserved.

Figure 7.3b shows the same collision in a frame S' that moves at velocity $-v$ relative to S; in S', body B is initially at rest. According to Galilean relativity, the initial velocity of A is $2v$ and the initial momentum of the system is $2mv$. After the collision both bodies have velocity v; the total final momentum is again $2mv$ and momentum is conserved in S' as it is in S.

---

sults actually proved little, if anything, more than Kaufmann's. C. T. Zahn and A. H. Spees, "A Critical Analysis of the Classical Experiments on the Relativistic Variation of Electron Mass," *Physical Review* 53 (1938):511–521.

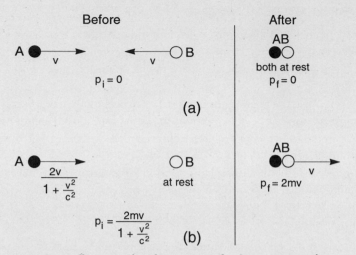

Fig. 7.3. Collision used to demonstrate that conservation of momentum is not covariant in special relativity if momentum is still defined as the product of mass and velocity. (a) Bodies A and B, of equal mass and moving in opposite directions with the same speed v, collide head-on. After the collision, both bodies are at rest. The total momentum is zero both before and after the collision; momentum is conserved. (b) The same collision is viewed in frame S', which moves from right to left at speed v relative to S. In S', B is initially at rest while the velocity of A is $2v/(1+v^2/c^2)$. The final velocity of both bodies is v. Momentum is not conserved.

According to special relativity, the initial velocity of body A in S' is not $2v$; using equation (4.15), we find instead

$$v'_A = \frac{2v}{1+v^2/c^2} \tag{7.23}$$

With momentum defined as mass times velocity, A's initial momentum in S' is

$$p'_A = \frac{2mv}{1+v^2/c^2} \tag{7.24}$$

Since the initial momentum of B is zero, (7.24) is also the total initial momentum of the system.

The final velocity of both A and B in S' is v, just as in Galilean relativity. Hence the final momentum of the system is $2mv$, which differs from the initial momentum, given by (7.24). Thus momentum is not conserved in S'.

The principle of relativity demands that all physical laws be covariant. We can draw one of two conclusions:

(i)   Momentum conservation is not a general law of nature. It is only an approximate law, valid when all velocities are small compared to *c;* or

(ii)   Momentum conservation is a strictly valid law, but momentum is not the product of mass and velocity. It takes that form only for small values of *v/c*.

Alternative (i) is unattractive; the conservation law has great appeal because of its simplicity and generality. We therefore pursue alternative (ii) and look for a new definition of momentum, one that makes the conservation law fully covariant under a Lorentz transformation. The new definition should reduce to the Newtonian form *mv* when *v/c* is very small.

A note of caution must be sounded at the outset. Although all laws of nature must be relativistically covariant, a covariant relation is not necessarily a law of nature. Covariance implies that if a certain relation holds in one inertial frame, it holds also in any other. But the relation might not hold in *any* frame. One can construct many "pseudo-laws" that are mathematically covariant but are not satisfied in nature. Even after we exhibit a definition of momentum that leads to a covariant conservation law, therefore, we cannot be sure that the new conservation law is actually valid. Only experimental evidence can provide that assurance.

A simple way to derive the expression for relativistic momentum is to return to the thought experiment of section 7.1, in which a body emits light waves of equal energies in opposite directions. This is the experiment that Einstein used to deduce the mass-energy relation. Let us examine the special case $\phi = 0$: the light waves are emitted in the $\pm x$ direction.

It is well known in electromagnetic theory that the momentum carried by an electromagnetic wave is $1/c$ times the energy flux. With the energies given by equation (7.1), the momenta of the two waves in frame S' are

$$\gamma(L/2c)\,(1 + v/c) \quad \text{and} \quad \gamma(L/2c)\,(1 - v/c) \tag{7.25}$$

where $L/2$ is the energy of each wave in frame S, and v is the velocity of the radiating body in S'.

Since the two waves are emitted in opposite directions, their total momentum p_{rad} is the difference between the two expressions in (7.25):

$$p_{\text{rad}} = \gamma\, L v/c^2$$

Let p_0 and p_1 denote the momentum of the radiating body in S' before and after the emission, respectively. Conservation of momentum gives the relation

$$p_0 = p_1 + p_{rad} = p_1 + \gamma \, Lv/c^2 \qquad (7.26)$$

According to equation (7.9), L/c^2 is equal to the body's decrease in mass as a result of the decay, $m_0 - m_1$. Substituting in equation (7.26), we can rewrite the equation in the form

$$p_0 - m_0 \gamma v = p_1 - m_1 \gamma v \qquad (7.27)$$

Equation (7.27) shows that the expression $p - \gamma mv$ is independent of the body's mass. It can, however, depend on velocity. The general solution of equation (7.27) is

$$p = \gamma \, mv + f(v) \qquad (7.28)$$

where $f(v)$ is an arbitrary function of velocity.

We can easily show that $f(v)$ must be identically zero. Imagine a system consisting of two noninteracting pieces of mass m_1 and m_2, each moving with velocity v. The momentum of each piece is given by equation (7.28):

$$p_1 = \gamma m_1 v + f(v) \qquad p_2 = \gamma m_2 v + f(v)$$

and their total momentum is

$$p = \gamma(m_1 + m_2)v + 2f(v) \qquad (7.29)$$

The system can also be considered a single body whose mass is $m_1 + m_2$; equation (7.28) gives for the momentum of that body

$$p = \gamma(m_1 + m_2)v + f(v) \qquad (7.29')$$

Equations (7.29) and (7.29') imply that

$$2 f(v) = f(v)$$

which in turn implies that $f(v) = 0$; hence

$$p = \gamma \, mv \qquad (7.30)$$

Equation (7.30) is the desired definition of relativistic momentum. If energy is measured in MeV, a convenient unit for momentum is MeV/c.

Notice that relativistic momentum can be expressed as the product of relativistic mass and velocity:

$$p = Mv \qquad (7.31)$$

where M is defined by equation (7.19).

We still have to verify that our new definition of momentum satisfies the stated requirements. It is clear that the nonrelativistic limit is correct: for small v, γ approaches unity and equation (7.30) reduces to the Newtonian definition $p = mv$. To verify that momentum conservation is relativistically covariant, we rewrite the conservation equation (7.20) using our new definition of momentum:

$$\gamma_A\, m_A\, v_A + \gamma_B\, m_B\, v_B = \gamma_C\, m_C\, v_C + \gamma_D\, m_D\, v_D \qquad (7.32)$$

Since momentum is a vector quantity, equation (7.32) stands for three equations, one for each component. We assume that these relations hold in inertial frame S, transform everything to another frame S', and check whether the same relations hold in S'.

The transformation properties of γ and of γv are derived in the appendix to this chapter. For the transverse components, the result is very simple: the product γv_y has the same value in both frames, as does γv_z. (The relative velocity between S' and S is, as usual, taken to be in the x direction.) Hence the transverse components of momentum are necessarily conserved in S' if they are conserved in S.

The x-momentum transforms in a more complicated way; we show in the appendix that it too is conserved in S' if it is conserved in S, provided the auxiliary condition

$$m_A\, \gamma(v_A) + m_B\, \gamma(v_B) = m_C\, \gamma(v_C) + m_D\, \gamma(v_D) \qquad (7.33)$$

is satisfied. Equation (7.33) is the relativistic counterpart of the conservation of mass, to which it reduces in the low-velocity limit ($\gamma \to 1$).

In Galilean relativity, conservation of mass is a necessary condition if momentum conservation is to be covariant. In special relativity, rest mass is not in general conserved; instead, the sum of γm for all the bodies present is conserved. We recognize this condition as the conservation of relativistic mass or (on multiplication by c^2) of total energy.

It is interesting that conservation of energy emerges as a necessary condition to make conservation of momentum covariant. The converse is also true: momentum must be conserved if energy conservation is to be covariant. Momentum and energy, which in classical mechanics are independent quantities, are closely linked in special relativity. We will show in section 7.6 that they can be considered a single entity, energy-momentum, analogous to space-time.

The validity of the relativistic momentum conservation law, with momentum defined by equation (7.30), is confirmed by experimental data from high-energy physics. As in the nonrelativistic case, conservation of momentum alone does not determine the outcome of a collision. In a two-body scattering problem, for example, if the masses and the initial velocities are given, there are six unknowns (the three components of each of the final velocities) and only three equations. Hence the equations have an infinite number of solutions. But if one of the final velocities is measured, the other is determined by the conservation law. In every instance, the measured velocity agrees with the value calculated from equation (7.32).

Finally, we return to the collision of figure 7.3 to tie up one loose end in the argument. We found in the earlier discussion that with the Newtonian definition $p = mv$, momentum is not conserved in this collision in frame S'. Let us now recalculate the initial and final momenta using the relativistic formula $p = \gamma mv$.

In frame S, the total momentum is still zero both before and after the collision, so momentum is conserved. In S', the initial velocity of A is given by equation (7.23); the corresponding momentum is

$$p'_A = \frac{2mv}{1 + v^2/c^2} \bigg/ \sqrt{1 - \frac{1}{c^2}\left(\frac{2v}{1 + v^2/c^2}\right)^2}$$

After some algebra this reduces to the form

$$p'_A = \frac{2mv}{1 - v^2/c^2} \tag{7.34}$$

which is also the total initial momentum of the system, since $p'_B = 0$.

After the collision, bodies A and B are both moving at velocity v; their total momentum is therefore

$$2mv/\sqrt{1 - v^2/c^2} \tag{7.35}$$

which differs from the initial momentum (7.34). Momentum still appears not to be conserved, even when the correct formula for relativistic momentum is employed. What is wrong?

The clue to the resolution of the puzzle comes from the realization that kinetic energy is not conserved in this collision. (This is most readily seen in frame S, in which the final kinetic energy is zero.) A collision in which kinetic energy is not conserved is called *inelastic*. Since total energy is always conserved, the lost kinetic energy in an inelastic collison must have been converted into some other form of energy—heat or internal excitation or perhaps sound or radiation. Our description of the collision is incomplete unless that other energy is taken into account.

Whatever form the lost kinetic energy takes, it contributes to the mass of the system. Hence our implicit assumption that the final state consists of two bodies of rest mass m (or a single body of mass $2m$) cannot be right.

Let μ denote the rest mass of the combined body AB. If all the energy remains within AB, we can calculate μ from the conservation of energy in frame S. The energy of each of the colliding bodies was γmc^2, so the total initial energy is $2\gamma mc^2$. After the collision AB is at rest; its entire energy is rest energy. Conservation of energy gives

$$\mu c^2 = 2\gamma mc^2 \quad \text{or} \quad \mu = 2\gamma m$$

With this value for the mass, the final momentum in S' is

$$p'_{AB} = \mu v\gamma = 2\gamma^2 mv = \frac{2mv}{1 - v^2/c^2}$$

which is the same as (7.33). Momentum is (at last!) conserved.

If any energy leaves the scene, for example, in the form of radiation, the energy and momentum of the radiation must be included in the conservation equations; the rest mass of AB in that case would be less than $2\gamma m$.

7.4. RELATIVITY IN NUCLEAR AND PARTICLE PHYSICS

Expressions (7.13) and (7.30) for relativistic energy and momentum provide the basis for relativistic dynamics. They must be used whenever a body travels at close to the speed of light, as happens routinely in elementary-particle physics. In most nuclear reactions, however, the bodies involved travel fairly slowly; in that sense low-energy nuclear physics is not relativistic. Relativity nonetheless plays an important role because it accounts for the energy balance through Einstein's mass-energy relation. Nuclear reactions in fact provided the first convincing proof of mass-energy equivalence.

Consider once again the general two-body reaction

$$A + B \rightarrow C + D$$

We define a parameter

$$\Delta m = m_C + m_D - (m_A + m_B) \tag{7.36}$$

which measures the difference between the total rest mass of the reacting bodies, A and B, and that of the bodies in the final state, C and D. (Positive Δm corresponds to an increase in rest mass.) We also define an energy

$$Q = -(\Delta m)c^2 \tag{7.37}$$

called the Q value for the reaction.

Equation (7.33), when multiplied by c^2, represents the conservation of energy for the reaction. Expressing each energy as the sum of rest energy and kinetic energy, we can rewrite the equation as

$$K_f = K_i + Q \tag{7.38}$$

where K_i and K_f are the total kinetic energies of the initial and final states, respectively.

If Δm and Q are zero, kinetic energy is conserved. Such a collision is said to be *elastic*. If Q is negative (Δm is positive), the kinetic energy is diminished by $|Q|$; such a reaction is called *endoergic*. Since kinetic energy cannot be negative, the reaction cannot take place at all unless K_i exceeds $|Q|$.

If Q is positive (Δm is negative), the kinetic energy increases by Q; such a reaction is called *exoergic*. (A decay process can be considered an exoergic reaction in which the initial state consists of only one body.)

An example of an exoergic nuclear reaction is

$$p + {}^7\text{Li} \rightarrow {}^4\text{He} + {}^4\text{He} \tag{7.39}$$

The rest masses of the isotopes involved have the following values:[18]

$$p \ ({}^1\text{H}): \ m = 1.007825 \ \text{u}$$
$${}^7\text{Li}: \ m = 7.016004 \ \text{u}$$
$${}^4\text{He}: \ m = 4.002603 \ \text{u}$$

The total mass of the colliding bodies is 8.02383 u, whereas that of the emergent ones is only 8.00521 u. The difference, $\Delta m = -.01862$ u, is about 0.2 percent of the total mass of the reactants. This is typical of nuclear reactions: only a very small fraction of the rest energy is converted to kinetic energy. The Q value for the reaction is 17.34 MeV.

Reaction (7.39) was studied by J. D. Cockroft and E. T. S. Walton in 1932 in one of the first experiments carried out with an accelerator they had invented. The measured increase in kinetic energy was 16.95 MeV, in good agreement with the value of Q.

---

18. The isotopic masses given in tables are generally those of neutral atoms, not of the bare nuclei. For an isotope with atomic number Z, this includes the mass of Z electrons. Since the total number of electrons is the same on both sides of the reaction, we can use atomic masses in calculating Δm and Q; the electron masses cancel.

Nuclear Power

In principle, any reaction with a positive Q value constitutes a potential source of usable energy. The kinetic energy of the reaction products can be used to boil water and run a turbine. The practical problems involved in doing this on a commercial scale are, however, immense.

For a long time there was widespread skepticism that nuclear energy could actually be harnessed as a practical source of power. In 1933, Ernest Rutherford, the discoverer of the nucleus, remarked in a public lecture that anyone looking to nuclear reactions for useful energy was "talking moonshine." Less than ten years after that pessimistic assessment, the first nuclear reactor became operational.

Nuclear reactors employ a particular reaction called a *fission* reaction; the most commonly used fuel is uranium. When a slow neutron strikes a uranium nucleus, it occasionally causes it to split into two large fragments; two or three neutrons are emitted as well. A typical fission reaction is

$$n + {}^{235}\text{U} \rightarrow {}^{141}\text{Ba} + {}^{92}\text{Kr} + 3\,n \tag{7.40}$$

The Q value for this reaction is 175 MeV; a great deal of energy is released. Moreover, the emitted neutrons can strike other uranium nuclei and cause them to react in turn. The result, called a chain reaction, is the basis for the operation of reactors.

With the help of control rods that absorb some of the neutrons, the reaction can be made to proceed at a steady rate. If the chain reaction is not controlled, it can run away, releasing energy too rapidly. If all the energy is released over a very short time, an explosion results; this is what happens in a nuclear weapon.

Another reaction that is a potential source of power is the *fusion* reaction, of which an example is

$$^{2}\text{H} + {}^{3}\text{H} \rightarrow {}^{4}\text{He} + n \tag{7.41}$$

The Q value for reaction (7.41) is 17.6 MeV. Although this is much less than the Q value for reaction (7.40), the energy release per gram of fuel in fusion reactions is more than three times greater than in fission. Moreover, the fuel is plentiful and readily available and the reactions leave no long-lived radioactive waste products, as fission reactions do.

Unfortunately, the technical problems involved in achieving a controlled fusion reaction have not yet been solved. The major difficulty is that the reactants must get very close to one another if the reaction is to proceed. (The nuclear force responsible for the reaction has a very short range.) Since the reactants in fusion reactions are charged, their electrical repulsion keeps them apart unless they have a fairly high velocity. By contrast, neutrons have no charge and encounter no difficulty in entering a uranium nucleus to initiate fission.

At very high temperatures, the thermal motion of the particles is sufficient to initiate fusion reactions. (For this reason the reactions are called thermonuclear.) Temperatures of hundreds of millions of degrees are required if the reactions are to proceed in sufficient number to make a reactor work; the problems of achieving such temperatures and keeping the reacting material confined within a small volume for sufficient time are daunting. Other methods, such as laser-initiated fusion, are under investigation.

Temperatures high enough to initiate thermonuclear reactions are known to occur in stellar interiors. This is the principal mechanism for the generation of the energy radiated by the sun.[19]

Binding Energies

All nuclei are composed of neutrons and protons: an isotope of atomic number Z and mass number A contains Z protons and $(A-Z)$ neutrons. One might therefore expect its rest mass to be Z times the proton mass plus $(A-Z)$ times the neutron mass. The measured mass of every isotope is, however, smaller than the value given by that prescription. The mass $M(A, Z)$ can be written in the form

$$M(A, Z) = Zm_p + (A - Z)m_n - \delta \tag{7.42}$$

where δ is a positive number called the *mass defect*.

For example, the deuteron (nucleus of heavy hydrogen) is composed of one neutron and one proton: $Z = 1$, $A = 2$. The relevant masses are

$$p \ (^1\text{H}): \ m = 1.007825 \text{ u}$$
$$n: \ m = 1.008665 \text{ u}$$
$$p + n: \ m = 2.016490 \text{ u}$$
$$deuteron \ (^2\text{H}): \ m = 2.014102 \text{ u}$$

The mass defect of the deuteron is 0.00239 u. When a proton and a neutron are brought together to form a deuteron, their combined mass decreases by about one part in a thousand.

19. Before the discovery of nuclear energy, the sun's energy was believed to come from gravitational potential energy released as the sun contracts. According to that model, the sun could emit at its present rate for only a few million years before collapsing.

Mass-energy equivalence accounts in a natural way for the mass defect. The deuteron, like any bound system, is held together by an attractive force between its constituents. The energy of the neutron and proton when bound in a deuteron is less than their energy when they are far apart and at rest; the latter energy is just the sum of the rest energies.

The difference between the energy of the separated particles and that of the bound state is called the *binding energy, E_b*:

$$E(\text{bound state}) = E(\text{separated particles}) - E_b \qquad (7.43)$$

The value of E_b can be determined experimentally by measuring how much energy must be provided to dissociate the nucleus.

According to equation (7.9), the negative contribution to the energy, $-E_b$, should result in a mass defect δ given by

$$\delta = E_b/c^2 \qquad (7.44)$$

Experimental data on nuclear masses and binding energies are in complete agreement with the relation (7.44). The binding energy of the deuteron, for example, is 2.22 MeV, exactly the predicted value.

Equation (7.44) should apply to any bound system, including atoms and molecules and even the solar system. In each case the mass of the bound system should be slightly smaller than the sum of the masses of the constituents. Typical chemical binding energies are, however, much smaller than those of nuclei; the associated mass defects are correspondingly small. In 1907, Planck estimated the decrease in mass of a mole of water due to its chemical binding. The result, about 10^{-8} g, is too small to be detected.

Elementary Particle Reactions

Although relativity accounts for the energy balance in nuclear reactions through equation (7.38), it has only a minor effect on the kinematics because Δm is generally much smaller than the masses of all the nuclei involved. Unless one of the reactants has a very high kinetic energy to begin with, the kinetic energies of all the particles involved are much smaller than their rest energies. All velocities are then small compared to c and the kinematics is essentially nonrelativistic. In writing the conservation of energy and momentum for reaction (7.39) at low energies, for example, it is unnecessary to use the relativistic forms γmv for momentum and $(\gamma - 1)mc^2$ for kinetic energy; the Newtonian expressions mv and $\frac{1}{2} mv^2$ give practically the same answers.

Dramatic manifestation of relativistic kinematics is found in the study of the unstable particles discovered during the past half century in high-energy physics. In reactions involving such particles, Δm is frequently greater than the rest masses of some of the particles involved. Wholesale conversion of rest energy to kinetic energy can take place. We have already encountered an extreme example of this in the decay of the π^0.

As an example of relativistic kinematics, consider the decay of an unstable particle called the K^+. Its decay scheme is the following:

$$K^+ \rightarrow \pi^+ + \pi^0 \tag{7.45}$$

The rest masses of the particles involved are as follows:

K^+: $m = 0.530$ u   (call this mass M)
π^+: $m = 0.150$ u   (call this m_1)
π^0: $m = 0.145$ u   (call this m_2)

Δm is -0.235 u, more than the pion rest mass. The Q value is 219 MeV.

Suppose a K^+ at rest decays. Momentum conservation demands that the two decay products travel in opposite directions, so the problem is essentially one-dimensional. The conservation equations are

$$m_1 v_1 \gamma_1 + m_2 v_2 \gamma_2 = 0 \quad \textit{conservation of momentum} \tag{7.46}$$
$$m_1 \gamma_1 c^2 + m_2 \gamma_2 c^2 = Mc^2 \quad \textit{conservation of energy} \tag{7.47}$$

Since each γ involves the corresponding velocity through a square root, the algebra involved in solving equations (7.46) and (7.47) is quite laborious. This is typical of relativistic kinematics. The result is

$$m_1 \gamma_1 = \frac{M}{2} + \frac{m_1{}^2 - m_2{}^2}{2M} \tag{7.48a}$$

$$m_2 \gamma_2 = \frac{M}{2} + \frac{m_2{}^2 - m_1{}^2}{2M} \tag{7.48b}$$

For K^+ decay, these equations give $\gamma_1 = 1.818$, $\gamma_2 = 1.776$. (The kinetic energies differ by very little because m_1 and m_2 are nearly equal.)

If the K^+ that decays is moving, the easiest approach is to analyze the problem first in the rest frame of the K^+, in which the solution is given by equations (7.48a,b). A Lorentz transformation gives the solution in the laboratory frame. In this case the results depend on the direction in which the decay products are emitted.

Production of Particles

Thus far we have discussed only cases in which mass-energy is converted into kinetic energy. The reverse process can occur as well: in a collision between energetic particles, new particles can be created at the expense of kinetic energy.

For example, pions are produced in high energy proton-proton collisions. One possible reaction is

$$p + p \rightarrow p + n + \pi^0 \tag{7.49}$$

In a typical experiment an energetic proton from an accelerator strikes a target proton at rest. The kinetic energy of the incident proton must of course exceed the rest energy of the pion, $m_\pi c^2$, if reaction (7.49) is to proceed.[20] Conservation of energy and momentum is described by equations like (7.48a,b). One has to write a separate equation for each component of momentum; there are four simultaneous equations, each containing square roots. The algebra is cumbersome, but it is straightforward to obtain numerical solutions with the help of a computer.

Special relativity accounts for the kinematics of particle production and decay. This is only the first step toward an understanding of the complex phenomena. Special relativity does not explain, for example, why only about 20 percent of K^+ decays proceed via the reaction (7.45), or why the half-life for the decay has the particular value 1.2×10^{-8} seconds. Our understanding of elementary particle phenomena is still very rudimentary. But at least the kinematics is understood.

7.5. Beta Decay and the Neutrino

An interesting application of relativistic kinematics and of the conservation laws is provided by the phenomenon of beta decay. Many radioactive nuclei decay by emitting a beta particle, either a positive or a negative electron. A typical example is the isotope ^{14}C, which emits a negative electron; its decay scheme is (or so it was at first believed)

$$^{14}C \rightarrow ^{14}N + e^- \tag{7.50}$$

This decay appears to be analogous to that of the K^+, equation (7.45); in each case one particle decays into two. Conservation of momentum and

---

20. The threshold energy for reaction (7.49) is somewhat greater than $m_\pi c^2$ because momentum conservation does not allow all the final-state particles to be at rest.

energy should therefore be described by equations (7.46) and (7.47), and the energies of the decay products should be given by equations (7.48a,b). Those equations imply that if one of the decay products is much lighter than the other, say, $m_1 << m_2$, its kinetic energy $(\gamma_1 - 1)m_1 c^2$ is much greater than that of the other. In the present case the electron is much lighter than the nitrogen nucleus; hence it should receive practically all the available energy.

The experimental data on beta decays were perplexing. According to equation (7.46), all the emitted electrons should have the same energy. Instead, electrons were detected with a continuous spectrum of energies, ranging from nearly zero to a maximum value E_m. Since the kinetic energy carried off by the ^{14}N is very small, energy appeared not to be conserved in the decay.

The electron energy spectrum was not the only puzzling feature of the results. The electron and the daughter nucleus were not always observed to be emitted in opposite directions, as they should be if their total momentum is zero. Thus momentum appeared not to be conserved. Finally, angular momentum appeared also not to be conserved.

Wolfgang Pauli came up with an ingenious idea that accounted for all the experimental findings. Pauli postulated that an additional neutral particle, which acquired the name neutrino, is emitted in every beta decay. Instead of (7.50) the decay reaction is

$$^{14}\text{C} \rightarrow {}^{14}\text{N} + e^- + \bar{\nu} \tag{7.51}$$

where $\bar{\nu}$ stands for the (anti)neutrino.

Having no charge, the neutrino could easily escape detection; it would not, for example, leave a track in an emulsion or in a cloud chamber photograph.

Pauli's neutrino hypothesis salvaged the conservation laws; the neutrino carries off the "missing" momentum and energy. When three particles are emitted in a decay, the energy that any one particle receives is not uniquely determined by the conservation laws. A spectrum of electron energies is therefore to be expected.

The neutrino hypothesis also explains why the decay products are not collinear. When a body at rest decays into three or more, conservation of momentum does not demand that any two of them be collinear. (Three vectors can add up to zero even if they are all in different directions.)

Physicists' confidence in the conservation laws was so strong that the existence of the neutrino was accepted for many years just on the basis of

Pauli's argument, without any direct evidence. Frederick Reines and Clyde Cowan finally confirmed the existence of the neutrino in 1956 by detecting inverse beta decay reactions initiated by neutrinos.

The neutrino story has one other interesting aspect. The value of E_m in every known beta decay is (within the experimental uncertainty) equal to the available energy $(\Delta m)c^2$. The neutrino that accompanies an electron with the maximum energy E_m therefore carries off little or no energy. But the energy of any particle must be at least its rest energy mc^2; it follows that the rest mass of the neutrino must be extremely small, even compared to that of the electron, the lightest known particle. It might even be zero.[21]

Our formalism at this stage is not prepared to deal with a particle whose rest mass is zero; equations (7.13) and (7.30) imply that both the energy and the momentum of such a particle are zero, no matter what its velocity. As we shall see in a later section, however, special relativity does admit the possibility of zero-mass particles, provided they travel at the speed of light. The photon is one such particle; if the neutrino indeed has zero rest mass, it is another.

7.6. The Transformation Law for Energy and Momentum; Four-vectors

In this section, we examine the transformation properties of relativistic energy and momentum under a Lorentz transformation. The results demonstrate the close relation between the two quantities, of which we have already seen some hint.

The necessary transformation relations are found in the appendix. Equations (7.A5) and (7.A7) express $\gamma(v')$ and $v'\gamma(v')$ in terms of $\gamma(v)$ and $v\gamma(v)$. On multiplying (7.A5) by mc^2 we can identify each term as either an energy or momentum and write the equation as

$$E' = \gamma(V)(E - Vp_x) \qquad (7.52a)$$

Similarly, multiplying (7.A7) by m gives

$$p'_x = \gamma(V)\left(p_x - \frac{VE}{c^2}\right) \qquad (7.52b)$$

According to (7.A9), the transverse components of momentum are invariant:

21. Experiment can only set an upper limit on the rest mass of the neutrino. The upper limit is very low (≤ 18 eV).

$$p'_y = p_y \qquad (7.52c)$$
$$p'_z = p_z \qquad (7.52d)$$

Equations (7.52a–d) constitute the desired transformation law. These equations have a familiar appearance: they have the same form as do the Lorentz transformations. If in equation (4.1) we replace x, y, z, and t by p_x, p_y, p_z, and E/c^2, respectively, we obtain precisely equation (7.52).

From the point of view of special relativity, space and time are parts of a single entity called space-time. An interval that is purely spatial in one frame appears as a combination of space and time intervals in another frame. In the same sense, relativistic energy and momentum are parts of a single four-dimensional entity that we may call energy-momentum. A Lorentz transformation turns energy and momentum into linear combinations of one another.

Four-vectors

According to equation (7.52) the three components of momentum, together with $(1/c)$ times the energy, obey the same transformation law as do the three components of position together with the time. Any set of four quantities that displays that transformation property is called a four-vector. In a particular frame of reference the first three (spatial) components of a four-vector form an ordinary three-vector; the fourth component is the temporal component.

Many quantities of physical interest turn out to be expressible as four-vectors. In electricity, for example, the three components of current density form the spatial part of a four-vector whose temporal component is the charge density. If a physical law can be expressed as a relation between four-vectors, its relativistic covariance is automatically assured.

We showed in chapter 4 that the combination $x^2 - c^2t^2$ (in one spatial dimension) or $x^2 + y^2 + z^2 - c^2t^2$ (in three dimensions) is a relativistic invariant: it has the same value in all inertial frames. An analogous invariant must exist for any four-vector: if (V_1, V_2, V_3, V_4) are the components of a four-vector, the quantity

$$V_1^2 + V_2^2 + V_3^2 - V_4^2 \qquad (7.53)$$

is invariant. Its square root can be viewed as a generalized "length" of the four-vector.[22]

---

22. In some formal treatments a four-vector is defined as a set of four quantities (V_0, V_1, V_2, V_3), where $V_0 = i V_4$. In this notation the space-time four-vector is written as *(ict, x, y, z)*. The invariant (7.53) is then $V_0^2 + V_1^2 + V_2^2 + V_3^2$, which is

The invariant associated with the energy-momentum four-vector is

$$p_x{}^2 + p_y{}^2 + p_z{}^2 - c^2 (E/c^2)^2 = p^2 - E^2/c^2 \qquad (7.54)$$

If a quantity is invariant, its value can be calculated in any convenient frame. We know that in the rest frame of a body, $p = 0$ and $E = mc^2$. Hence the value of the invariant must be $-m^2 c^2$. We have derived the very useful relation

$$p^2 - \frac{E^2}{c^2} = -m^2 c^2$$

or equivalently,

$$E^2 = p^2 c^2 + m^2 c^4 \qquad (7.55)$$

Equation (7.55) is often useful in carrying out relativistic calculations. By using it, one can avoid tedious application of the transformation laws. The relation can of course be derived directly from the definitions of E and p, equations (7.13) and (7.30).

Since the Lorentz transformations are linear, the sum of two four-vectors is another four-vector. Thus the quantity $p^2 - (E/c)^2$ is invariant even when p and E represent the total momentum and energy of a system of particles. For applications of this technique, see the problems.

Another useful relation can be derived directly from the defining relations (7.13) and (7.30) for momentum and energy. Dividing one equation by the other, one gets

$$\frac{pc}{E} = \frac{v}{c} \qquad (7.56)$$

Equation (7.56) shows that in the ultrarelativistic limit, when the velocity of a body is nearly c, its energy and momentum become directly proportional to one another. When $v = c$, (7.56) gives the energy-momentum relation for a photon, discussed in the next section.

7.7. PHOTONS

As we noted in chapter 2, the long-standing debate over the nature of light appeared by the middle of the nineteenth century to have been settled in

---

consistent with the usual geometric definition of the length of a vector as the square root of the sum of the squares of its components. However, the "zeroth" component of any four-vector is imaginary. The Lorentz transformation can be viewed formally as a rotation through an imaginary angle in four-dimensional space.

favor of the wave picture. After Maxwell's work, it was universally accepted that light is an electromagnetic wave.

At the end of the century, however, the issue was reopened with the discovery of phenomena that could not be explained on the basis of the wave theory. In 1900, Planck, to explain certain features of the so-called blackbody radiation spectrum, proposed the idea that electromagnetic energy is emitted and absorbed in the form of discrete bundles or "quanta." The energy of each quantum depends only on the frequency of the radiation and is given by the simple formula

$$E = hf \tag{7.57}$$

The constant of proportionality h in this equation is called Planck's constant. Its numerical value in Mks units is 6.7×10^{-34}. The quanta, which came to be known as photons, behave like particles. The intensity of a beam measures the number of photons that pass through a unit area in unit time.

A single photon of visible light carries a very small amount of energy, a couple of electron volts. Ordinary light beams contain so many photons that for the description of many phenomena the energy can be regarded as being continuously distributed, as in the wave picture.

In one of his three great papers of 1905, Einstein employed the photon model to provide a full explanation of the photoelectric effect, which had been discovered by Heinrich Hertz in 1887 and studied in detail by Philipp Lenard and others.

When ultraviolet light strikes the surface of a metal, electrons are sometimes emitted. A surprising feature of the phenomenon was the existence of a "frequency threshold": unless the frequency of the light exceeds a certain minimum, no photoelectrons are emitted, no matter how intense the beam. The threshold frequency is different for different metals.

Although the wave theory does provide a mechanism for transferring energy from light to an electron, it cannot account for the frequency threshold. Several other features of the effect are very hard to understand according to the wave picture.

Einstein's explanation of the photoeffect was based on the premise that an electron receives the energy it needs to leave the metal by absorbing a single photon. Unless the photon's energy exceeds the required amount (called the work function of the metal, ϕ) the electron cannot leave. Equation (7.57) therefore implies that no photoelectrons can be liberated unless the frequency of the light is greater than ϕ/h. With the photon model Einstein was able to account for every feature of the experimental results.

Our concern here is not with the photoeffect but with how the concept of the photon as a relativistic particle fits in with the results of this chapter. The photon's speed must be c in every inertial frame of reference. As has been noted on several occasions here, serious difficulties arise in relativity if the velocity of a particle is set equal to c. The parameter γ is infinite and the energy defined by equation (7.13) is likewise infinite. That conclusion follows, however, only if the particle's rest mass is finite.

The key to the description of the photon as a relativistic particle is that it has zero rest mass. According to equation (7.11), the rest energy of a particle with zero rest mass is zero. Thus a photon at rest would have zero energy. This causes no problems because a photon is never at rest in any frame.

Although the photon has no rest mass, we can use equation (7.19) to assign it a relativistic mass based on its energy:

$$M_{ph} = hf/c^2 \qquad (7.58)$$

With $m = 0$, $v = c$, and $\gamma = \infty$, the expression γmc^2 for particle energy is mathematically undefined. (The product of zero and infinity can have any value.) Notice, however, that γ does not appear in equation (7.55), which relates a particle's energy and its momentum. In that expression we can sensibly set m equal to zero, obtaining the simple relation

$$E = pc \qquad (7.59)$$

The energy and momentum of a photon are directly proportional to one another. The same result (7.59) is obtained by putting $v = c$ in equation (7.56).[23]

Combining equations (7.57) and (7.59), we find that the momentum of a photon of frequency f is

$$p = hf/c \qquad (7.60)$$

Equation (7.59) provides a bridge between the photon picture and the wave description of light. Although the energy and momentum in an

23. Eq. (7.56) was obtained by dividing (7.30) by (7.13). One could object that when we put $m = 0$ the result contains the factor $0/0$, which is undefined. We can, however, consider a limiting process in which $m \rightarrow 0$. At every step the mass cancels; hence in the limit, we obtain the result (7.59).

electromagnetic wave are not localized but are continuously distributed, the energy and momentum *densities* are related by equation (7.59). We have already made use of this property in deriving the relativistic expression for momentum, equation (7.30).

According to equation (7.59), the x momentum of a photon is

$$p_x = p \cos \theta = (E \cos \theta)/c \tag{7.61}$$

where θ is the angle between the direction of the photon and the x axis. The general energy transformation (7.52a) takes the form

$$E' = \gamma(V) \, E\left(1 - \frac{V}{c} \cos \theta\right) \tag{7.62}$$

Since the frequency of a photon is proportional to its energy, it must obey the same transformation law:

$$f' = \gamma(V) \, f\left(1 - \frac{V}{c} \cos \theta\right) \tag{7.63}$$

When $\theta = 0$, equation (7.63) reduces to the simpler form

$$f' = f \sqrt{\frac{1 - V/c}{1 + V/c}} \tag{7.64}$$

Equations (7.63) and (7.64) are nothing but the Doppler effect formulas, equations (4.41) and (4.38).[24]

Equation (7.62) is the same as equation (7.1), which Einstein had derived in his first paper from the wave theory of light. The energy transformation law for light is the same whether the light is treated as a classical wave or a collection of photons.

In the 1905 paper, Einstein derived the Doppler effect formula (7.63) independently and remarked that "it is noteworthy that the energy and the frequency of a light complex vary with the observer's state of motion according to the same law." Strangely enough, Einstein did not refer to his own paper on the photoeffect, written only three months earlier, which provides a natural explanation for that result. He apparently did not have complete confidence in the photon picture at the time.[25]

The photon hypothesis leads to a consistent kinematic picture of the photon as a relativistic particle that always moves at speed c along a well-

24. Eq. (7.63) looks different from eq. (4.41) because the angle θ in (7.63) is measured in the source frame S, whereas in (4.41) the angle θ is measured in the observer's frame. The two expressions are in fact equivalent.
25. See Pais, *Subtle Is the Lord,* 462, for additional comments on this point.

defined path. In phenomena such as interference and diffraction, in contrast, light definitely behaves as a wave. Young's two-slit experiment, for example, is readily explained by the wave picture: the part of the wave that passes through one slit interferes with the part that passes through the other and forms the familiar interference pattern. A classical particle, following a well-defined path, cannot pass through both slits simultaneously.

A complete description of light is provided only by quantum field theory, in which the particle and wave aspects are synthesized and account for all phenomena. For the analysis of many phenomena, the particle aspect dominates. This is particularly true at high frequencies, when one is dealing not with visible light but with X rays or gamma rays. In elementary particle reactions, photons are emitted and absorbed and can be treated just like any other particle; their energies and momenta, given by equations (7.57) and (7.60), must be included in the conservation equations.

Since a photon has no rest mass and no rest energy, its energy is entirely kinetic. The ultimate in mass-energy conversion takes place in reactions in which the only end products are photons. We have already encountered one such reaction, the decay of the neutral pion. Another example is a collision between an electron and a positron, in which both particles are annihilated and two gamma rays are produced:[26]

$$e^+ + e^- \rightarrow \gamma + \gamma \qquad (7.65)$$

In reaction (7.65) the right side has no rest mass; the entire mass energy of both electron and positron has been converted into radiant energy. Such a result would be unintelligible without relativity.

The energies of the gamma rays in reaction (7.65) are easily calculated. Suppose first that the electron and positron approach one another head-on with equal speeds. The total momentum of the system is zero; conservation of momentum requires that the two photons be emitted in opposite directions with equal momenta, hence with equal frequencies. The direction of the photons is arbitrary.

Conservation of energy gives us the energy of each photon:

$$E_i = 2\gamma mc^2 = E_f = 2hf$$

26. The positron is the antiparticle of the electron; it has the same mass and a positive charge equal numerically to that of the electron. The symbol γ in reaction (7.65) stands for a gamma ray and is not to be confused with the relativistic parameter γ.

where γmc^2 is the energy of each of the annihilating electrons. The photon frequencies are therefore given by

$$hf = \gamma mc^2 \qquad (7.66)$$

In the most common experimental arrangement, a positron in a beam collides with an electron which is essentially at rest. To find the energies of the photons in that case, we argue as follows. There must exist a frame of reference in which the electron and positron have equal and opposite speeds. Such a frame is sometimes called the center-of-mass frame; zero-momentum frame is a better name for it.

In the zero-momentum frame, the energy and frequency of each photon are given by equation (7.66). The zero-momentum frame is related to the laboratory frame by a Lorentz transformation, in which the velocity V is just the velocity of the electron in the zero-momentum frame. The frequencies of the gamma rays in the laboratory are obtained from equation (7.63.)

Incidentally, conservation of momentum explains why a positron-electron pair cannot annihilate into just one photon. In the zero-momentum frame, the total final momentum must be zero. But a single photon cannot have zero momentum. Hence there must be at least two. Arguments of this type, based on momentum conservation, are often useful in the analysis of complicated experiments. They cannot specify what *will* happen, but they can sometimes imply that certain reactions are impossible.

In reaction (7.65), rest mass is converted entirely into photon energy. The opposite type of reaction, in which photons materialize into electron pairs, also occurs. In principle, the reaction inverse to (7.65) could take place; this, however, would entail aiming two photon beams directly at each other and is too difficult to carry out in practice.

The momentum argument presented above implies that the reaction

$$\gamma \rightarrow e^+ + e^- \qquad (7.67)$$

cannot take place in free space. (This is best seen by viewing the reaction in the zero-momentum frame of the two electrons.) If, however, a gamma ray photon passes near an atom, the atom can recoil, absorbing some momentum, and reaction (7.67) can proceed. Because the mass of the atom is so much greater than that of the electrons, the atom receives practically no energy. The reaction, called pair production, is routinely observed in experiments.

As noted in section 7.5, the data on beta decay indicate that the mass of the neutrino is very small, perhaps zero. If the mass is indeed zero, it follows that neutrinos must travel at the speed of light; otherwise, their energy and momentum would be zero. The energy and momentum of a neutrino would, in that case, be related by equation (7.58); from a kinematic point of view, there would be nothing to distinguish a neutrino from a photon.

We have already noted that in the extreme relativistic limit, when the energy of a particle is very large compared to its rest mass, the energy-momentum relation, (7.55), approaches (7.58). The photon is always in the extreme relativistic limit. The same is true of the neutrino, if its mass is truly zero.

One final example of the role of photons in elementary particle reactions may be mentioned. Suppose we want to make deuterons by firing neutrons at protons. According to the data on page 222, the reaction

$$n + p \rightarrow d \qquad (7.68)$$

would be exoergic. (The total mass of neutron + proton exceeds that of the deuteron.) However, an analysis similar to the one we applied to electron annihilation shows that (7.68) is inconsistent with momentum conservation. (Once again, the easiest way to see that is to look in the zero-momentum frame.) If, however, we add a photon to the right side, making the reaction

$$n + p \rightarrow d + \gamma \qquad (7.69)$$

the conservation laws can be satisfied and the reaction can proceed. This process is called radiative capture. Its inverse, called photo-disintegration, is also observed.

7.8. Tachyons

I referred briefly in section 4.4 to the possible existence of tachyons, particles that move faster than light. I noted that this property is invariant: if a particle's speed is greater than c in one inertial frame of reference, it is greater than c in any other (provided the relative velocity between frames is less than c).

There is an immediate problem with energy and momentum, since the parameter $\sqrt{1 - (v/c)^2}$ that appears in their definition is imaginary when v is greater than c. Energy and momentum, being measurable quantities, must be real.

The key to the theory of tachyons is to assign them imaginary rest masses.[27] If we put

$$m = i\mu \tag{7.70}$$

with $i = \sqrt{-1}$ and μ a real number, the expressions for momentum and energy (7.30) and (7.13), become

$$p = \frac{\mu v}{\sqrt{(v/c)^2 - 1}} \tag{7.71}$$

$$E = \frac{\mu c^2}{\sqrt{(v/c)^2 - 1}} \tag{7.72}$$

both of which are real. The energy-momentum relation, equation (7.55), becomes

$$E^2 = p^2 c^2 - (\mu c^2)^2 \tag{7.73}$$

What sense does it make for a particle to have an imaginary rest mass? The rest energy of such a particle, mc^2, is imaginary. This is not a disaster for a tachyon, however, because it is never observed to be at rest. Its rest mass is therefore not a measurable quantity. Just as we can assign zero rest mass to the photon because it is never at rest, so we can formally assign an imaginary rest mass to the tachyon.

Equations (7.71) and (7.72) have an intriguing property: the energy and momentum are decreasing functions of v. As a tachyon gains energy, its velocity decreases! Infinite energy would be required to slow it down to the speed of light, just as infinite energy is required to accelerate an ordinary particle to speed c. A particle produced at a speed greater than c is therefore destined to spend its entire life in that state. The realms of tachyons and of ordinary particles are entirely disjoint, with the speed of light forming an impassable barrier between them.

As noted in chapter 5, the existence of tachyons poses a problem for causality. Suppose a tachyon is emitted at time t_1 at position x_1 (event E_1) and absorbed at time t_2 at position x_2 (event E_2). Equation (4.9) specifies the time interval $\Delta t'$ between the two events in a frame S' in terms of Δx and Δt. If v is the tachyon's velocity in frame S, we can put $\Delta x = v\Delta t$ and the equation becomes

$$\Delta t' = \gamma\left(\Delta t - \frac{V}{c^2}\Delta x\right) = \gamma\Delta t\left(1 - \frac{Vv}{c^2}\right) \tag{7.74}$$

27. O. M. P. Bilaniuk, V. K. Deshpande, and E. C. G. Sudarshan, " 'Meta'-Relativity," *American Journal of Physics* 30 (1967):718–723. G. Feinberg, "Possibility of Faster-than-Light Particles," *Physical Review* 159 (1967):1089–1105.

If the relative velocity V between S' and S satisfies

$$\frac{c}{v} < \frac{V}{c} < 1 \qquad (7.75)$$

which is possible when $v > c$, equation (7.74) shows that $\Delta t'$ is negative: an observer in S' sees the tachyon absorbed before it is emitted. Causality seems to be violated.

The situation is saved by another interesting property of tachyons. Equation (7.52a) gives the energy of the tachyon in S':

$$E' = \gamma(V)\,(E - Vp) \qquad (7.76)$$

For ordinary particles, the value of E' determined by equation (7.76) is necessarily positive since the magnitude of pV is less than that of E. For a tachyon, however, equation (7.73) shows that the magnitude of pc is greater than that of E. It is not hard to show that if V is in the range specified by equation (7.75), E' is negative.

If a particle absorbs a negative-energy tachyon, its energy decreases. The process can be viewed sensibly as the *emission* of a positive-energy tachyon in the opposite direction. This idea is used in quantum field theory. S' observers see the tachyon emitted at event E_2 and absorbed at E_1, which according to them occurs later. By means of this interpretation, causality is salvaged.

Several attempts to find evidence of tachyon production in high-energy particle interactions are described in an article by Feinberg.[28] The results of all the searches were negative: no candidates have been identified. In all probability, tachyons are destined to remain an amusing footnote in relativity theory.

APPENDIX: TRANSFORMATION OF MOMENTUM AND ENERGY

In this appendix, we derive several mathematical results needed in sections 7.1 and 7.6.

First, we require the transformation property of the parameter

$$\gamma(v) = \frac{1}{\sqrt{1 - \dfrac{v^2}{c^2}}} \qquad (7.A1)$$

which appears in the definition of relativistic energy and momentum.

28. G. Feinberg, "Particles that Go Faster than Light," *Scientific American*, February 1970, 69–72.

The transformation law for the components of v was derived in chapter 4. Equation (4.19) read

$$v'_x = \frac{v_x - V}{1 - \dfrac{v_x V}{c^2}} \tag{7.A2a}$$

$$v'_y = \frac{v_y}{\gamma(V)\left(1 - \dfrac{v_x V}{c^2}\right)} \tag{7.A2b}$$

$$v'_z = \frac{v_z}{\gamma(V)\left(1 - \dfrac{v_x V}{c^2}\right)} \tag{7.A2c}$$

V, the velocity of frame S' relative to frame S, is as usual assumed to be in the x direction, and

$$\gamma(V) = \frac{1}{\sqrt{1 - V^2/c^2}} \tag{7.A3}$$

Using equation (7.A2) we can express $\gamma(v')$ as

$$\gamma(v') = \frac{1}{\sqrt{1 - v'^2/c^2}}$$

$$= \frac{1}{\left\{1 - \dfrac{1}{c^2}\left[\left(\dfrac{v_x - V}{1 - \dfrac{v_x V}{c^2}}\right)^2 + \left(\dfrac{v_y}{\gamma(V)\left(1 - \dfrac{v_x V}{c^2}\right)}\right)^2 + \left(\dfrac{v_z}{\gamma(V)\left(1 - \dfrac{v_x V}{c^2}\right)}\right)^2\right]\right\}} \tag{7.A4}$$

(7.A4) is a formidable-looking expression, but after some algebra it reduces to the much simpler form

$$\gamma(v') = \gamma(V)\gamma(v)\left(1 - \frac{v_x V}{c^2}\right) \tag{7.A5}$$

The analogous expression for $\gamma(v)$ in terms of $\gamma(v')$ is

$$\gamma(v) = \gamma(V)\gamma(v')\left(1 + \frac{v'_x V}{c^2}\right) \tag{7.A6}$$

Consider next the transformation properties of the products γv_x, γv_y, and γv_z. Multiplying equation (7.A5) by (7.A2) we obtain

$$v'_x \gamma(v') = \gamma(V)\gamma(v)(v_x - V) \tag{7.A7a}$$

$$v_y' \gamma(v') = v_y \gamma(v) \tag{7.A7b}$$
$$v_z' \gamma(v') = v_z \gamma(v) \tag{7.A7c}$$

The inverse form of equation (7.A7a) is

$$v_x \gamma(v) = \gamma(V) \gamma(v')(v_x' + V) \tag{7.A8}$$

Notice that although the transverse components of velocity, v_y and v_z, transform in a complicated way, the products γv_y and γv_z transform much more simply: they are invariant. Multiplying equations (7.A7b,c) by m gives directly:

$$p_y' = p_y \qquad p_z' = p_z \tag{7.A9}$$

The transverse components of momentum are invariant. If they are conserved in one frame, they are automatically conserved in any other.

Multiplying equation (7.A8) by m gives the transformation law for the x component of momentum:

$$p_x = \gamma(V) \, [p_x' + mV\gamma(v')] \tag{7.A10}$$

We are now prepared to investigate the covariance of the momentum conservation law. In the collision discussed in section 7.3, assume that x momentum is conserved in frame S:

$$p_A + p_B = p_C + p_D \tag{7.A11}$$

and transform each term to S' using equation (7.A10). The result is

$$\gamma(V)\{p_A' + p_B' + V[m_A\gamma(v_A') + m_B\gamma(v_B')]\}$$
$$= \gamma(V)\{p_C' + p_D' + V[m_C\gamma(v_C') + m_D\gamma(v_D')]\} \tag{7.A12}$$

The factor $\gamma(V)$ appears on each side in equation (7.A12) and cancels. The equation now represents conservation of x momentum in S', provided the terms that multiply V cancel, that is, provided that

$$m_A\gamma(v_A') + m_B\gamma(v_B') = m_C\gamma(v_C') + m_D\gamma(v_D') \tag{7.A13}$$

(7.A13) is the auxiliary condition that must be satisfied if conservation of momentum is to be a covariant law. It asserts that the quantity γm is conserved in S'. Since the designation of S and S' is arbitrary, the same relation must hold in frame S, that is

$$m_A\gamma(v_A) + m_B\gamma(v_B) = m_C\gamma(v_C) + m_D\gamma(v_D) \tag{7.A14}$$

PROBLEMS

7.1. The rest mass of the proton is 1.007277 u.
   (a) What is the rest energy of a proton in MeV?

(b) A proton moves with velocity $0.1c$. What is its (total) energy in MeV? What is its relativistic mass? What is its kinetic energy in MeV? What is its momentum in MeV/c?

(c) Calculate the kinetic energy and momentum using the Newtonian relations $K = \frac{1}{2} mv^2$ and $p = mv$; compare with the results of (b).

(d) A proton has kinetic energy of 2,000 MeV. What is its relativistic mass? What is its velocity? Calculate the Newtonian velocity and compare with the relativistic result.

7.2. The nucleus ^{7}Be contains 4 protons and 3 neutrons. The mass of the neutral ^{7}Be atom is 7.016928 u.

(a) Find the binding energy of ^{7}Be in MeV.

(b) Consider the reaction $n + {}^7\text{Be} \rightarrow {}^4\text{He} + {}^4\text{He}$. If the incident neutron has negligible kinetic energy, what will be the total kinetic energy of the two alpha particles? (The rest mass of the neutron is 1.008665 u.)

7.3. A hypothetical particle, the Q meson, has a rest energy 7.5 GeV (which equals 7,500 MeV). It decays into two Y particles; the rest energy of a Y is 3.0 GeV. Suppose a Q meson decays at rest.

(a) Find the (total) energy of each Y.

(b) Find the kinetic energy of each Y.

(c) Find the momentum of each Y.

(d) Find the velocity of each Y.

7.4. A Q meson moving at $0.8c$ in the x direction decays. (See problem 7.3.) The decay products are emitted along the x axis. Find the energy, momentum, and velocity of each Y.

7.5. Particle A (rest mass m_A) decays into particle B (rest mass m_B) and a photon.

(a) An A particle decays at rest. Write the equations for conservation of energy and momentum in this process.

(b) Solve the equations. Find the energies of the B and the photon in terms of $m_A c^2$ and $m_B c^2$. Compare your results with eq. (7.48).

(c) The Σ^0 hyperon decays into a Λ hyperon and a photon. The rest energies of Σ^0 and Λ are

$$m_\Sigma c^2 = 1{,}139 \text{ MeV} \qquad m_\Lambda c^2 = 1{,}116 \text{ MeV}$$

Find the energies of the Λ and the photon when a Σ^0 decays at rest.

(d) A Σ^0 moving at $0.6c$ in the x direction decays. The Λ is emitted in the $+x$ direction. Find the energy and momentum of the Λ and the photon. Verify that energy and momentum are conserved.

7.6. A photon of energy 3 GeV moving in the $+x$ direction collides with a particle B at rest. The rest energy of B is 2 GeV. After the collision a new particle, D, moves with speed v. (Conservation of momentum demands that the direction of D must be along the x axis.) Let m denote the rest mass of D.

(a) Write the equations that describe conservation of energy and momentum for this reaction.

(b) Solve the equations for v/c and mc^2. (You should find it easier to solve for v/c first.)

(c) Find the energy and momentum of particle D.

(d) Suppose this reaction is observed in a frame S' that moves in the $-x$ direction at $0.4c$ relative to the laboratory. Find the energy of the incident photon and the energy and momentum of B and D in S'. Verify that energy and momentum are conserved.

7.7. A useful concept for the analysis of particle reactions is the frame of reference in which the total momentum is zero. This is generally called the center-of-mass (CM) frame because the center of mass of the system is at rest. For a two-particle system, the momenta of the particles in the CM frame are equal and opposite.

The relation between the laboratory frame and CM frame is most easily established by exploiting the invariance of the quantity $E^2 - p^2 c^2$, discussed in section 7.6. E and p can be the energy and momentum of a single particle or of any system of particles.

Consider a reaction in which a particle of rest mass m and energy E_0 collides with another particle, of equal mass, at rest. Let E^* denote the energy of each particle in the CM frame. (For the case of equal masses, the CM energies must be equal.)

(a) Express the total energy and momentum of the system in the laboratory in terms of mc^2 and E_0 and in the CM frame in terms of mc^2 and E^*.

(b) Using the invariance argument, solve for E^* in terms of mc^2 and E_0. (The total energy in the CM frame is of course $2E^*$.)

(c) A proton of kinetic energy 100 MeV collides with a proton at rest. Find the total kinetic energy of the system in the CM frame.

(d) Analyze the problem using Galilean relativity. Show that for equal-mass particles, the total CM kinetic energy is $E_0/2$. Compare with the result of (c).

(e) Repeat (c) for a proton with kinetic energy 5 GeV.

7.8. When an energetic proton collides with a proton at rest, an X particle (rest energy = 2 GeV) can be created via the reaction

$$p + p \rightarrow p + p + X$$

The threshold energy for a reaction is defined as the minimum kinetic energy that allows the reaction to proceed. Obviously the threshold energy for the above reaction must be at least equal to the rest energy of the created particle, 2 GeV.

(a) Give a qualitative argument that explains why the threshold energy must in fact be higher than 2 GeV. (Hint: If all the kinetic energy of the incident proton is converted into rest energy, the particles in the final state must all be at rest. Why is this impossible?)

(b) The threshold energy measured in the CM frame is 2 GeV. Why? Using the result of problem 7.7, find the threshold energy in the laboratory.

(c) High-energy physicists nowadays often perform colliding-beam experiments. In such an experiment, instead of the target being at rest, the colliding protons (or protons and antiprotons) have equal energies. What advantage does this arrangement provide?

7.9. Energy from the sun arrives at the earth's atmosphere at the rate of 1.36×10^3 joule/m^2/sec. The (mean) distance between earth and sun is 1.5×10^{11} m. The sun's mass is 2×10^{30} kg.

(a) Calculate the rate at which energy is radiated by the sun in joule/sec.

(b) At what rate is the sun's mass decreasing as a result of its radiation?

(c) The radiated energy is produced by nuclear reactions in the interior of the sun. Assuming that one-third of 1 percent of the reactant rest energy is converted to radiant energy in the reactions, estimate how long the sun can continue to radiate at its present rate.

8   General Relativity

8.1. INTRODUCTION

In special relativity, as in Newtonian mechanics, inertial frames enjoy a preferential status. The principle of relativity applies only to them. The laws of nature are the same in all inertial frames and no experiment can distinguish between one and another. An inertial frame is one in which Newton's law of inertia holds: a body subject to no net external force remains at rest or moves in a straight line with uniform velocity.

All inertial frames move uniformly relative to one another. If K is an inertial frame and K' is accelerated relative to K, then K' is noninertial. The laws of nature are more complicated in K' than in K. As noted in chapter 1, in a noninertial frame bodies experience pseudo-forces, called inertial forces, for which no agent can be identified; in a rotating frame, for example, every body is pushed away from the axis of rotation by a "centrifugal" force.

Einstein was dissatisfied with this state of affairs. In his popular book he posed his objection as follows:

> How does it come that certain reference systems (or their states of motion) are given priority over other reference systems (or their states of motion)? *What is the reason for this preference? . . .* I seek in vain for a real something in classical mechanics (or in the special theory of relativity) to which I can attribute the different behavior of bodies considered with respect to the reference systems K and K'.[1]

For Newton, absolute space was the "something" that provides a definition of absolute acceleration and distinguishes between inertial and non-

1. Einstein, *Relativity*, 71.

inertial frames. In support of this view Newton described an experiment in which a bucket half-filled with water hangs from a long rope. If the rope is twisted many times and then released, the bucket begins to rotate. The surface of the water at first remains level. Gradually, however, as the rotation of the bucket is transmitted to the water, the water becomes depressed in the middle and rises on the sides of the bucket, forming a concave figure. If the bucket is suddenly stopped, the water continues to rotate for a while and retains its concave form.

According to Newton, the bucket experiment (which he actually performed) demonstrates the existence of absolute space and of absolute acceleration. Inertial forces act on the water when it accelerates relative to absolute space; its motion relative to the bucket is irrelevant. A similar argument accounts for the equatorial bulge in the rotating earth.

The nineteenth-century physicist/philosopher Ernst Mach rejected Newton's conception of absolute space. In Mach's view, space without matter has no properties; the mean velocity of all the matter in the universe (the "celestial bodies") establishes the standard with respect to which acceleration may be defined.[2]

Mach attributed the inertia of every body to the rest of the matter in the universe. According to this view, the water in Newton's experiment experiences centrifugal forces when it rotates relative to the distant matter. If the bucket and the water were the only objects in the universe, the outcome of the experiment would be different: the water would remain level at all times. This hypothesis is obviously difficult to test.

Mach's ideas had a profound influence on Einstein's early thinking; he coined the term "Mach's principle" to describe these ideas. Although Einstein eventually abandoned Mach's principle, he retained its central premise—that absolute acceleration has no meaning.

Elsewhere, Einstein pointed out that the law of inertia involves a circular argument.[3] A body moves without acceleration if no forces act on it, but how do we know that no forces are acting? Contact forces can be excluded by inspection, but what about forces that act at a distance? In particular, what about gravity? We infer that such forces are absent (or negligibly small) only from the fact that the body is observed to move without acceleration. This line of reasoning leads to the principle of equivalence, discussed in the next section.

2. Similar ideas had been expressed in 1721 by Bishop Berkeley.
3. Einstein, *The Meaning of Relativity*, 58.

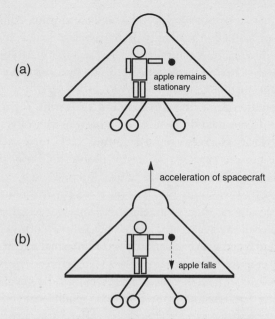

Fig. 8.1.   Testing a reference frame for acceleration by releasing an apple. (a) The apple remains
stationary when released. The laboratory must be
moving with constant velocity. (b) The apple
"falls" when released. The laboratory must be accelerating in a direction opposite to that of the
apple's fall.

"Every intellect which strives after generalization," said Einstein,
"must feel the temptation to venture the step towards a general principle
of relativity."[4] Guided by that conviction, he sought to extend the principle of relativity to encompass *all* frames of reference.

At first sight such a quest seems doomed to fail, for the very reasons
we have been discussing. Our everyday experience indicates that the laws
of physics are *not* the same in all frames. How then can a general principle
of relativity possibly hold?

Suppose a spacecraft is cruising in outer space, far from any astronomical body. According to the principle of relativity, no experiment performed inside the spacecraft can distinguish between a state of rest and a
state of uniform motion. It is easy enough, however, for an astronaut to

4. Einstein, *Relativity*, 61.

determine whether or not the spacecraft is accelerating. All he has to do is hold out an apple and release it. If the apple "floats" (fig. 8.1a), the spacecraft is moving with constant velocity and its rest frame is inertial; if the apple "falls" in some direction (fig. 8.1b), the craft must be accelerating in the opposite direction.

In the reference frame of an accelerating spacecraft, the ordinary laws of mechanics do not hold. In particular, the law of inertia is not satisfied: the apple's velocity changes even though no "real" forces act on it. How then can one hope to express the laws of physics in a way that does not distinguish between inertial and accelerated frames?

After grappling with this problem for several years, Einstein finally arrived at the general theory of relativity, an intellectual tour de force that accomplished the apparently impossible task of putting all frames of reference on an equal footing. The solution is a formal one; it involves the introduction of abstract coordinates called Gaussian coordinates, in terms of which the laws of physics can be expressed in fully covariant fashion.

As we shall see, Einstein's quest led him along unexpected paths. In pursuing a general theory of relativity he arrived at a theory of gravitation. The mathematical formulation of general relativity, unlike that of the special theory, is quite complicated, employing the techniques of tensor calculus and differential geometry. A fully quantitative treatment is not possible here. We can, however, describe the basic concepts of the theory and its most important implications.

8.2. The Principle of Equivalence

The last section of Einstein's 1907 review article on special relativity is entitled "Principle of Relativity and Gravitation."[5] It was here that Einstein put forward the idea that later came to be known as the principle of equivalence and that constitutes the first step toward a general theory of relativity. He developed the idea more fully in a paper published in 1911.[6]

To introduce the principle of equivalence, we return to the problem of the spacecraft discussed above. Suppose the spacecraft is not accelerating but happens to be parked on the surface of a planet. (See fig. 8.2.) The

5. A. Einstein, "On the Relativity Principle and the Conclusions Drawn from It," *Jahrbuch der Radioaktivität und Elektronik* 4 (1907):411–462; also *Collected Papers*, doc. 47.
6. A. Einstein, "On the Influence of Gravitation on the Propagation of Light," *Annalen der Physik* 35 (1911):898–908. An English translation appears in *The Principle of Relativity*, 99–108. This paper is quite readable.

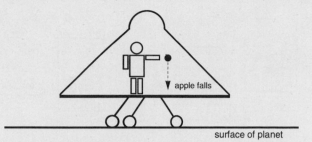

Fig. 8.2. A spacecraft parked on the surface of a planet. A released apple falls under the influence of the planet's gravity. This situation cannot be distinguished from that of fig. 8.1.

gravitational pull of the planet attracts a released apple and causes it to fall, just as it does when the spacecraft is accelerating. If the direction of the gravitational force in figure 8.2 is opposite to that of the acceleration of the spacecraft in figure 8.1b, the apple falls in the same direction in both cases.

The magnitude of the apple's acceleration in figure 8.2 depends on the strength of the planet's gravity, which is determined by its mass and its radius. By assigning appropriate values to those quantities, we can make the apple's acceleration in the planet's gravitational field equal to that in the accelerating spacecraft in figure 8.1. All other mechanical effects are likewise the same. If, for example, the astronaut is standing on a scale, the reading of the scale is the same in either case.

In this problem, then, an acceleration of the frame of reference in a particular direction and a gravitational force field acting in the opposite direction produce identical effects. Observers within the spacecraft cannot distinguish between the two alternative explanations for their observations.

Figure 8.3 shows another example. A passenger in a freely falling elevator releases a tennis ball. As viewed from an inertial frame fixed on the ground, both the elevator and the tennis ball are acted upon by the earth's gravity and both fall with the same acceleration (fig. 8.3a). The ball remains at a constant height above the elevator's floor.

Figure 8.3b shows the situation as seen in an accelerated frame of reference that moves with the elevator. Here the ball floats without falling. In the falling elevator, as in the accelerated spacecraft of figure 8.1b, the laws of Newtonian mechanics do not hold. In this case the tennis ball is subject to an unbalanced force (the earth's gravity), yet it does not accelerate.

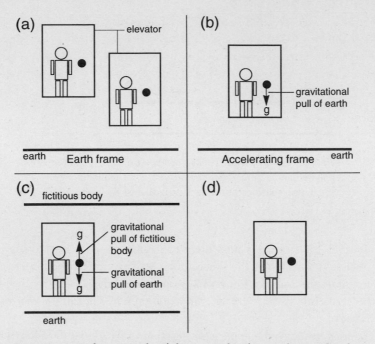

Fig. 8.3. Another example of the principle of equivalence. The elevator is in free fall in the earth's gravity. (a) As seen in an inertial frame fixed on earth, the elevator, passenger, and tennis ball are all falling with the same acceleration. Their relative position does not change. (b) As seen in an accelerating frame fixed on the elevator, the tennis ball is subject to the earth's gravity but does not fall. Newton's laws of motion do not apply in this accelerated frame. (c) Here the acceleration of the elevator has been replaced by a massive object above it. That object produces a gravitational upward pull that cancels the earth's downward pull. The tennis ball remains at rest. (d) Here the earth's gravity has been replaced by an upward acceleration of the frame. The net acceleration is zero: the elevator frame is inertial. The tennis ball remains at rest because no forces act on it.

In figure 8.3c, the downward acceleration of the elevator frame has been replaced by a massive body, located above the elevator, whose gravity is such that the upward force it exerts on the tennis ball equals the downward pull of the earth. The elevator frame in this case is inertial. Since the net force on the ball is zero, it remains at rest after being released, just as in fig. 8.3b.

Instead of replacing the downward acceleration of the elevator frame with a gravitating body, we could have substituted for the earth's gravity an upward acceleration of the elevator frame. When this is added to the "actual" downward acceleration, the net acceleration is zero: the elevator moves with constant velocity. The elevator frame is again an inertial frame, this time with no gravitating bodies whatever in its vicinity. A released ball floats freely because no forces act on it (fig. 8.3d).

Observers inside the elevator cannot distinguish among the last three pictures in figure 8.3. No mechanical experiment can determine whether their laboratory is in free fall under the influence of the earth's gravity (fig. 8.3b), is at rest and sandwiched between the earth and another massive body (fig. 8.3c), or is coasting in outer space, far from any gravitating matter (fig. 8.3d).

The principle of equivalence is a bold generalization of these findings. Einstein posited that an accelerated frame of reference and a gravitational force field that points in the direction opposite to the acceleration are equivalent *in every respect*. Einstein's hypothesis goes far beyond the observations described thus far. The equivalence principle identifies gravity as a unique force in nature; it asserts that by studying motion in accelerated frames, one is also studying the effects of gravity, and vice versa. The equivalence principle leads to several predictions that can be tested by experiment; they are discussed in the following sections. In each case, an effect calculated in an accelerated frame is used to predict the outcome of an experiment involving gravitation.

Just as special relativity forbids us to speak of absolute velocity, the principle of equivalence makes it impossible to speak of the absolute acceleration of a frame of reference. No experiment of any kind can distinguish between acceleration and a (uniform) gravitational field.

Gravitational and Inertial Mass

An important assumption is implicit in the very statement of the principle of equivalence. In a noninertial frame any body placed at a given point and subject to no real forces is observed to accelerate at precisely the same rate (which is just equal to the acceleration of the frame and points in the opposite direction). If an acceleration of the reference frame is indeed equivalent to a gravitational force field, the effect of gravity must likewise be the same on all bodies, irrespective of their composition or their mass. This property of gravity is subject to experimental test.

Early evidence was provided by Galileo, who noted that in a fall of 100 braccia (about 46 m) in air, "a ball of gold will not have outrun one of copper by four fingers," a difference of about one part in a thousand.

"This seen," Galileo added, "I came to the opinion that if one were to remove entirely the resistance of the medium, all materials would descend with equal speed."[7]

The principle of equivalence demands that the rate of acceleration under gravity be *exactly* the same for all bodies. Even a tiny difference would distinguish between the effects of gravity and of acceleration and thereby invalidate the principle.

The requirement can be expressed in a different way by noting that mass plays two distinct roles in mechanics. The mass that appears in Newton's law of motion, $F = ma$, is a measure of a body's inertia—its resistance to a change in its motion. We call it *inertial* mass, m_I, and write Newton's law as

$$F = m_I a \qquad (8.1)$$

Mass appears also in the equation $W = mg$, which relates a body's weight, W (the force of gravity on the body), to its mass. (The constant of proportionality g has the dimensions of acceleration.) The mass in this equation has nothing to do with inertia; it is a measure of the "quantity of matter" contained in the body. We therefore call it *gravitational* mass, m_G, and write

$$W = m_G g \qquad (8.2)$$

There is no reason a priori why a body's gravitational mass need be related to its inertial mass.

Consider a freely falling body, that is, one subject only to the earth's gravity. We put $F = W$ in equation (8.1) and obtain

$$a = W/m_I \qquad (8.3)$$

Substituting for W from equation (8.2) gives

$$a = (m_G/m_I)g \qquad (8.4)$$

Equation (8.4) implies that the acceleration due to gravity has the same value for all objects, provided that *the ratio of gravitational mass to inertial mass is the same for every object*. (By an appropriate choice of units, the two masses can be made numerically equal.) This requirement must be strictly satisfied if the principle of equivalence is valid.

7. Galileo Galilei, *Two New Sciences*, translated by Stillman Drake (Madison: University of Wisconsin Press, 1974), 75.

Newton recognized the distinction between the two types of mass and carried out experiments to investigate whether their ratio is indeed the same for all materials. He measured the periods of pendulums of equal length, made of a variety of materials, and found them to be the same within one part in a thousand. He thus proved that the ratio m_G/m_I is the same for all the materials he tested, within the accuracy of his experiment.[8] In 1827, Friedrich Bessel carried out an improved version of the same experiment and verified the equality of inertial and gravitational mass to within two parts in 10^5.

Much more accurate confirmation was provided by a series of experiments carried out over many years by Roland Eötvös, who employed a sensitive torsion balance from whose ends he suspended bars made of different materials. Eötvös's experimental arrangement was such that the rotation of the earth would cause the balance to twist if the ratio of gravitational to inertial mass for the two materials was different. In 1890, Eötvös was able to show that the ratios for a variety of materials are equal to within a few parts in 10^9.[9]

In 1964, R. H. Dicke improved the precision of the result still further and confirmed the equality of m_G/m_I for aluminum and gold to within one part in 10^{11}, one of the most precise experimental results ever obtained. Dicke's result placed Einstein's principle of equivalence on a very firm experimental foundation.[10] The equality of inertial and gravitational mass is sometimes called the *weak* principle of equivalence; it is a necessary but not a sufficient condition for the validity of Einstein's postulate, which is called the *strong* principle.

One important aspect of the equivalence principle remains to be discussed. In a uniformly accelerated frame, every force-free body has the same acceleration. The equivalent gravity field is likewise uniform: it has the same strength and points in the same direction everywhere. But *no such gravity field exists in nature;* real gravitational forces point toward the bodies that create them and are stronger at points near the source than at points farther away. Hence the accelerated spacecraft frame in figure

8. The force that acts on a pendulum is produced by gravity and is proportional to the pendulum's gravitational mass. The acceleration this force produces is inversely proportional to the inertial mass. A standard calculation shows that the period of oscillation is proportional to the square root of the ratio m_I/m_G.
9. A good description of the Eötvös and the Dicke experiments is found in Hans Ohanian, *Gravitation and Spacetime* (New York: W. W. Norton, 1976).
10. In 1971, V. B. Braginsky improved the precision by another order of magnitude.

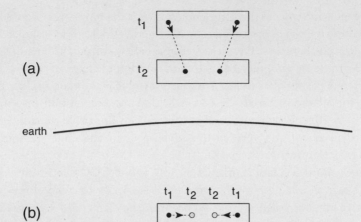

Fig. 8.4. Departure from the principle of equivalence caused by nonuniformity of the earth's gravity field. (a) Two balls are released at equal heights at the same time. As viewed from a reference frame fixed on the ground, the balls draw closer together as they fall. (b) As viewed in the (accelerated) reference frame that falls with the elevator, the balls stay at the same height but draw closer together as time passes.

8.1b and the gravitational field of the planet in figure 8.2 are not strictly equivalent. For the same reason, the uniform gravitational field that replaces the acceleration of the elevator frame in figure 8.3d does not quite cancel the nonuniform field of the earth.

These differences give rise to effects that are in principle detectable. Suppose two tennis balls are released simultaneously at the same height in a falling elevator. Ground-based observers see the balls draw gradually closer as they fall (fig. 8.4a) because the force of gravity, which points toward the center of the earth, acts in a slightly different direction on each ball. Observers in an accelerated frame that falls with the elevator see the balls move slowly toward each other (fig. 8.4b). In the frame of figure 8.3d, obtained by replacing the earth's gravity by an upward acceleration of the elevator frame, the balls remain at rest and their separation remains constant.

A similar effect occurs if two balls are released simultaneously at different heights. The balls fall at slightly different rates because the lower one, being closer to the center of the earth, experiences a slightly stronger gravitational force and therefore accelerates more rapidly. Observers in the accelerated elevator frame find that the distance between the balls increases gradually with time, whereas in the frame of figure 8.3d their separation remains constant.

These remarks do not invalidate the principle of equivalence, but they do restrict its applicability. A real gravitational field is equivalent to an accelerated reference frame only *locally;* that is, over a region small enough that the gravity within it can be considered approximately uniform. If experiments are performed in an elevator of normal size that falls only a short distance, the effects of the earth's nonuniform gravity are extremely small. (But see the remarks below concerning tidal effects.)

Freely Falling Frames

A freely falling laboratory accelerates in the direction of the gravity field at its location. According to the equivalence principle, we can replace that acceleration with a gravitational force field that (locally) cancels the one that produced the acceleration. The freely falling frame is therefore equivalent to a frame that moves without acceleration and with no gravity; it plays in general relativity the same role as the inertial frame does in special relativity. The (local) laws of physics are the same in all freely falling frames. An observer in any freely falling frame experiences weightlessness.

The frames in figures 8.3b and 8.3c are both freely falling; so is a frame fixed in a space vehicle in orbit. Unlike the inertial frames of special relativity, which extend throughout all space, a freely falling frame is defined only locally.

Dicke has phrased the strong principle of equivalence as follows:

> In a freely falling, nonrotating laboratory the local laws of physics take on some standard form, including a standard numerical content, independent of the position of the laboratory in space and time.[11]

This statement implies that by going into a freely falling frame, one "transforms away" any gravitational field that exists in a small region. (But see the remarks below.)

A frame of reference fixed on earth is freely falling in the gravitational field of the sun (and the moon).[12] In an earth-based frame the sun's gravity has been transformed away, but only locally, of course. It would be

11. R. H. Dicke, *The Theoretical Significance of Experimental Relativity* (New York: Gordon and Breach, 1964), 4.
12. The earth-based frame is not freely falling with respect to the earth's own gravity.

absurd to claim that the sun's gravity is not detectable from earth; the observed motion of the planets is its most obvious manifestation. But in first approximation the sun's gravity does not affect the motion of a body *near the earth's surface*, as viewed in a reference frame fixed on earth.

Tidal Forces. Even in a freely falling frame, the nonuniformity of gravitational fields gives rise to detectable effects. The best known of these is the phenomenon of ocean tides, caused by the difference in the strength of the moon's gravitational field on opposite sides of the earth.[13] For this reason the variation in the strength of a gravitational field is called a *tidal* force.

By detecting tidal forces, an observer can determine that he is in a real gravitational field even though his frame is freely falling. This conclusion seems inconsistent with the assertion that the local laws of physics are the same in all freely falling frames. If tidal forces are included, the assertion is false.

Dicke's version of the principle of equivalence, quoted above, contains the disclaimer, "It is of course implicit in this statement that the effects of gradients in the gravitational field strength are negligibly small, i.e., tidal interaction effects are negligible." Hans Ohanian has shown, however, that tidal effects persist even when the object in question is arbitrarily small.[14] If the radius of the earth were to shrink to zero, its density remaining constant, the shape of the tidal bulges would remain unchanged. In view of this result, it is hard to argue that tidal effects are nonlocal. In principle, an observer in a freely falling elevator could deduce that he is in a gravitational field by detecting tidal bulges in a liquid drop.

Many subtleties are associated with the principle of equivalence. A recent review article by John Norton contains a useful discussion.[15]

The Genesis of the Principle of Equivalence

Einstein referred to his discovery of the principle of equivalence as "the happiest thought of my life".[16] In an address given in Japan in 1922, he

13. The sun also contributes to tide formation. Although the moon's gravity is much weaker than that of the sun at the position of earth, its fractional variation is greater because the moon is so much closer than the sun. Hence tides on earth are due primarily to the influence of the moon.
14. H. C. Ohanian, "What Is the Principle of Equivalence?" *American Journal of Physics* 45 (1977):903–909.
15. John Norton, "What Was Einstein's Principle of Equivalence?" in *Einstein and the History of General Relativity*, ed. Don Howard and John Stachel (Boston: Birkhäuser, 1989), 5–47.
16. Unpublished manuscript by Einstein, quoted in Pais, *Subtle Is the Lord*, 178.

described the circumstances under which the happy thought occurred to him.

> I was sitting in a chair in the patent office at Bern when all of a sudden a thought occurred to me: "If a person falls freely he will not feel his own weight." I was startled. This simple thought made a deep impression on me. It impelled me toward a theory of gravitation.[17]

The episode must have occurred sometime in 1907.

There can be no doubt that the equality of gravitational and inertial mass was the principal experimental fact on which the principle of equivalence (and ultimately the general theory of relativity) was based. In an article written in 1934, Einstein says:

> The equality of inertial and gravitational mass was now brought home to me in all its significance. I was in the highest degree amazed at its existence and guessed that in it must lie the key to a deeper understanding of inertia and gravitation. I had no serious doubts about its strict validity even without knowing the results of the admirable experiment of Eötvös, which—if my memory is right—I only came to know later.[18]

The remark about the Eötvös experiment is fascinating. According to Pais, Einstein learned about the experiment only in 1912 and first referred to it in a paper written in 1913. (Eötvös's paper had been published in 1890.) As in the case of special relativity and the Michelson-Morley experiment, Einstein apparently had no need to know the up-to-date experimental situation. His intuition stood him in good stead.

8.3. THE GRAVITATIONAL RED SHIFT

In his 1911 paper, Einstein derived two consequences of the principle of equivalence that are subject to experimental test. Both predictions have been confirmed, providing strong support for the principle. In this section we calculate the change in the frequency of light measured in an accelerated frame and apply the principle of equivalence to infer that a similar effect takes place in a gravitational field. The second effect, the deflection of light in a gravitational field, is treated in section 8.4.

Consider the following thought experiment. Light of frequency f is emitted at the bottom of an elevator of height L that is moving in free

17. J. Ishiwara, *Einstein Koen-Ruku* (Tokyo: Tokyo-Tosho, 1977). Cited in Pais, *Subtle Is the Lord,* 179.
18. Albert Einstein, "Notes on the Origin of the General Theory of Relativity," reprinted in Albert Einstein, *Ideas and Opinions* (New York: Dell, 1954), 279.

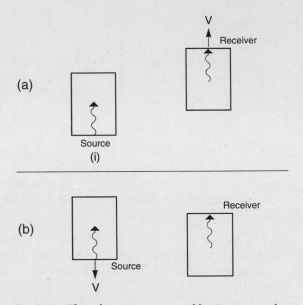

Fig. 8.5. Thought experiment used by Einstein to derive the gravitational red shift. Light is emitted at the bottom of an accelerating elevator and detected at the top. (a) The experiment is shown in frame S, in which the elevator is at rest when the light is emitted. When the light arrives at the top the elevator is moving with speed V. (b) The same experiment is shown in frame S', in which the elevator is at rest when the light arrives. The source was moving downward when the light was emitted. In either case, a Doppler shift is detected.

space with a constant "upward" acceleration g. The light is detected by an observer O stationed at the top of the elevator. What frequency does O measure?

We face a serious problem at the outset. O is clearly not an inertial observer. How can we say anything about his measurements?

Einstein assumed that the result of every measurement made by accelerated observer O is the same as that obtained by a co-moving inertial observer, that is, an inertial observer, at the same location, who has the same velocity as O at the time the measurement is made. This assumption may not be rigorously valid but it should be satisfactory, at least as a first approximation, for moderate accelerations.

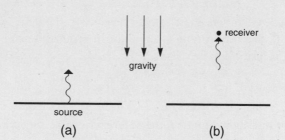

Fig. 8.6.   The gravitational red shift. Light emitted
by source (a) travels against a gravitational field
and is detected by a receiver at a higher elevation
(b). By the principle of equivalence, this experiment
is equivalent to that of fig. 8.5; hence the detected
light is red-shifted.

Let S be the inertial frame in which the source is (momentarily) at rest
when the light is emitted and S' the inertial frame in which the receiver
is at rest when the light is detected. Figures 8.5a and 8.5b show the experiment as it looks in each of those frames.

From either figure it is clear that the source and receiver are receding
from one another. Hence, according to the theory of the Doppler effect
discussed in section 4.8, the frequency f' measured at the receiver is lower
than f: observer O detects the light as red-shifted.

According to the principle of equivalence, the same effect must be observed if the elevator is not accelerating but is instead subject to a gravitational force field directed downward (fig. 8.6). Since O in this case is always at rest relative to the source, he cannot attribute the red shift to the
Doppler effect but must interpret it as an effect of gravity. We conclude
that *light is red-shifted when it moves opposite to the direction of a gravitational field.* Light that moves in the direction of a gravitational field is
similarly blue-shifted. Because the most interesting astronomical applications involve red shifts, the effect is known as the gravitational red shift.

An exact calculation of the frequency shift is quite complicated. In either frame S or S', the velocity of the top of the elevator at any given time
differs from that of the bottom, and the height of the elevator changes as
it accelerates. (Recall the discussion of the accelerated pole in sec. 6.3.) We
can greatly simplify the problem by considering only the case in which
the relative velocity V between source and receiver is much less than c,
neglecting all but the lowest power of V/c in every expression.

In this approximation, the distance traveled by the light in either S or S' is just L.[19] The transit time of the light in either frame is $t = L/c$, and the relative velocity between S and S' is

$$V = gt = \frac{gL}{c}$$

In view of our approximation, the results are valid only when

$$\frac{gL}{c^2} << 1 \tag{8.5}$$

The fractional frequency shift is given by the first-order Doppler formula

$$\frac{\Delta f}{f} = -\frac{V}{c} = -\frac{gL}{c^2} \tag{8.6}$$

By virtue of condition (8.5), the result applies only when the fractional frequency shift is very small. This condition is satisfied in all the applications to be discussed.

According to the principle of equivalence, the frequency shift of light when it moves a distance L away from a gravitating body is also given by equation (8.6), in which g now denotes the acceleration due to the gravity of the body. If light moves toward a massive body, the minus sign in equation (8.6) is replaced by a plus sign: the light is blue-shifted.

In a freely falling elevator there are two effects, a Doppler shift due to the acceleration and a gravitational shift of equal magnitude and opposite sign. The two effects cancel, and no frequency shift should be detected.

Equation (8.6) applies only if g, the strength of the gravitational field, can be regarded as constant in the region through which the light moves. If that condition is not satisfied, we can subdivide the path of the light into short segments. In each segment equation (8.6) applies, with g the strength of the gravity in that segment. The total shift is the sum of the contributions from all the segments.

The quantity gL represents the difference in gravitational potential (potential energy per unit mass) between the two ends of the segment. The total frequency shift is therefore determined by the difference in potential between the end points of the light path. Instead of equation (8.6), we have (to first order)

19. The actual distance traveled by the light, as measured in frame S, is greater than L; in S' the distance traveled is less than L. (See figs. 8.5a and 8.5b.) This correction would change the calculated frequency shift but only in second order.

$$\frac{\Delta f}{f} = -\frac{\Delta \Phi}{c^2} = \frac{\Phi_1 - \Phi_2}{c^2} \qquad (8.7)$$

where Φ_1 is the potential at the point of emission and Φ_2 the potential where the light is measured. Positive $\Delta \Phi$ implies a red shift.

An alternative derivation of the gravitational red shift makes use of the photon model of light. Suppose a photon is emitted by an atom at point 1 in figure 8.6 and absorbed by another atom at point 2. As we learned in chapter 7, the mass of the emitting atom changes by the amount

$$\Delta m_1 = -hf/c^2 \qquad (8.8a)$$

while that of the absorbing atom changes by

$$\Delta m_2 = +hf'/c^2 \qquad (8.8b)$$

where f and f' are the frequencies of the emitted and the absorbed photons.

If the process takes place in a gravitational field, the changes in mass are accompanied by changes in gravitational potential energy:

$$\Delta PE_1 = \Phi_1 \Delta m_1 = -hf\Phi_1/c^2 \qquad (8.9a)$$
$$\Delta PE_2 = \Phi_2 \Delta m_2 = +hf'\Phi_2/c^2 \qquad (8.9b)$$

The total changes in energy for the two atoms are:

$$\Delta E_1 = -hf + \Delta PE_1 = -hf(1 + \Phi_1/c^2) \qquad (8.10a)$$
$$\Delta E_2 = hf' + \Delta PE_2 = +hf'(1 + \Phi_2/c^2) \qquad (8.10b)$$

Conservation of energy requires that the change in the total energy of the system, $\Delta E_1 + \Delta E_2$, be zero. Equations (8.10a,b) give

$$f' - f = \frac{1}{c^2}(f\Phi_1 - f'\Phi_2) \qquad (8.11)$$

If the fractional change in frequency is small, we can approximate f' by f on the right side of equation (8.11). (The error will be of second order.) In this approximation we recover the result (8.7).

The principle of equivalence appears implicitly in this derivation because we have used the same expressions for Δm in equations (8.8a,b), which involve the atom's inertial mass, and in equations (8.9a,b), which involve its gravitational mass. (Notice, however, that only the weak principle is required.) The equivalence principle assigns to a photon a gravitational mass equal to its effective inertial mass hf/c^2. As a result, the photon can be viewed as having a potential energy when it is in a gravitational field. Its "kinetic" energy hf changes when it moves in a gravitational field, just as the kinetic energy of any other particle does.

Observational Tests

The principal applications of the gravitational red shift are in astronomy. Light emitted at the surface of a star moves against the gravitational attraction of the star and should therefore be detected on earth as red-shifted. Equation (8.7) applies. The gravitational potential outside a spherical star of mass M is

$$\Phi(r) = -GM/r \qquad (8.12)$$

where G is the universal gravitational constant, 6.67×10^{-11} N m^2/kg^2, and r is the distance from the center of the star. For this problem r_1 is R, the radius of the star, and r_2 is the earth-star distance, which is much greater than R. Hence we can set $\Phi_2 = 0$ and obtain

$$\frac{\Delta f}{f} = -\frac{GM}{Rc^2} \qquad (8.13)$$

In his 1911 paper, Einstein derived equation (8.13) and applied it to the sun. The theory predicts that each line in the solar spectrum should be red-shifted by about two parts in a million. So small a shift is hard to detect because the solar lines are subject to a number of other effects, including Doppler broadening due to the thermal motion of the atoms that emit the light.

At the surface temperature of the sun, the Doppler width is about ten times the predicted gravitational red shift. In addition, convection currents in the solar atmosphere give rise to shifts whose magnitude and direction are unpredictable. Electric fields can also cause the lines to shift. Nonetheless, in 1962, James Brault was able to confirm the presence of the gravitational shift in a strong sodium line in the solar spectrum.

More convincing evidence is provided by observations on white dwarfs, dense stars whose masses are about equal to that of the sun but whose radii are about a hundred times smaller. The red shifts predicted by equation (8.13) are therefore about one hundred times greater, or about one part in ten thousand. The best candidates to study are white dwarfs that are members of binary systems, which orbit about one another; the mass of such a star can be accurately calculated from its observed orbital period. In 1971, Jesse Greenstein measured a fractional red shift of 2.7×10^{-4} for the spectrum of the white dwarf Sirius B, in excellent agreement with the value predicted by equation (8.13).

For light confined to the neighborhood of the earth, equation (8.6) applies. The earth's gravity is so weak that the predicted frequency shift is minute. The effect was nonetheless detected in a brilliant experiment carried out by Robert Pound and G. A. Rebka, who sent a beam of gamma rays down a 20-meter shaft.[20] According to equation (8.6), the radiation should be detected as blue-shifted, but by only two parts in 10^{15}. By exploiting the newly discovered Mossbauer effect, which makes it possible to measure frequencies with extremely high precision, Pound and Rebka were able to detect the tiny blue shift. The result was in full agreement with the prediction based on the equivalence principle.

In the 1960s, astronomers identified very large red shifts in the spectra of faint sources called quasars. Those red shifts are generally interpreted as being cosmological in origin, associated with the expansion of the universe. (See the discussion in chap. 9.) If that explanation is correct, quasars are the most distant as well as the most luminous objects yet observed. An alternative interpretation is that quasars are extremely dense objects with very high surface gravity and the observed red shifts are in fact gravitational in origin. In such a model, the quasars could be relatively nearby and not nearly so luminous.[21] That interpretation is currently not in favor among astronomers.

Gravitational Time Dilation

As Einstein himself pointed out, the change in frequency caused by gravitation seems on superficial consideration to be absurd. Consider the experiment in figure 8.6. If the source emits a given number of waves each second, how can a different number per second arrive at a receiver at rest with respect to the source?

The result makes sense only if the source clock and the receiver's clock keep time at different rates. The source emits N waves during a time interval Δt, as measured by a clock stationed there. (Since f is the number of waves per second emitted, $N = f\Delta t$.) According to a clock at the location of the receiver, those same N waves arrive over a time interval $\Delta t'$, where $f'\Delta t' = N = f\Delta t$. Since f' is smaller than f, $\Delta t'$ must be greater than Δt. Because the source and receiver are at rest relative to one another, all the wave fronts travel the same distance at the same speed. Hence an observer

---

20. R. V. Pound and G. A. Rebka, "Apparent Weight of Photons," *Physical Review Letters* 4 (1960):337–341.
21. The first-order result (8.13) is not applicable when the red shift is large; a more accurate expression is required.

at O concludes that the N wave fronts, which arrived over the time interval $\Delta t'$, must have been emitted over an equal time interval.

Inasmuch as the source clock recorded the shorter interval Δt, it must be running slow relative to the clock at O. In our example, the source is close to the gravitating body and the receiver is farther away. The general result can be expressed as follows: *clocks close to a massive body run slow compared to ones that are farther away*. The effect is known as gravitational time dilation.

Clocks at different points in an accelerating elevator also keep time at different rates. In any inertial frame the top and the bottom move with different velocities at a given time. Hence the ordinary time dilation of special relativity implies that clocks at the top and bottom must run at different rates. If the acceleration is upward, a clock at the bottom of the elevator runs slow relative to one at the top. (In a freely falling frame, however, all clocks keep time at the same rate.)

In section 6.5, we described an experiment in which atomic clocks were carried around the world on jet planes to test the time dilation predicted by special relativity. In analyzing their results, the experimenters had to take into account the effect of gravitational time dilation as well as that of special relativity. Because the plane clocks were at a higher elevation than the ground clocks, gravitational time dilation caused the plane clocks to gain a little on the ground clocks.

The magnitude of the gravitational effect can be calculated from equation (8.6). The time difference is very small but is comparable in magnitude to that of special relativity. For an altitude of 10,000 meters the theory predicts that plane clocks should gain about 1.6×10^{-7} sec relative to earth clocks in one trip around the world; this gain must be combined with the time difference due to special relativity. The measured differences between clock readings agreed, within the experimental errors, with the calculated values, which took into account both special relativity and gravitational time dilation.

Gravitational time dilation complicates the task of assigning time coordinates to events in the presence of gravitational fields. In special relativity the time of an event in a particular inertial frame is defined by the reading of a clock that is at rest in that frame and is present at the event. Because all clocks in a given inertial frame can be synchronized, this prescription assigns a unique time to each event for the frame in question. But if clocks at different locations in a gravitational field keep time at different rates, they cannot be synchronized. How, then, can the times of events that occur at different places be compared?

In his 1911 paper, Einstein proposed a rather awkward solution to this problem. If we measure time at some reference point with a standard clock U, he suggested, then at any other point we should use a clock that runs at a rate different from that of U when the two are at the same location. In other words, at every point of space we must station a clock of different construction. Although such a procedure is in principle possible, the idea is not an attractive one. Alternatively, we could use identically constructed clocks at all points but correct all clock readings for gravitational time dilation. That solution is not very satisfying either.

As Einstein pointed out, the difference in clock rates has another important implication: observers at two different points in a gravitational field measure different values for the speed of light. When gravity is taken into account, c is no longer a universal constant. It follows that special relativity is strictly valid only in the absence of gravity. Only because gravity is such a weak force has special relativity proven to be successful.

8.4. Bending of Light in a Gravitational Field

For the second application of the principle of equivalence, we again refer to the accelerating elevator of the preceding section. Suppose a pulse of light enters the elevator at point P, in a direction perpendicular to the motion of the elevator, as seen in an inertial frame S (fig. 8.7a). Let S be the frame in which the elevator is at rest when the light enters.

By the time the light reaches the far wall, the elevator has "risen" some distance. The light strikes the far wall at point Q, which is lower than the entry point P. The difference in elevation between P and Q is just the distance the elevator has traveled while the light was in transit.

Figure 8.7b shows how the same experiment looks in the (noninertial) elevator frame. Here the light "falls" in a direction opposite to that of the elevator's acceleration; its path is a parabola. To first order, the distance through which the light falls in the elevator frame is the same as the vertical distance between P and Q in frame S.

According to the equivalence principle, the upward acceleration of the elevator is equivalent to a gravitational field directed downward. If a horizontal light beam enters an elevator at rest on the surface of the earth, the light must strike the far wall at a lower elevation than it entered (fig. 8.7c). The light "falls" in the gravitational field of the earth, just as any material particle would.

In our hypothetical elevator experiment, the light falls only a minute distance because it takes a very short time to cross the elevator and the

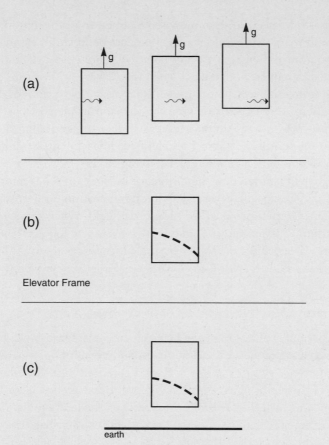

Fig. 8.7. Hypothetical experiment to demonstrate the bending of light in a gravitational field. (a) An elevator accelerates upward, as seen in an inertial frame. Light travels in a straight horizontal line. (b) As seen in the (accelerated) elevator frame, the light follows a parabolic path. (c) The elevator is at rest in a gravitational field. By the principle of equivalence, the path must look the same as in (b).

gravitational pull of the earth is very weak. The deflection of light caused by the earth's gravity is much too small to be detected. A stronger gravity field is needed to bring about a measurable deflection.

The strongest gravitational field in our neighborhood is that in the vicinity of the sun. Einstein pointed out that if the line of sight to a star happens to pass close to the sun, light coming from the star should be

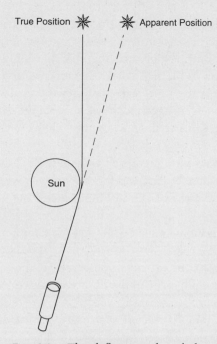

True Position · Apparent Position

Sun

Fig. 8.8. The deflection of starlight by the gravitational pull of the sun, predicted by the principle of equivalence. The effect is much exaggerated.

deflected by the sun's gravitational pull, as shown in figure 8.8.[22] The apparent position of the star in the sky is determined by the direction of the light when it enters our telescope. Hence when the line of sight to a star passes close to the sun, the star's apparent position relative to all the other stars in the sky should shift.

Einstein calculated the deflection and found it to be about 0.83 seconds of angle when the line of sight to the star just grazes the surface of the sun. If the line of sight passes farther from the sun, the deflection is correspondingly less. Although the predicted deflection is small, its detection is within the capability of modern telescopes. The full theory of general relativity, which Einstein published in 1916, predicts a deflection just

22. In 1801, Johann Soldner had predicted that light should be deflected toward the sun, on the basis of a particle theory of light. His result was the same as that obtained by Einstein from the principle of equivalence. The "orbit" of a light beam would be the same as that of a comet traveling at speed *c*.

twice the value obtained from the principle of equivalence; the additional deflection is due to the curvature of space. (See sec. 8.6.)

A practical problem complicated the task of testing the prediction: a star whose position in the sky is near the sun has to compete with very intense sunlight. At the time, astronomers were unable to observe stars in the daytime. Einstein suggested, however, that the deflection should be observable during a solar eclipse, when the light of the sun is blocked out by the moon.

Efforts to test Einstein's prediction were beset by bad luck. An attempt during the solar eclipse of 1912 was foiled by cloudy weather at the observation site. The outbreak of World War I caused the next attempt to be abandoned.

A group of astronomers led by Sir Arthur Eddington planned a new attempt for the total solar eclipse of May 29, 1919. Because the region of totality did not pass over any major observatory, the astronomers had to transport all their equipment to the observation site. As a hedge against possible bad weather, two separate expeditions were mounted, one to Brazil and the other to an island off the coast of West Africa. Both teams observed the deflection just as Einstein had predicted, although the uncertainties in the measured results were sizable.[23] The expeditions were highly publicized and attracted a great deal of attention worldwide; their success made Einstein an overnight celebrity.[24]

More precise results have recently been obtained using radio waves instead of optical light; the predicted deflection is the same. With radio sources the observer does not have to wait for an eclipse; even though the sun is a copious source of radio waves, the radiation from a strong source is readily distinguished from the solar radio emission.

A single radio receiver cannot determine the direction of a source with the required precision, but the directional sensitivity can be greatly improved by employing two or more separated receivers and using interferometry. Observations have been carried out on the quasar 3C279, which is occulted by the sun once a year; the data are in full accord with the prediction of general relativity. In 1974, C. C. Counselman et al. found a deflection of 1.73 ± 0.05 angular seconds.

23. One team measured a deflection of 1.98 ± 0.96 sec, the other 1.61 ± 0.40. Subsequent measurements yielded similar results.
24. For an interesting account of the expeditions and the problems they encountered, see Sir Arthur Eddington, *Space, Time, and Gravitation* (Cambridge: Cambridge University Press, 1920), 114–122.

8.5. CURVED SPACE

The key to general relativity is Einstein's revolutionary idea that space becomes curved in the presence of gravitating matter. This is a difficult concept to grasp. We are accustomed to thinking of space as a shapeless matrix within which material objects are located. Anyone can visualize a curved object, but the notion that *space itself* can be curved is a foreign one.

Our intuitive picture of curvature is two-dimensional: a surface can be either flat or curved.[25] The curvature of a surface is apparent, however, only when it is viewed from the perspective of the three-dimensional space in which it is embedded. Early thinkers deduced that the earth is round from observations such as the following: (i) when a ship sails away, the top of the mast remains visible for some time after the bottom has disappeared below the horizon; (ii) as one travels westward, sunrise occurs later and later; (iii) during a lunar eclipse, the earth casts a round shadow on the moon. Today, the view from a space vehicle constitutes a direct demonstration that the earth is round. All these observations are intrinsically three-dimensional.

If two-dimensional curvature is visible only from a third dimension, we cannot visualize curved three-dimensional space because no fourth spatial dimension from which to view it is available. We can, however, characterize the properties of a curved surface in terms of measurements that can be carried out entirely on the surface, with no reference to a third dimension. Curvature in three (or more) dimensions is defined by generalizing those properties.

Imagine a two-dimensional surveyor, whose observations are confined to a surface. She is not aware that a third dimension even exists. For her, space has two dimensions; objects have length and width but not height. She is equipped with instruments to measure lengths, areas, and angles on the surface, but the term "volume" has no meaning for her. Can any measurements enable the surveyor to determine whether her "space" is flat or curved?

A straightforward way to verify that a sphere is curved is to take a trip along a fixed heading. Eventually one will return to the starting point; that could not happen on a flat surface. Magellan's journey was convincing proof that the earth is not flat. This method, however, works only for

25. Curvature in one dimension is also familiar. A line can be straight or curved.

curved surfaces that are closed and sufficiently symmetric. On the saddle-shaped surface of figure 8.11, for example, one would never return to the starting point even though the surface is, by any reasonable standard, curved.

Another disadvantage of the circumnavigation method is that it requires exploration of the entire surface. A test that requires us to travel across the entire universe to discover whether space is curved is obviously impractical. We seek a test that involves only *local* measurements, that is, measurements confined to a small neighborhood.

The solution is to investigate the geometry of the surface. The familiar geometry of Euclid, which is taught in high school, is the geometry of a flat plane. On a curved surface, many of Euclid's postulates and theorems do not hold.[26] By testing whether or not geometry is Euclidian, therefore, our two-dimensional surveyor can determine whether the space she inhabits is flat or curved. Non-Euclidian geometry is interesting in its own right and was studied by several mathematicians during the nineteenth century.

Geodesics

On a curved surface there are no straight lines, but one can define a curve that shares their principal property: it is the shortest curve that connects two given points. A curve with this property is called a geodesic; it is the "straightest" possible line. On a sphere, for example, a geodesic is a great circle (a circle whose plane contains the center of the sphere; see fig. 8.9).[27]

The strict definition of a geodesic is a curve whose length is an *extremum*, which can be either a minimum or a maximum. In ordinary Euclidian space the extremum is a minimum. In some spaces, however, a geodesic is the longest curve that joins two given points; length in such a case has a generalized definition. An example is discussed in the next section.

Geometric figures on curved surfaces are generally defined in terms of geodesics. A polygon is a figure whose boundaries are geodesics; a circle is the locus of points connected to a fixed point by geodesics of equal length; and so on.

26. There are two exceptions: on a cylinder and on a cone, Euclidian geometry is valid. Either of those surfaces can be flattened into a plane; for our purposes they can be considered flat. A sphere, however, is intrinsically curved: it cannot be rolled flat without leaving any bulges.
27. Sometimes, more than one geodesic exists between two points. On a sphere, infinitely many great circles connect two opposite poles; all have the same length.

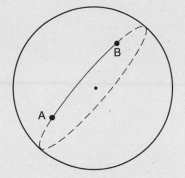

Fig. 8.9. A great circle is a geodesic
on a spherical surface. It is the shortest
line joining points *A* and *B*.

Table 8.1. Contrast between properties of flat and curved surfaces

Flat *Euclidian Geometry Holds*	*Curved* *Non-Euclidian Geometry Holds*
1. Lines perpendicular to the same line are parallel.	1. Lines perpendicular to the same line can intersect. (For example, meridians on a sphere.)
2. Sum of the angles of a triangle = 180°.	2. Sum of the angles of a triangle ≠180° (on a sphere, the sum is more than 180°).
3. Pythagorean theorem holds: $$a^2 + b^2 = c^2$$ where a and b are the sides of a right triangle and c is the hypotenuse.	3. Pythagorean theorem does not hold.
4. Circumference of a circle $= 2\pi r$.	4. On a sphere, $C < 2\pi r$.
5. A finite area must have a boundary.	5. A surface can have a finite area without any boundary (e.g., a sphere). But a saddle-shaped surface is infinite.

Table 8.1 lists some properties of a flat surface that are not true on a curved one. We focus our attention on two:

(i)   the sum of the angles of a triangle is 180°; and
(ii) the circumference of a circle is 2π times its radius.

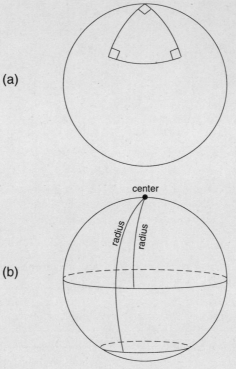

(a)

center

radius   radius

(b)

R = radius of sphere

Fig. 8.10. Geometry on a sphere is non-Euclidian.
(a) A triangle with three right angles. (b) The radius
of a circle along the equator is ½ πR and its circumference is 2πR, where R is the radius of the sphere;
C = 4r in this case. Also shown is a circle in the
Southern Hemisphere; its radius is greater than that
of the equatorial circle, but its circumference is less.

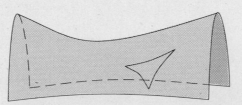

Fig. 8.11.   A surface of negative curvature (saddle-
shaped). The sum of the angles of a triangle on this
surface is less than 180°, while the circumference of
a circle is more than 2πr.

On a spherical surface, the sum of the angles of a triangle is more than 180°, and the circumference of a circle is less than 2π times its radius (fig. 8.10). The equator, for example, is a circle whose circumference is only four times its radius.[28] On the saddle-shaped surface of figure 8.11, the angles of a triangle add up to less than 180° and the circumference of a circle is more than 2π times its radius.

The departure from flatness of a surface can be measured by a quantity called the Gaussian curvature, defined as follows. Construct circles of various radii centered at a given point on the surface, and measure their circumferences. For each value of the radius, calculate the quantity

$$K_r = \frac{3(2\pi r - C_r)}{\pi r^3} \tag{8.14}$$

where C_r is the circumference of a circle with radius r.

Now let the circles become smaller and smaller. If K_r approaches some limit K as r approaches zero, that limiting value is defined to be the curvature of the surface at the point in question. K can be positive or negative and in general varies from one point to another on the surface.

On a plane, the circumference of any circle is $2\pi r$; hence K_r is zero for all values of r and its limit is zero: a plane is a surface of zero curvature.

On a spherical surface, K must have the same value at every point because of symmetry. We show in the appendix to this chapter that the circumference of a small circle on the surface of a sphere is approximately

$$C_r \approx 2\pi r\left(1 - \frac{r^2}{6R^2}\right) \tag{8.15}$$

where R is the radius of the sphere and $r \ll R$. Substituting this expression in equation (8.14) gives $K = 1/R^2$: a spherical surface has constant positive curvature. A small sphere has greater curvature than a large one, in accord with our intuitive notion of curvature. The saddle-shaped surface of figure 8.11 has constant negative curvature.

The preceding paragraph refers to the radius of the sphere, R. From our three-dimensional perspective, we can actually see the sphere and locate its center; for us, R is the distance between the center and the surface. *That statement has no meaning for the two-dimensional surveyor.* Although she can deduce from her measurements that space is curved and can calculate the value of R, she cannot identify any point in her two-

28. Circles on a sphere have other unusual properties. For radii greater than a quarter of a great circle, the circumference decreases as the radius increases. As the radius approaches half of a great circle, the circumference shrinks to zero.

dimensional space as the "center" of the sphere; for her, R is just a parameter that characterizes the (non-Euclidean) geometry of space.

Locally Flat Space

An important property of curved surfaces can be inferred by examining the behavior of equation (8.14) as r goes to zero. The left side approaches the curvature K. Since the denominator on the right side goes to zero, the numerator must also go to zero (provided K is finite). Hence the circumference of a small circle on *any* surface of finite curvature approaches the Euclidian form $2\pi r$ as r approaches zero.

The sum of the angles of a small triangle can similarly be expressed in terms of the curvature of the surface. The following relation holds for a small triangle:[29]

$$\text{sum of angles} = 180° \left(1 + \frac{K \text{ (Area of } \triangle)}{\pi}\right) \qquad (8.16)$$

Equation (8.16) implies that for any finite K, the sum of the angles approaches 180° as the area of the triangle goes to zero.

These results illustrate a general property: *any surface with finite curvature is locally flat:* the geometry in a small enough region is very nearly Euclidian.[30] This conclusion should come as no surprise, for we know that Euclidean geometry works quite well on our spherical earth in a sufficiently small region.

A simple test for curvature requires only the measurement of the distances between any four points on a surface. If the surface is flat, the six distances are not independent; given any five, one can calculate the value of the sixth.

We illustrate the method using cities on the earth. The following are airline distances in miles between the designated cities:

Melbourne–Chicago	9,673
Melbourne–Rio de Janeiro	8,226
Melbourne–Moscow	8,950
Chicago–Rio de Janeiro	5,282
Chicago–Moscow	4,987
Rio de Janeiro–Moscow	7,170

29. On a uniformly curved surface, eq. (8.16) holds for a triangle of any size.
30. One can construct mathematical spaces that are not locally flat; such spaces have singular points or "cusps." At a singularity the curvature is either infinite or is undefined.

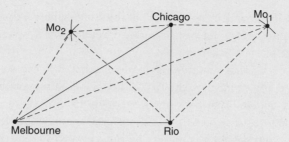

Fig. 8.12. Proof that the earth is not flat. The triangle Melbourne-Chicago-Rio is constructed from the known intercity distances. Two possible locations for Moscow are determined by the distances Rio-Moscow and Chicago-Moscow. The distance Moscow-Melbourne for either of those locations does not agree with the measured distance. Hence the geometry cannot be Euclidian.

In figure 8.12, the triangle Melbourne-Chicago-Rio has been drawn to scale. (The shape of a triangle on a plane is completely determined by the lengths of its sides.) Arcs whose radii are the distances Rio-Moscow and Chicago-Moscow have been drawn with their centers at Rio and Chicago. Moscow must lie at an intersection of those arcs. There are two intersections, marked Mo_1 and Mo_2; on our flat map their distances from Melbourne are 14,100 and 5,700 miles, respectively. Since neither of these values is close to the actual Melbourne-Moscow distance, we conclude that the earth is not flat. Its radius can in fact be calculated from the six intercity distances.

Another property of curved surfaces will prove useful in the discussion of cosmology in chapter 9: a curved surface can have a finite area even though it has no boundary. A sphere has that property. A flat surface without boundaries must be infinite in extent.

The preceding discussion can be generalized to spaces of three or more dimensions. By definition, *a curved space in any number of dimensions is one whose geometry is not Euclidian.* Three-dimensional Euclidian space is the space of ordinary solid geometry, in which the surface area of a sphere of radius r is $4\pi r^2$ and the volume of a cube of side h is h^3. In a curved three-dimensional space those relations do not hold.

The simplest example of curved three-dimensional space is a region of constant positive curvature. Such a region can be regarded as the "surface" of a hypersphere, the four-dimensional analogue of a sphere. It is the locus of points equidistant (in four dimensions) from a given point.

That common distance is the "radius" of the hypersphere. A formula analogous to equation (8.15) gives the surface area of a small sphere of radius r in a hypersphere of radius R.

Three-dimensional curvature is an abstraction. We cannot visualize a hypersphere any more than the two-dimensional surveyor can visualize an ordinary sphere. That does not prevent us from defining the mathematical properties of curved three-dimensional space and (at least in principle) testing our space for curvature by carrying out measurements analogous to the ones described above for two dimensions. For example, we could measure the surface areas of small spheres. If the area of every sphere is $4\pi r^2$, space is Euclidean or "flat"; if not, it is curved. Three-dimensional curvature can be defined by a procedure analogous to the one based on equation (8.12).[31] The curvature of a hypersphere is $1/R^2$.

The procedure illustrated in figure 8.12 can also be generalized to provide a test for three-dimensional curvature. Five cities are required in this case, with ten intercity distances. If space is Euclidian, those distances are not independent.

Legend has it that the great mathematician Karl Gauss tried to determine whether space is curved, long before Einstein had proposed general relativity, by measuring the angles of the triangle formed by three mountaintops in Germany. That story has been refuted by Arthur Miller,[32] but a similar test was in fact carried out in 1900 by Karl Schwarzschild, using the triangle defined by the position of a star and that of the earth at two points in its orbit.[33] Schwarzschild did not succeed in demonstrating any curvature; he was able to show only that if space is curved, its "radius" in our neighborhood is greater than 1,600 light-years. That lower bound is many orders of magnitude less than the value suggested by general relativity. As we shall see, the curvature of space is very small except in the vicinity of extremely dense objects like black holes.

---

31. In two dimensions, a single number, the curvature, suffices to fix the properties of a curved surface at a given point. Spaces of greater dimensionality are more complicated. The curvature becomes a tensor (a mathematical quantity with several components). Six numbers are needed to describe fully the curvature of a three-dimensional space, 20 in four dimensions. In a space of n dimensions, the curvature tensor has $n^2(n^2-1)/12$ independent components. In Euclidian space, all components of the curvature tensor vanish.

32. A. Miller, "The Myth of Gauss' Experiment in the Euclidean Nature of Physical Space," *Isis* 63 (1972):345–348. Gauss did measure the three mountaintops but for a different purpose.

33. Schwarzschild's "experiment" is described by Robertson in *Einstein, Philosopher-Scientist*, 323.

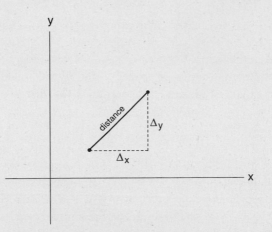

Fig. 8.13. Distance in a two-dimensional Euclid-ian space. The dependence of distance on Δx and Δy is given by eq. (8.17).

The Geometry of Space-Time

An additional complication arises when curvature is introduced in relativity theory. We already know from special relativity that space and time are intermixed: a space coordinate in one frame of reference corresponds to a combination of space and time coordinates in another frame. Curvature therefore cannot be confined to spatial dimensions. *It is not three-dimensional space but rather four-dimensional space-time that is curved according to general relativity.*

Curved space-time is an even more abstruse concept than is curved space. On the basis of the earlier discussion, one may surmise that the geometry of flat space-time is Euclidian, whereas that of curved space-time is non-Euclidian. But what is the meaning of that statement? The term "geometry" has a purely spatial connotation; how does geometry apply to time?

In addressing these questions, it is helpful to introduce a new concept, the *metric* of a space. The metric specifies how the distance between two neighboring points depends on the differences in their coordinates. If we know the metric at every point, we know everything about the space, including its curvature.

We illustrate with the familiar case of plane geometry. In rectangular coordinates, the distance between the points (x, y) and $(x + \Delta x, y + \Delta y)$ (see fig. 8.13) is given by

$$(\text{distance})^2 = (\Delta x)^2 + (\Delta y)^2 \tag{8.17}$$

Equation (8.17) is the metric for rectangular coordinates in two dimensions.[34]

The form of the metric for a given space is not unique. When we transform to a different coordinate system, the metric relation in general changes. In polar coordinates, for example, the distance between the points (r, θ) and $(r + \triangle r, \theta + \triangle \theta)$ is

$$(\text{distance})^2 = (\triangle r)^2 + r^2 (\triangle \theta)^2 \qquad (8.18)$$

which is not of the simple form (8.17). The numerical distance between two given points is, however, the same in any coordinate system.

The quadratic form (8.17) is characteristic of Euclidian (flat) space. On a curved surface, the metric never takes that simple form, no matter what coordinate system one chooses.

These ideas can be extended to abstract vector spaces in any number of dimensions. A "point" in an n-dimensional vector space is simply a collection of n numbers $x_1, x_2, \ldots, x_n$ called the coordinates; the "distance" between the points $(x_1, x_2, \ldots, x_n)$ and $(x_1 + \triangle x_1, x_2 + \triangle x_2, \ldots, x_n + \triangle x_n)$ is some scalar function of the infinitesimal differences $\triangle x_1, \triangle x_2, \ldots, \triangle x_n$. This is a generalization of the concept of distance; it is not the sort of distance that can be read on a tape measure.

The form of the metric determines the properties of the space. If in some set of coordinates the metric has the form

$$(\text{distance})^2 = (\triangle x_1)^2 + (\triangle x_2)^2 + \ldots + (\triangle x_n)^2 \qquad (8.19)$$

the space in question is said to be Euclidean. Equation (8.19) is the obvious generalization of (8.17).

The space-time of relativity is a four-dimensional vector space in which the "points" are events; can a "distance" be defined in this space?

We are guided by the fact that the subspace of space-time that corresponds to a fixed value of t is ordinary three-dimensional space. Hence the four-dimensional "distance" between two events that occur at the same time should be just the three-dimensional spatial distance between them. Furthermore, the distance between two specified events should be independent of the frame of reference in which their coordinates are measured. These are the only conditions we can impose.

In special relativity a quantity with the desired properties is the invariant interval $(\triangle s)^2$, discussed in chapters 4 and 5. Accordingly, we define

34. Eq. (8.17) happens to be valid for any value of $\triangle x$ and $\triangle y$. In general, however, the metric relation gives only the distance between points whose coordinates differ by infinitesimal amounts.

the distance between the events whose coordinates are (x, y, z, ct) and $(x + \triangle x, y + \triangle y, z + \triangle z, ct + c\triangle t)$ by the relation

$$(\text{distance})^2 = (c\triangle t)^2 - (\triangle x)^2 - (\triangle y)^2 - (\triangle z)^2 \qquad (8.20)$$

The metric (8.20) is almost, but not quite, of the form (8.19). The difference is the presence of the minus signs in (8.20). Because of those minus signs, the space-time of special relativity is not quite Euclidian; it is sometimes called pseudo-Euclidian. The minus signs have an important consequence, as we shall see.

I showed in section 5.3 that if the interval between two events is timelike, $\triangle s$ is just (c times) the proper time interval between them. (There must exist a frame in which the events occur at the same place; $\triangle s$ in that frame is just $c\triangle t$ and $\triangle t$ is by definition a proper time interval.) Since the world line of any material particle consists of a succession of events separated by timelike intervals, the length of a world line as defined by equation (8.20) is the total proper time for a body that follows that world line, that is, the elapsed time shown by a clock that moves with the body.

If two events occur at the same time ($\triangle t = 0$), $\triangle s$ is, except for a multiplicative constant $i = \sqrt{-1}$, the ordinary spatial separation $\sqrt{(\triangle x)^2 + (\triangle y)^2 + (\triangle z)^2}$ between the events.[35] If two events are separated by a spacelike interval, we can always find a frame in which they occur simultaneously and calculate the metric distance in that frame. With the metric defined by equation (8.20), the distance between any two events separated by a spacelike interval is imaginary.

8.6 General Relativity: Gravity as Geometry

By 1912, Einstein had realized that the principle of equivalence implies that space cannot remain flat in the presence of matter. He was apparently led to this conclusion by thinking about measurements carried out in a rotating frame.[36]

35. The metric for special relativity is sometimes defined as

$$(\text{distance})^2 = (\triangle x)^2 + (\triangle y)^2 + (\triangle z)^2 - (c\triangle t)^2$$

instead of as in (8.20). With this definition the distance between two events with the same time coordinate is real and is exactly equal to their spatial separation. Multiplying the metric by a constant does not change the properties of a space; the form (8.20) has the advantage that it makes the lengths of all world lines real.
36. For a detailed discussion, see J. Stachel, "The Rigidly Rotating Disk as the 'Missing Link' in the History of General Relativity," in *Einstein and the History of General Relativity*, 48–62.

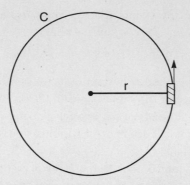

Fig. 8.14.   Geometry on a rotating turntable. Observers in the inertial frame find the length of the short segment of the rim to be shorter than the length measured by observers on the turntable. Since this is true for every piece of the rim, the circumference according to the inertial observers is less. The two sets of observers agree as to the length of the radius, which is transverse to the relative motion. Hence if $C = 2\pi r$ according to inertial observers, it must be greater than $2\pi r$ according to observers in the rotating frame. The geometry is non-Euclidian.

A horizontal platform rotates uniformly about a vertical axis through its center (fig. 8.14). Each point on the platform moves in a circular path and therefore accelerates toward the center; a frame of reference fixed on the platform is an accelerated frame in which both the magnitude and the direction of the acceleration vary from point to point. In this respect it differs from the uniformly accelerated frames considered earlier, in which every point has the same acceleration.[37]

We want to compare length measurements carried out by observers riding on the platform with those of observers in an inertial frame fixed on the ground. We assume as before that the result of every measurement

37. Notice that a uniformly rotating disk must be limited in size. Since the tangential velocity is proportional to the radius, for a large enough radius the velocity would exceed c.

carried out by a platform observer is the same as that obtained by a co-moving inertial observer, that is, an observer at the same point who moves in a tangential direction with a constant speed equal to the platform's rotational velocity.

A platform observer located on the rim measures the length of the small segment marked with cross-hatching in figure 8.14. Since the segment is at rest relative to that observer, the value she obtains is a proper length. Ground-based observers who measure the length of the same segment obtain a Lorentz-contracted result, since for them the segment is moving and is aligned in the direction of its motion.

The same argument applies to each segment of the rim. Platform observers therefore measure a greater value for the circumference of the rim than do ground-based observers; the ratio of the two lengths is just the value of γ appropriate to the tangential velocity of the rim.

When the two sets of observers measure the radius of the platform, they obtain equal results because the radius is transverse to the relative motion. Hence if ground observers find the circumference to be 2π times the radius (as they must, since their frame is inertial), platform observers must find it to be *more* than 2π times the radius. Geometry on the rotating platform is not Euclidian!

According to the equivalence principle, the accelerating platform frame is equivalent to a gravitational field. It follows that geometry in the presence of a gravitational field must likewise be non-Euclidian.

Another thought experiment that illustrates the intimate connection between gravity and geometry has been described by Edwin Taylor and John Wheeler; they call it "the parable of the two travelers."[38] Two travelers start out from the equator, 20 km apart. (See fig. 8.15) Each heads due north, along a line of constant longitude, at the same speed. After the travelers have gone 200 km, their separation is only 19.99 km. As they proceed on their journeys, they continue to draw closer together, at an ever-increasing rate; they are accelerating toward one another.

From a geometric point of view, the travelers are simply force-free bodies following geodesic paths on a curved earth; their apparent acceleration is a consequence of the geometry. An alternative explanation is available, however. The travelers are not conscious of being on a curved surface. Believing in Newton's laws, they attribute their relative acceleration

38. E. F. Taylor and J. A. Wheeler, *Spacetime Physics,* 2d ed. (New York: W. H. Freeman, 1992), 184.

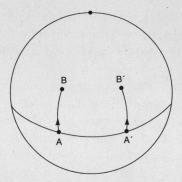

Fig. 8.15. Two particles on a spherical surface at points A and A' move off in parallel directions. Some time later they are at B and B' and are found to be closer together than at the start. The curved geometry simulates the effect of an attractive force.

to some mysterious force that acts on them. Since the acceleration is independent of the travelers' composition or mass, the force can be identified as being gravitational in nature. It is similar to the tidal force responsible for the decreasing separation of the falling tennis balls in figure 8.4b.

The two descriptions (in terms of gravity and in terms of geometry) must be equivalent. The lesson of the parable is that gravitational effects can be represented by changes in the geometry of space. This is the basic idea of general relativity.

Having recognized that gravity and geometry are related, Einstein pursued the daring hypothesis that the *entire* effect of gravitating matter is manifested as a distortion of the space-time in its vicinity. The problem is then to find the relation between the distribution of mass and the geometry. After several false starts, Einstein finally arrived at a satisfactory solution, which he presented to the Prussian Academy of Sciences in late 1915.[39] This was the general theory of relativity.

General relativity is based on two hypotheses:

39. A. Einstein, *Proceedings of the Prussian Academy of Sciences* (Nov. 11, 1915):778–786; (Nov. 18, 1915):799–801, 831–839; (Nov. 25, 1915):831–839; and (Dec. 2, 1915):844–847. A comprehensive paper describing the entire theory was published under the title, "The Foundation of the General Theory of Relativity," *Annalen der Physik* 49 (1916):769–822. An English translation of this paper appears in *The Principle of Relativity*, 111–164.

(i)   Gravitating matter distorts the space-time in its vicinity, causing it to become curved. The curvature depends on the distribution of mass; far from any matter, special relativity is valid and space-time is (pseudo-) Euclidian: the metric approaches the form (8.20).

(ii)   The world lines of all freely falling bodies (including light rays) are geodesics in space-time.

It can be shown that hypothesis (ii) is not independent but is actually a consequence of (i).

The Geodesic Law

I illustrate Einstein's geodesic law by applying it to the simplest possible problem—the motion of a body on which no forces act. The solution is well known: the body moves with constant velocity. I will show that the geodesic law reproduces this solution.

In the absence of gravity the space-time is that of special relativity, with a metric given by equation (8.20). Let event *A* denote the position of the body at a given time and event *B* its position at some later time. According to the geodesic law, the body's world line is the curve that connects events *A* and *B* whose length is an extremum. Since motion with constant velocity is described by a straight world line, we have to prove that of all possible world lines that connect *A* and *B*, the straight line is either the shortest or the longest.

If space-time were strictly Euclidian, the proof would be trivial; a straight line is the shortest distance between two points in Euclidian space. With distance defined by the pseudo-Euclidian metric (8.20), however, the nature of the geodesic path is not at all obvious.

If events *A* and *B* can be connected by a world line, the interval between them must be timelike. We have seen that in such a case there must exist a frame in which the events occur at the same location. It is simplest to carry out the calculation in that frame. (Since Δs is invariant, we are free to calculate its value in any frame.)

Figure 8.16 shows the two events and the straight world line *AB* that joins them. The coordinates of the events in this frame are:

$$\text{event } A: x = a, \ t = 0$$
$$\text{event } B: x = a, \ t = T$$

World line *AB* describes a body at rest at the point $x = a$. Applying equation (8.20) with $\Delta x = 0$, $\Delta t = T$, we find for the length of *AB*:

$$\text{length of world line } AB = cT \qquad (8.21)$$

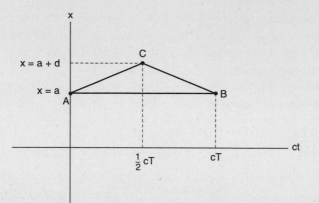

Fig. 8.16. This diagram illustrates the peculiar geometry of space-time. The "length" of world line *ACB* is less than that of *AB*. If $d = cT/2$, the length of *ACB* is zero.

Also shown in the figure is another world line, *ACB,* that connects the same two events. *AC,* the first segment of that world line, describes a body that moves to the right at constant velocity, arriving at the point $x = a + d$ at time $T/2$. The body then reverses its direction (segment *CB*) and returns to the starting point $x = a$ at time T.

The length of world line *ACB* is the sum of the lengths of the segments *AC* and *CB.* Those lengths are equal; each one has the value

$$\sqrt{\left(\frac{1}{2}cT\right)^2 - d^2}$$

Hence we find

$$\text{length of world line } ACB = 2\sqrt{\left(\frac{1}{2}cT\right)^2 - d^2}$$
$$= \sqrt{(cT)^2 - 4d^2} \qquad (8.22)$$

Clearly, (8.22) is less than (8.21) for any value of d.

Many other world lines connect events *A* and *B;* it can be verified, however, that their lengths are all less than cT. The straight world line *AB* is therefore a geodesic, but its four-dimensional length is a *maximum,* not a minimum. This surprising result is a consequence of the minus signs in the metric (8.20).

If $d = cT/2$, the speed of the body along paths *AC* and *CB* is c: the "body" in this case must be a light ray. The length of its world line, according to equation (8.22), is zero. For this reason the path followed by a light ray is called a *null geodesic.*

When d is greater than $cT/2$, the metric distance given by equation (8.22) becomes imaginary. Such a path cannot be the world line of any material body, however, since it corresponds to a speed greater than c. The lengths of all physically permitted world lines are real.

As noted in the preceding section, the length of any world line represents proper time for a body whose motion is described by that world line. *A body follows the world line that maximizes the proper time between its end points.* Although this result has been demonstrated only for a body free of any forces, it in fact applies even when gravitational forces are present.

The Field Equations of General Relativity

In the presence of matter, the metric is no longer the simple one of special relativity. The central problem of general relativity is to find the metric that corresponds to a given distribution of mass. In 1916, Einstein derived his famous field equations, which specify the relation between the metric and the mass distribution. The equations employ the language of tensor calculus; I shall not attempt to write them out.

Symbolically, the Einstein field equations are of the form [40]

$$\text{(a tensor related to the curvature of space-time)} = \text{constant} \times \text{(a tensor related to the mass-energy distribution)} \quad (8.23)$$

The curvature is a geometric quantity; it depends on the spatial variation of the metric. The right side of equation (8.23) is gravitational.

Once the metric has been determined for a given mass distribution, the motion of a body can be found from the geodesic law. In curved space-time, geodesics are not straight lines. A curved world line describes accelerated motion, caused by gravitational forces.

Tests of General Relativity

In his 1916 paper, Einstein proposed three observational tests of general relativity, all of which involve gravitational effects of the sun. Because gravity is such a weak force, all three effects are very small and their calculation does not require exact solution of the field equations; an approximate solution suffices.

The first of Einstein's proposed tests was the deflection of starlight, already discussed in section 8.4 in connection with the principle of equiva-

40. The curvature is not simply proportional to matter density; that would imply that space-time is curved only at points where mass is located. The actual curvature extends to the surrounding region.

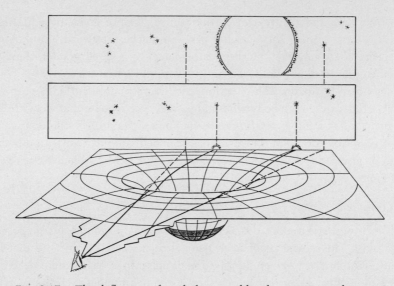

Fig. 8.17.　The deflection of starlight caused by the curvature of space in the vicinity of the sun. Light rays from the two stars follow the solid curves. The apparent positions of the stars are shown in the upper rectangle. The lower rectangle shows their positions when the sun is not near the line of sight. From *Gravitation* by Misner, Thorne, and Wheeler. Copyright © 1973 by W. H. Freeman and Company. Used with permission.

lence. General relativity predicts an additional deflection caused by the curvature of space-time. A light ray, like anything else with inertia, follows a geodesic path through the curved space-time in the neighborhood of the sun.

Figure 8.17 is an artist's attempt to depict the curved world line of a light ray that passes close to the sun. Since the actual geodesic is in four-dimensional space-time, it cannot be drawn on a two-dimensional graph. The figure is intended to be suggestive only.

The magnitude of the deflection calculated by Einstein turned out to be exactly twice the value he had earlier obtained from the equivalence principle alone. The predicted deflection is still very small: 1.75 angular seconds for a light ray that just grazes the edge of the sun. The observations that confirmed this prediction have been described in section 8.4.

A second observational test of general relativity involves planetary orbits. According to Newtonian mechanics, the orbit of a planet under the influence of the sun's gravitational attraction alone is an ellipse. As a result of the perturbations caused by the other planets, the ellipse does not

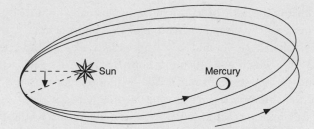

Fig. 8.18. Precession of the orbit of Mercury (schematic). Most of the precession is due to perturbations caused by the gravitational attraction of Jupiter and the other planets. The last 43 seconds of arc per century is accounted for by general relativity.

quite close on itself after each revolution; the orbit precesses very slowly. (See fig. 8.18.)

The precession of planetary orbits has been observed. The magnitude of the effect is greatest for Mercury, the planet whose orbit is the most eccentric. When the precession was calculated using Newtonian mechanics, there remained a difference of 43 seconds of angle per century between theory and observation. Although the difference is minute, it is greater than the uncertainty in the observations. The discrepancy puzzled astronomers for many years. Several unsatisfying explanations were proposed, including the presence of an unseen small planet near Mercury as well as a slight departure from the inverse-square dependence in Newton's law of gravity.[41]

Einstein pointed out that the curvature of space causes an additional precession. He calculated the effect and found that it had just the right magnitude to account for the missing 43 seconds. This result gave Einstein great satisfaction. He later wrote, "For a few days I was beside myself with excitement," and told a colleague that the discovery had given him palpitations of the heart.

In recent years, the same effect has been detected in a case in which its magnitude is much greater—in a pulsar. Pulsars, discovered in the 1960s, emit radio pulses at very regular intervals. A pulsar is believed to be a rotating neutron star, a collapsed object of stellar mass and radius about

41. Simon Newcomb showed that the discrepancy could be accounted for if gravity, instead of being an inverse square force, depended on the distance to the inverse power 2.0000001574. This was not an attractive solution to the problem. Besides, it led to a discrepancy in the calculated orbit of the moon.

10 km which is the relic of a supernova explosion. The measured pulsar period represents the period of rotation of the neutron star.

In 1974, John H. Taylor and R. A. Hulse discovered a pulsar, designated PSR 1913 + 16, which exhibits anomalous behavior. Its period (about 59 milliseconds) is not constant but oscillates in a regular manner. The amplitude of the variation in arrival times of the pulses is about 4 minutes.

These data indicate that PSR 1913 + 16 is a member of a binary system. The regular oscillation in its period is due to the changing travel time for pulses as the pulsar traverses its orbit. Detailed analysis indicates that the companion is also a compact object, probably another neutron star.

After observing the pulsar for nearly twenty years, Taylor has been able to establish its orbit with great precision. The orbit is very small and highly eccentric: the semimajor axis is only about 1 solar radius and the eccentricity is .617. The period of the orbit is about 8 hours.

This combination of circumstances makes the binary pulsar an ideal laboratory for testing general relativity. The precession of the orbit predicted by general relativity is about 4° of arc per year, some 30,000 times greater than for Mercury. The observations confirm that the orbit precesses at just the predicted rate. The orbit is also small enough to make the gravitational red shift detectable. Finally, the secular decrease in the pulsar's rotation period due to the emission of gravitational radiation is also observed as predicted by general relativity.

The third prediction of general relativity, the gravitational red shift, is not really a test of the theory because it can be calculated from the equivalence principle alone, as we saw in section 8.2. Unlike the case of the deflection of light, the curvature of space causes no additional shift.

A fourth test was proposed by Irwin Shapiro in 1964 and carried out by him in 1968 and 1971. Shapiro, a radio astronomer, was one of the pioneers in the technique of radar ranging, determining the orbits of the planets Mercury, Venus, and Mars by reflecting radar pulses from them. The reflected signal is extremely weak but is detectable. By measuring the round-trip time of the pulse, which can be done with great precision, one can measure the distance of the planets from earth to within 1 km, much more precise than any previous determination.

Since radar waves are electromagnetic radiation, they are affected by gravity in the same way light is. In particular, their speed is diminished when they pass through a strong gravitational field. Shapiro realized that this relativistic effect should cause an additional time delay in the radar pulse whenever the line of sight to the planet passes close to the sun. The predicted extra time delay for Venus is about 200 microseconds.

Shapiro's data for Venus agreed with the prediction of general relativity to better than 5 percent. As in the case of the light deflection effect, some time delay can be inferred from the principle of equivalence alone, but there is an additional contribution due to the curvature of space-time.

Similar time-delay observations were carried out using the spacecrafts Mariner 6 and Mariner 7, which were part of the Mars landing expeditions. The spacecrafts passed close to the sun on several occasions; the measured delays in the return of radio signals reflected from them were again in agreement with general relativity.

Because the classic tests described above require only approximate solution of the field equations, the agreement with observation does not prove that general relativity is correct. In recent years several alternative theories of gravitation have been put forward, the most noteworthy being the one proposed by R. H. Dicke and Carl Brans. All the theories involve curved space-time, but the field equations differ from Einstein's; the Brans-Dicke theory, for example, is a scalar-tensor theory.

The predictions of the alternative theories for the classic tests differ from those of general relativity, but not by very much. Within the observational uncertainties, they too agree with observation. Some observational tests are possible which will discriminate among the various candidate theories, but these tests are very difficult and have not yet been carried out with the accuracy required to make a definite choice. At present, general relativity remains the favored theory.

Exact solutions of Einstein's field equations have been found for only a few problems. In 1916, a few months after Einstein had published his paper, Schwarzschild found the solution for the metric in the presence of a point mass. Shortly afterward he solved the more interesting problem of a spherically symmetric star. The Schwarzschild solution is the basis for a number of applications.

Black Holes

Even without looking at the detailed mathematical solution, we can predict that unusual effects should take place in the vicinity of an extremely dense body. The gravity near the surface of such a body is very strong, and the deflection of light it causes is correspondingly large. If the gravity is strong enough, a light ray moving tangentially to the body will be bent into a circular "orbit," similar to that of a planet.

Figure 8.19 shows a still denser body. The circular light orbit is now outside the body. Consider light rays that leave the surface of the body. Those that leave in an approximately radial direction are deflected as

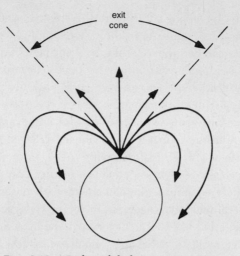

Fig. 8.19. Paths of light rays emanating from the surface of a very dense body that is nearly a black hole. Only the rays whose directions lie within the exit cone escape; the others are attracted back by the gravity of the source. If the body is dense enough to be a black hole, the exit cone disappears and no light can escape. From Harrison, *Cosmology: The Science of the Universe*, copyright © 1981 by Cambridge University Press. Reprinted with the permission of Cambridge University Press.

shown. At some angle the deflection is so great that the light ray "falls" back into the source. Only rays emitted in a cone around the radial direction escape; this is known as the *exit cone*.

As the density increases, the size of the exit cone decreases. At some critical density the exit cone entirely disappears: gravity is so strong that no light whatever can escape. An object with this property is called a black hole; the term was coined by Wheeler.

All the properties that we have inferred heuristically are confirmed by exact solutions of the field equations. The possible existence of black holes is a definite prediction of general relativity.

If an object of mass M and uniform density is to be a black hole, its radius must be less than

$$R = 2GM/c^2 \qquad (8.23)$$

This quantity is called the Schwarzschild radius, R_s. It is obtained from the Schwarzschild solution mentioned above.

The value of R_s for $M = 1$ solar mass is about 3 km; the sun would have to have a radius of 3 km to be a black hole. For a typical galaxy ($M \approx 10^{11}$ solar masses), R_s is about .01 light-year.

The gravitational red shift for light emitted at the surface of a body that is nearly a black hole is very large. As the radius approaches R_s, the red shift goes to infinity.

If a black holes exist, can they be observed? By definition, we cannot see a black hole inasmuch as no light can escape from its surface. The presence of a black hole is manifested by its very strong gravitational field, which acts on the surrounding matter; external electromagnetic fields could also exist.

The most plausible mechanism for the creation of a black hole is gravitational collapse. According to conventional theories of stellar evolution, stars contract as they burn their fuel, becoming more dense. The contraction generally comes to an end long before the star reaches black hole density, but under the right conditions the contraction can continue and can eventually lead to the formation of a black hole.

Another possibility is the gravitational collapse of an object of quasi-galactic mass, for example, a large star cluster. It has been speculated that such objects exist at the centers of active galaxies or quasars, perhaps even at the center of our own galaxy. According to equation (8.23), the radius of a black hole is proportional to its mass. The density, proportional to M/R^3, therefore decreases with increasing mass. A black hole of galactic mass is much less dense than one of stellar mass.

The search for black holes is one of the most active areas of current astrophysical research. The most promising candidates are members of binary systems. If one star in a binary is a black hole, it will accrete matter from its companion. The matter will radiate copiously as it falls in; much of the radiation is in the form of X rays. Several possible black holes have been identified among strong X-ray sources; the strongest candidate is the source Cygnus X-1.

APPENDIX: CURVATURE OF A SPHERICAL SPACE

We give here a proof that the curvature of a sphere defined by equation (8.8) is $1/R^2$ where R is the radius. The figure below shows a circle of radius r on a sphere of radius R. (r is measured on the surface of the sphere, of course.)

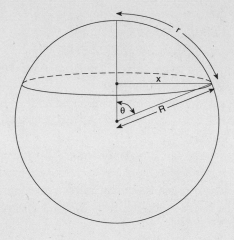

Let θ be the angle subtended at the center of the sphere by the length r. The relation between r and θ is

$$r = R\theta \qquad (8.A1)$$

where θ is measured in radians. If we construct the plane of the circle, we see that its radius in the plane is

$$x = R \sin \theta \qquad (8.A2)$$

and the circumference is

$$C_r = 2\pi x = 2\pi R \sin \theta = 2\pi r \frac{\sin \theta}{\theta} \qquad (8.A3)$$

where (8.A1) has been used to obtain the last equality.

For small θ, we can use the power series expansion for the sine function:

$$\sin \theta = \theta - \frac{1}{6}\theta^3 + \dots \qquad (8.A4)$$

and keep only the first two terms. Substituting this for $\sin \theta$ in equation (8.A3), we get

$$C_r = 2\pi r \left(1 - \frac{1}{6}\theta^2 + \dots\right)$$
$$= 2\pi r \left(1 - \frac{1}{6}\frac{r^2}{R^2} + \dots\right) \qquad (8.A5)$$

Hence

$$C_r - 2\pi r = -\frac{\pi}{3}\frac{r^3}{R^2} + \ldots$$

and

$$K = \lim_{r \to 0} \frac{3(2\pi r - C_r)}{\pi r^3} = \frac{1}{R^2}$$

as claimed.

PROBLEMS

8.1. A pendulum bob hangs vertically from the roof of a railroad car that is at rest.

(a) Use the principle of equivalence to deduce what happens to the pendulum when the car accelerates in the $+x$ direction. Draw a sketch.

(b) What happens to the pendulum when the car moves at uniform velocity? What law did you use to answer this question?

8.2. Suppose the ratio of inertial mass to gravitational mass were greater for lead than for iron. Would the period of a pendulum with a lead bob be longer or shorter than that of a pendulum of equal length with an iron bob? Explain.

8.3. Identical clocks are carried on airliners that circle the globe in easterly and westerly directions at the equator. The airliners fly at an altitude of 12 km at a speed of 0.25 km/sec. On their return the readings of the airliner clocks are compared with one another and with the reading of a clock that remained at the starting point. The rotational speed of the earth at the equator is about 0.5 km/sec.

(a) Find the differences between the readings of the airliner clocks and the ground clock at the end of the journey, attributable to special relativity. (Hints: All three clocks are traveling on circular paths, at different velocities relative to a hypothetical clock that remains stationary at the starting point. We can neglect the orbital motion of the earth around the sun.) Imagine the circular paths as being straightened out, and calculate the clock readings in the rest frame of the "stationary" clock.

(b) Find the differences in clock readings attributable to gravitational time dilation. Combine with the results of (a) to obtain the net differences in clock readings.

8.4. The mass of the sun is 2×10^{30} kg and its radius is 7×10^8 m. Calculate the fractional gravitational red shift of a spectral line from the sun detected on earth.

8.5. Consider the following three possible motions:

(a) body starts at $x = 0$ and moves with constant speed $0.2c$ for 10 sec;

(b) body starts at $x = 0$, moves with constant speed $0.4c$ for 5 sec and remains stopped for the next 5 sec;

(c) body starts at $x = 0$, moves with constant speed $0.6c$ for 5 sec and then moves in the opposite direction at $0.2c$ for the next 5 sec.

In each motion the body's position at $t = 10$ sec is $x = 2$ light-seconds.

Show all three motions on a space-time diagram. Using the space-time metric (8.20), calculate the four-dimensional "distance" in light-seconds between the body's initial and final positions for each path. Verify that the distance for the geodesic path (a) is the longest.

9 Cosmology

9.1. Basic Facts: The Cosmological Principle

Cosmology is concerned with the nature and the history of the universe. Although men and women have pondered these questions for millennia, early cosmologies were little more than myths. Scientific cosmology began with the Greeks some 2,500 years ago but was for a long time largely speculative. Only in the present century have observational data become available that bear directly on the questions posed by cosmology. The evidence points strongly toward the world picture currently in favor—the "big bang" and expanding universe. Einstein's conception of curved space-time, described in chapter 8, provides the framework for all modern cosmological models.

Stars and Galaxies

One of the oldest cosmological problems is the nature of the stars. The notion that they are distant suns goes back to antiquity. If that is so, their small apparent size and low apparent brightness indicate that they must be very far away. Newton estimated that the stars must be at least a hundred thousand times as distant as the sun. His estimate turned out to be quite accurate: the nearest star, Alpha Centauri, is 4.3 light-years away, about 270,000 times the earth-sun distance.

A prominent feature of the night sky is the Milky Way, or Galaxy, long recognized as being a densely packed collection of stars. It contains some hundred billion stars, a few thousand of which are visible to the naked eye. The Galaxy has the form of a disk, about 100,000 light-years in diameter and 5,000 light-years in thickness. Our sun sits toward the outside of the disk, some 30,000 light-years from the galactic center.

The key to modern cosmology lies in the identity of the "nebulae," fuzzy patches of light a few of which are visible to the naked eye. Others are so faint that they can be seen only through the most powerful telescopes. By 1780, over one hundred nebulae had been cataloged by Charles Messier. Many exhibited a characteristic spiral structure and were called "spiral nebulae."

A nebula might be a diffuse cloud of weakly shining gas within the Galaxy; some of them turned out to be just that. In 1755, however, Immanuel Kant suggested that most of the nebulae are in fact "island universes" of stars—galaxies similar to the Milky Way but so distant that the star images merge into a continuum (at least when viewed through the low-resolution telescopes available at the time).

The island-universe hypothesis gained increasing acceptance during the nineteenth century. It was finally confirmed in 1923 when Edwin P. Hubble, using the recently completed 100-inch telescope on Mount Wilson, resolved the outer regions of M31 (the Andromeda nebula) and of M33 into what he described as "dense swarms of images which in no way differ from those of ordinary stars" (except for being much fainter).[1]

The mean distance between galaxies is large compared to their dimensions: galaxies occupy a very small fraction of the universe. They are not uniformly distributed; most are found in clusters that consist of as few as ten galaxies or as many as a thousand. The Milky Way belongs to a small cluster of about twenty called the "local group." Each cluster is held together by the gravitational attraction of its members.

Most clusters of galaxies are themselves grouped into larger aggregations called superclusters; the local group is part of a supercluster centered in the constellation Virgo. Some cosmologists have speculated that this process continues without end, that is, there exist clusters of superclusters, and so on ad infinitum. Such an arrangement, called a hierarchical universe, would have interesting implications; for one, the average density of matter would decrease with each level of clustering and could ultimately approach zero. However, no evidence of clustering beyond the scale of superclusters has been found.

Distance Indicators

The distances of nearby stars are determined by the method of trigonometric parallax, which is based on the change in a star's apparent direction

---

1. E. P. Hubble, "Cepheids in Spiral Nebulae," *Observatory* 48 (1925):140. The notation M31 refers to the number in Messier's catalog.

as the earth traverses its orbit around the sun. The parallax angle measures the ratio between the radius of the earth's orbit and the distance to the star. This method can be applied only to stars within about 100 light-years of earth; at that distance the parallax angle is 0.03 seconds, about the smallest that can be reliably measured with existing telescopes.

For more distant objects astronomers have to resort to less direct methods, which make use of so-called standard candles or distance indicators. The intensity of the light from a given source falls off with the square of the distance from the source. Hence one can calculate the ratio of the distances of two identical sources by comparing their apparent brightnesses. Suppose, for example, we observe two stars whose characteristics are identical and find that one looks a hundred times brighter than the other. The dimmer one must be ten times as distant. If we know the distance of the brighter one, we can infer the distance of the dimmer one.[2]

A key role in the determination of the extragalactic distance scale was played by Cepheid variables, stars whose brightness oscillates regularly with a period typically between three and fifty days. Study of Cepheids in the Small Magellanic Cloud by Henrietta Leavitt disclosed a strong correlation between their periods and their apparent brightnesses: the longer the period, the brighter the star looks.

Since all the stars in the Magellanic Cloud are at (nearly) the same distance, the apparent brightness of any star in the Cloud is a measure of its absolute luminosity. Hence the measured period of a Cepheid variable determines its absolute luminosity; they make excellent standard candles.[3]

Hubble identified some of the stars he had observed in M31 and M33 as being Cepheid variables. Using the Cepheids as distance indicators, he concluded that M31 and M33 are nearly a million light-years away. The discovery of external galaxies thus extended the distance scale of the observable universe by more than an order of magnitude.

2. The inverse-square dependence of light intensity on distance holds only in static Euclidian space; the argument must be modified when applied to very distant sources in an expanding universe. (See the discussion in secs. 9.2 and 9.7.) Absorption of light by interstellar gas or dust can also modify the variation of intensity with distance.

3. To calibrate the period-luminosity relation for Cepheids, one has to know the absolute luminosity of at least one; this requires knowing its distance. Unfortunately, no Cepheid is close enough to earth to allow a reliable distance determination by means of trigonometric parallax. The calibration was first carried out by Ejnar Hertzsprung in 1913, using thirteen galactic Cepheids. Although each one has a barely detectable parallax, by averaging the data Hertzsprung was able to obtain a fairly good calibration.

To be useful as a distance indicator, an object must be recognizable as belonging to a class whose members all have nearly the same absolute luminosity. Cepheids make the most reliable indicators, but they have been identified only in galaxies within about 20 million light-years of earth. For more distant galaxies, astronomers must fall back on indicators that are intrinsically more luminous and can therefore be identified at greater distances. H II regions (clouds of ionized hydrogen surrounding a bright O or B star), globular clusters of stars, supernovas, and the brightest star in a galaxy are all used as indicators.

At distances greater than about 80 million light-years, no distinct features within a galaxy can be resolved and the determination of distance becomes even more problematic. The best one can do is to use the brightest galaxy in a cluster as a standard candle. (Studies of nearby clusters show that although individual galaxies vary in brightness a great deal, the brightest galaxy in a cluster generally has the same absolute luminosity within a factor of about two.) With this technique, distances up to 3 billion light-years have been assigned. Such estimates are subject to considerable uncertainty, but no better way of estimating such great distances is available.

For cosmological purposes, the more distant an object, the more interesting it is; by observing a very faraway source we are probing the universe deeply not only in space but also in time: the light we receive from the source today was emitted a very long time ago. If the universe evolves, the very distant sources provide direct information about conditions during its early history.

Cosmological Principle

In ancient cosmologies the earth was the center of the universe. That notion was challenged by some Greek thinkers and was finally abandoned with the acceptance of the Copernican system, which relegates the earth to an inconspicuous spot in the solar system. The logical extension of Copernicus's idea is that neither the sun nor our galaxy nor any other object occupies a preferred place in the universe: apart from local irregularities, the universe is homogeneous and isotropic.[4] To observers in another galaxy, the heavens look about the same as they do to us; the average density of matter is the same everywhere, and the sky looks more or less the same in all directions. This set of assumptions, which underlies nearly all modern cosmologies, is called the cosmological principle.

4. It can be shown that isotropy about every point implies homogeneity.

The cosmological principle seems plausible and is weakly supported by observational evidence such as number counts of galaxies; the portion of the universe we can observe appears to be roughly homogeneous and isotropic. Most cosmologists adopt the cosmological principle as a working hypothesis, not necessarily because they are convinced it is correct but because it restricts the range of possible models to a manageable number. The observational data are quite limited; even with a cosmological principle, as we shall see, it is difficult to discriminate among the models. If nonhomogeneous or nonisotropic universes were included, the number of possible models would be vastly increased and the task of choosing among them would become even more daunting.

On a local scale the cosmological principle is obviously not valid. The sky looks very different in the direction of the Milky Way than in other directions. The principle applies only on the cosmological scale, that is, to regions large enough to contain many clusters of galaxies. Two such regions, of equal size, should contain approximately equal numbers of galaxies no matter where they are located. The diameter of the region must be of the order of 10^8 to 10^9 light-years.

A stronger statement is the perfect cosmological principle (PCP), which asserts that the universe is homogeneous not only in space but in time as well. According to the PCP, the universe has always looked the same as it does today. This assumption leads to the steady-state cosmology, which was in favor for some time; as we shall see, there is now strong evidence against it.

9.2. Hubble's Law and the Expansion of the Universe

The light from a star or a galaxy contains spectral lines of common elements—hydrogen, calcium, and so on—with the wavelength of every line shifted by the same fraction from the value measured in the laboratory. Astronomers use the symbol z to denote the fractional change in wavelength:

$$z = \frac{\Delta\lambda}{\lambda_0} = \frac{\lambda - \lambda_0}{\lambda_0} \qquad \text{or} \qquad 1 + z = \frac{\lambda}{\lambda_0} = \frac{f_0}{f} \qquad (9.1)$$

Here λ is the measured wavelength and λ_0 is the laboratory wavelength of the same line (f and f_0 are the corresponding frequencies). z is called the red shift because for positive z a line in the visible part of the spectrum is shifted toward the red. When z is negative, the spectrum is said to be blue-shifted.

In section 4.8 we discussed the Doppler effect, the change in frequency measured by an observer in motion relative to the source of radiation. When v/c is small, the fractional shift in wavelength is the same as the fractional shift in frequency and is given by equation (4.40):[5]

$$\frac{\Delta\lambda}{\lambda_0} = \pm\frac{v}{c} \tag{9.2}$$

where v is the radial velocity of the source relative to the observer. A receding source gives rise to a red shift, and an approaching source gives rise to a blue shift.

The observed wavelength shifts in stellar spectra are routinely interpreted as being Doppler shifts. According to equations (9.1) and (9.2), the value of z gives the radial velocity of the star through the relation

$$v = cz \tag{9.3}$$

A typical value of z for a star is 0.001, which corresponds to a radial velocity 300 km/sec.

By 1923, the spectra of 41 galaxies had been analyzed. Surprisingly, in 36 cases the spectrum was found to be red-shifted while only five galaxies exhibited blue shifts (all of them very small). According to the Doppler interpretation, the preponderance of red shifts implies that most of the galaxies are moving away from us. If the motions were random one would expect to observe approximately equal numbers of red and blue shifts, as is the case for stars within our galaxy.

In 1929, Hubble plotted red shift versus distance for 18 galaxies and discovered the law that bears his name and that is the key to modern cosmology: the red shift is proportional to distance. All the galaxies in Hubble's original plot are fairly close by and have small red shifts; the most distant ones are members of the Virgo cluster, with z about 0.03. From the data, which exhibited considerable scatter, Hubble optimistically inferred a linear relation between red shift and distance. Later measurements extended the sample to more distant galaxies and confirmed that the relation is indeed linear up to z around 0.2.

According to equation (9.3), the recession velocity is also proportional to distance, provided the velocity is low enough to justify the use of the first-order Doppler formula (9.2).

---

5. The signs are of course opposite. If the frequency decreases, the wavelength increases.

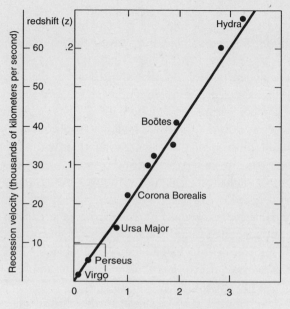

Fig. 9.1. Modern version of the Hubble plot. The gal-axies used by Hubble in his original plot are all in the box at the lower left. There are two vertical scales, one labeled by red shift and the other by recession velocity. The red shift is what is actually measured. The reces-sion velocity is inferred from the Doppler formula. (See discussion in the text.) From Harrison, *Cosmology: The Science of the Universe,* copyright © 1981 by Cambridge University Press. Reprinted with the per-mission of Cambridge University Press.

Figure 9.1 is a modern version of the Hubble diagram; the sample on which Hubble based his original conclusion is all found in the box at the lower left corner of the diagram. The ordinate axis is marked in units both of z and of v; keep in mind, however, that the red shift is the measured quantity.

The slope of the Hubble diagram (with velocity as ordinate) is known as Hubble's constant and is assigned the symbol H. Hubble's law thus takes the form either of

$$v = Hd \tag{9.4}$$

or, using relation (9.3),

$$z = (H/c)d \qquad (9.5)$$

I emphasize that equation (9.5), not (9.4), is the true statement of Hubble's law.

H has the dimensions of velocity divided by distance, or reciprocal time. From his original data, Hubble found the value of the constant to be about 150 km per second per million light-years. Subsequent study showed, however, that the galactic Cepheid variables used by Hubble as distance indicators are actually farther away than had at first been believed; as a result, the extragalactic distance scale had to be recalibrated and the value of H changed. The new value was substantially lower than the one first deduced by Hubble.

There remains some controversy among the experts as to the correct value of H; the suggested values range between 15 and 30 km/sec/million light-years.[6] For this discussion, I adopt a nominal value of 20.

Figure 9.2a shows the recession of distant galaxies implied by Hubble's law.[7] The picture seems to violate the cosmological principle since it is symmetric about point A, the location of the earth. We appear to be in a special position. This, however, is only an illusion; it is easy to show that if one plots the velocities relative to any other point, say, B, the picture looks exactly the same.

In Figure 9.2a, I have indicated the positions of point B and of an arbitrary galaxy G, and their velocities $\mathbf{v}_B$ and $\mathbf{v}$. According to equation (9.4),

$$\mathbf{v}_B = H\mathbf{d}_B \qquad (9.6a)$$

and

$$\mathbf{v} = H\mathbf{d} \qquad (9.6b)$$

(Vector notation is used to emphasize that each velocity points in the same direction as the corresponding displacement.)

6. The constant is generally quoted in these unusual units because it allows immediate conversion of distance to recession velocity. For example, a source 100 million light-years away is receding at 2,000 km/sec. In more conventional units, the value of H is 2×10^{-18} sec^{-1} or 6×10^{-11} yr^{-1}. See Michael Rowan-Robinson, *The Cosmological Distance Ladder* (New York: W. H. Freeman, 1985) for detailed discussion of the determination of Hubble's constant.
7. In the Doppler picture, the value of z determines only the radial component of velocity; each galaxy has, in addition, an unknown velocity perpendicular to the line of sight. Fig. 9.2a is based on the assumption that the transverse velocities are negligible. In the expanding universe model, the recession is in fact purely radial.

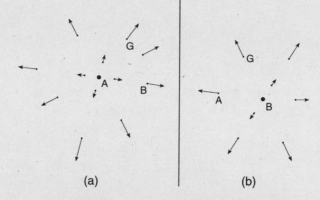

Fig. 9.2. Recession of galaxies inferred from Hubble's Law. (a) Galaxies are all receding from earth (point A) at speeds proportional to their distances. (b) When the motion relative to an arbitrary point B is calculated, it likewise represents a recession proportional to the distance from B. Thus the model is consistent with the cosmological principle. (G is an arbitrary galaxy.)

Figure 9.2b shows the motions of all the galaxies as they look to an observer at B. The displacement of G relative to B, labeled $\mathbf{d'}$, is

$$\mathbf{d'} = \mathbf{d} - \mathbf{d}_B \qquad (9.7\text{a})$$

while the velocity of G relative to B, labeled $\mathbf{v'}$, is

$$\mathbf{v'} = \mathbf{v} - \mathbf{v}_B \qquad (9.7\text{b})$$

From equations (9.6) and (9.7), we obtain

$$\mathbf{v'} = H(\mathbf{d} - \mathbf{d}_B) = H\mathbf{d'} \qquad (9.8)$$

which is the same as equation (9.4) referred to an origin at B. Figures 9.2a and 9.2b look the same. The radial recession picture is thus fully consistent with the cosmological principle.

The Expanding Universe. Hubble's law was originally interpreted as implying that all the galaxies are rushing away from us. An alternative interpretation, first proposed by Georges Lemaitre and Howard Robertson and now universally accepted, is that *space itself* is expanding. Hubble's law is a direct consequence of the expansion, as I shall show.

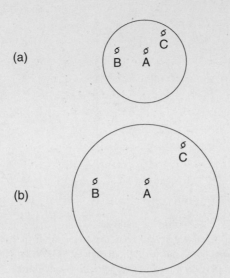

Fig. 9.3. An expanding two-dimensional universe. *A, B,* and *C* represent galaxies. Each one remains fixed on the surface of the sphere. No matter which galaxy is taken as a center, all the others appear to recede at velocities proportional to their distances. This model accounts for Hubble's Law.

To elucidate the idea of the expanding universe, let us return to the two-dimensional analogy discussed in chapter 8, in which space is a surface. The expansion of a surface is readily visualized; for concreteness let the surface be a sphere.

Figure 9.3a shows a spherical universe with several galaxies on its surface. Suppose the sphere expands but the position of each galaxy on the sphere remains unchanged. Figure 9.3b shows the situation a short time afterward. It is apparent that the distance between any two galaxies has increased by an amount proportional to their initial separation. If we interpret the rate of increase of the separation as a recessional velocity, that velocity is proportional to the separation, precisely as in equation (9.4).

When generalized to three-dimensional space, figure 9.3 is the model for the expanding universe. According to this picture *the galaxies are not actually moving;* their measured separation increases because of the expansion of space. If that is the case the red shifts cannot be true Doppler shifts, although they were so interpreted at first and continue to be so described in many textbooks.[8] How then are the red shifts accounted for?

It is convenient to define a "co-moving" coordinate system, which follows the expansion of space; on an expanding sphere, for example, latitude and longitude are a satisfactory set of co-moving coordinates. The co-moving coordinates of each galaxy remain constant as space expands.

We further define a time-dependent dimensionless parameter $R(t)$ called the *scale factor*. R increases as the universe expands but at any given time has the same value everywhere in space.[9] Its present value is denoted by $R(t_0)$ or simply R_0 (t_0 is the present time). Each cosmological model specifies how R varies with time.

As space expands, *all measured distances between co-moving points (points that are stationary with respect to co-moving coordinates) increase in direct proportion to R.* This fundamental relation can be written in the form

$$\frac{d(t)}{d_0} = \frac{R(t)}{R_0} \qquad (9.9)$$

where $d(t)$ is the measured distance at time t and d_0 is the present measured distance.

Quantities such as the radius of an atom or of a galaxy, which are determined by local physical laws, do *not* increase as the universe expands. (If all lengths, including those of metersticks, increased in the same proportion, there would be no observable consequences.) In figure 9.3 the galaxies can be represented by small stickers of constant size, affixed to the surface of the expanding sphere. This property plays an important role in the early history of the universe, as we shall see in section 9.8.

To illustrate what happens to the wavelength of a light wave in expanding space, imagine two athletes running at the same speed in the same lane of a track. If the track expands, the separation between the runners increases by the same fraction as the length of the track. Replacing the runners by two successive crests of a light wave, whose separation

8. For small values of z, the expansion picture and the Doppler picture turn out to be equivalent, but the distinction between them is crucial for cosmology. See Edward R. Harrison, *Cosmology: The Science of the Universe* (Cambridge: Cambridge University Press, 1981), 236–240, for a clear discussion.
9. R is sometimes referred to as the "radius of the universe." This nomenclature is somewhat misleading, however. In a spherical universe, the radius is indeed proportional to the scale factor, but a scale factor can be defined as well in a flat universe, which has no radius.

is one wavelength, we conclude that the wavelength increases in proportion to the scale factor. This argument is somewhat heuristic, but the result, equation 9.10 below, follows rigorously from the mathematical models.

If a light wave was emitted at time t_e, when the scale factor had the value $R(t_e)$, and is received now when it has the value R_0, the wavelength is multiplied by the factor $R_0/R(t_e)$. (In an expanding universe $R(t_e)$ is less than R_0.) Hence the formula for the cosmological (expansion) red shift is

$$1 + z = R_0/R(t_e) \qquad (9.10)$$

The red shift of a galaxy is a measure of how much the universe has expanded between the time light was emitted by the galaxy and the time the light reaches us.[10] The shape of the Hubble plot therefore implicitly determines how the universe has been expanding ever since the light from the most distant galaxy was emitted.

Although the galaxies are not moving, one can nonetheless define a recession velocity v_r for a galaxy as the rate of change of its measured distance d; call that quantity $\dot{d}$, and let $\dot{R}$ similarly denote the rate of change of the scale factor R. The ratio $\dot{R}/R$ is then the fractional rate of expansion of space. Since by hypothesis, d is proportional to R, the fractional rate of change of d must be the same as that of R; that is to say,

$$\frac{\dot{d}}{d} = \frac{\dot{R}}{R} \qquad (9.11)$$

Equation (9.11) implies that the recession velocity is

$$v_r = \dot{d} = \left(\frac{\dot{R}}{R}\right)d \qquad (9.12)$$

which is the same as equation (9.4) if we identify the Hubble constant as

$$H = \dot{R}/R \qquad (9.13)$$

According to equation (9.13), H measures the fractional rate of expansion of the universe. As this relation indicates, the Hubble "constant" is really not constant but varies with time, depending on how the expansion

10. Any true relative motion contributes a Doppler shift that should be subtracted to obtain the expansion red shift. The Doppler shift can be in either direction but is small compared to the expansion red shift except for very close galaxies. One source of proper motion is known: because of galactic rotation, the solar system has a velocity about 300 km/sec relative to the galactic center. After this velocity has been subtracted, only one nearby galaxy (NGC 253) exhibits a blue shift.

proceeds. Even if $\dot{R}$ were constant, H would decrease as R increases.[11] (In most models, as we shall see, the expansion slows down, which makes H decrease even faster.) We use H_0 to denote the present value of H. The measured value of H_0 implies that the universe is now expanding by about one part in 15 billion each year.

The expansion of the universe is described by specifying how the scale factor R varies with time. Over a short enough time period, the curve of R vs. t can be approximated by a straight line. The slope of that line is the rate of change $\dot{R}$, which according to equation (9.13) is equal to HR. A straight line fitted to the present rate of expansion has the slope H_0R_0. Its equation can be written in the form

$$R(t) = R_0[1 + H_0(t - t_0)] \qquad (9.14)$$

Equation (9.14) applies only for times close to t_0.

We can now show that the expanding universe picture leads to Hubble's law, provided the emission time t_e is close enough to t_0 to justify the use of the linear approximation (9.14). Putting $t = t_e$ in (9.14) and substituting in equation (9.10), we obtain

$$1 + z = \frac{1}{1 + H_0(t_e - t_0)}$$
$$\approx 1 + H_0(t_0 - t_e) \qquad (9.15)$$

To get the last form of the equation, we used the approximation $1/(1 + x) \approx 1 - x$, which is justified when x is small.

The quantity $t_0 - t_e$ is the look-back time—the time required for light to travel from the source to us. A rigorous calculation of that time must take into account the expansion of space while the light is in transit, as well as the curvature of space. The result is model-dependent. In first approximation, however, we can simply put

$$t_0 - t_e = d/c \qquad (9.16)$$

where d is the distance of the source. (To first order, it is unnecessary to distinguish between the present distance and the distance at the time of emission.)

With the substitution (9.16), equation (9.15) becomes Hubble's law, equation (9.5). As our derivation indicates, the linear relation between red shift and distance is valid only for small z.

---

11. If $\dot{R}$ is proportional to R, H is a true constant. This occurs only in the steady-state universe, described in section 9.6.

I emphasize again that the Doppler and expansion interpretations are not equivalent ways of explaining the observed red shifts but are fundamentally different. The difference has consequences that are in principle observable (though not for small values of z).

In the Doppler picture, velocity and red shift are directly related. The observed linear dependence of red shift on distance, equation (9.5), translates into a relation between velocity and distance, but that relation is linear only in first approximation. The exact relativistic Doppler formula (4.38), written in terms of z, is

$$1 + z = \left(\frac{1 + v/c}{1 - v/c}\right)^{1/2} \tag{9.17}$$

Combining equation (9.17) with (9.5), we obtain

$$\left(\frac{1 + v/c}{1 - v/c}\right)^{1/2} - 1 = \frac{H}{c}d \tag{9.18}$$

If the red shifts were due to Doppler effect, the precise velocity-distance relation would be given by equation (9.18). For small values of v/c this reduces to the linear relation (9.4), but a plot of velocity vs. distance would depart noticeably from linearity for v/c more than about 0.1. In figure 9.1 the velocity scale would have to be revised at the high end. For $z = 0.2$, for example, the exact relation (9.17) gives $v/c = 0.18$; the approximate form (9.3) is more than 10 percent high.

In the expansion picture, in contrast, the linear velocity-distance relation (9.12) is valid for *all* values of v_r: it simply expresses the fact that at any given time the entire universe is expanding at the same fractional rate, as required by the cosmological principle. A plot of recession velocity against distance is therefore necessarily linear. The observed dependence of red shift on distance leads to a derived relation between recession velocity and red shift.

Notice that although the Doppler velocity defined by equation (9.18) is always less than c, the expansion velocity defined by equation (9.12) clearly exceeds c at great enough distances. Special relativity is not violated because no object is actually traveling at that speed; the apparent recession "velocity" is simply a manifestation of the expansion of space.

One other conceptual difference between the two pictures should be noted. In an expanding universe, the red shift of a given source depends on the entire history of the expansion between the emission of the light and its arrival at the observer. The Doppler effect depends only on the

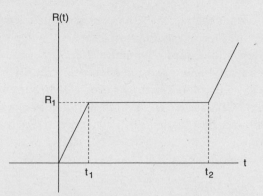

Fig. 9.4. "Dachshund" universe used to illustrate the difference between the expansion and the Doppler interpretations of cosmological red shifts.

relative velocity between the source at the time of emission and the observer at the time of reception. The motion of the source after it has emitted the light is irrelevant.

Figure 9.4 illustrates how this difference can affect observations. The expansion of this hypothetical universe halts at time t_1 and resumes at t_2; between t_1 and t_2 the scale factor has the constant value R_1.[12] Light emitted by a source during the resting stage and received on Earth after t_2 is detected as red-shifted by the factor R_0/R_1. If, instead, the same curve represented the position of a moving galaxy, the source would be at rest between t_1 and t_2. Light emitted during that period would be detected with no Doppler shift at all.

Cosmological Distances. When discussing a distant source, we must distinguish between the source's present distance, d_0, and its distance when it emitted the light we receive now, d_e. The relation between the two distances is

$$d_0/d_e = R_0/R_e = 1 + z \qquad (9.19)$$

In any expanding universe d_0 is greater than d_e; the difference is critical for the discussion of high-z sources. Observation provides information only on d_e; we know nothing about the motion of the source after it emitted the light that is now reaching us. d_0 is just a convenient theoretical

12. A model with such behavior is the Lemaitre universe, sometimes whimsically referred to as the "dachshund" universe.

number that tells us the present distance of the source if it has continued to follow the expansion of the universe. That assumption could be wrong; the source might have exploded a minute after t_e.

One can similarly define two recession velocities for each source: its present velocity, v_0, and its velocity at time t_e, when it emitted the light we now detect. Call the latter v_e. Each velocity is related to the corresponding distance by equation (9.4), evaluated at the appropriate time. That is,

$$v_0(z) = H_0 d_0(z) \tag{9.20a}$$

and

$$v_e(z) = H(t_e)\, d_e(z) \tag{9.20b}$$

Both velocities are model-dependent.

Luminosity Distance. Both d_0 and d_e are so-called proper distances; they are distances that would be measured on metersticks held by a chain of observers at the same instant of cosmic time. This, of course, is not a practical method of measuring the distance of a galaxy. Astronomers therefore define still another distance, called the luminosity distance.

As noted earlier, the distance of a faraway cluster of galaxies is deduced from the apparent brightness of its brightest member, under the assumptions that (i) the absolute luminosity of the brightest galaxy is the same for all clusters and (ii) the apparent brightness of a source diminishes as the square of the distance between source and observer. If proper distance is used, the second assumption is valid only in a static universe with Euclidian geometry. The luminosity distance d_L is defined by *postulating* an inverse-square variation. It differs from both d_0 and d_e.

The effect of expansion is not hard to calculate for Euclidian space. The luminosity of a source, L, is the radiant energy per unit time it emits. If the source were at rest at a distance d in a nonexpanding universe, the energy would be distributed over a sphere whose area is $4\pi d^2$. Hence the flux density (energy/unit area/unit time) reaching the observer's telescope would be

$$F = L/4\pi d^2 \qquad \text{(in static Euclidian space)} \tag{9.21}$$

The expansion of space affects the intensity in two ways. The effect is more easily described in terms of the photon picture of light, although a wave description must lead to the same result. First, each photon is redshifted. Since the energy of a photon is $hf = hc/\lambda$, the red shift reduces the energy by the factor $1+z$. In addition, the number of photons per second that reach the observer is reduced by the same factor $1+z$.[13] Hence the flux density at the telescope is smaller by a factor $(1+z)^2$ than it would be if the source were at the same distance in a nonexpanding space. We have instead of (9.21):

$$F = L/4\pi d^2 \, (1+z)^2 \qquad \text{(in expanding Euclidian space)} \quad (9.22)$$

The luminosity distance d_L is defined as the distance at which the flux density would have the value given by equation (9.22) if the apparent brightness obeyed an inverse-square law. That is,

$$F = L/4\pi d_L{}^2 \qquad (9.23)$$

Equations (9.22) and (9.23) yield the desired result:

$$d_L = d_0(1+z) = d_e(1+z)^2 \qquad (9.24)$$

The red shift makes the source look dimmer and therefore appear to be more distant than it actually is. In curved space, the relation between d_L and d_0 contains the same factor $1+z$ caused by expansion, multiplied by another z-dependent factor due to the non-Euclidian geometry.

The Distances of High-z Objects. The Hubble relation (9.5) can be used to assign a distance to any source for which a red shift has been measured and no other distance determination is possible, for example, a distant galaxy that cannot be identified with any cluster.

The nature of quasars—compact objects with faint starlike images and with high red shifts (some as high as 3)—is of great interest. If their red shifts are cosmological, high-z quasars are the most distant objects yet observed as well as the most luminous. To calculate a cosmological distance for such a source, one has to extend the red shift–distance relation far beyond the range in which it has been established observationally. As we shall see, each cosmological model predicts a unique form for the relation. For high values of z, the relations are quite different. Distance estimates for high-z quasars are therefore model-dependent.[14]

13. Imagine that N photons per second are emitted by the source at a steady rate. The spacing between successive photons is then c/N. The expansion of space causes the spacing to increase by the factor $1+z$ by the time the photons reach the observer. The arrival rate is correspondingly less.
14. Straightforward linear extrapolation of eq. (9.5) would yield a distance of 45 billion light-years for a quasar with $z=3$. The distances given by most models are substantially lower.

To be sure, quasar red shifts might not be cosmological. They might be gravitational, as noted in chapter 8, or they might even be Doppler shifts caused by true motion. In the latter case, the velocity would be very close to c. A quasar might be the result of a massive explosion that sends matter hurtling off at relativistic speeds.

If either of the alternative interpretations is correct, the red shifts of quasars tell us nothing about their distances. They could even be relatively nearby. Most experts, however, favor a cosmological origin. Several cases of quasars associated with clusters of galaxies have been discovered recently. In each case, the quasar has nearly the same red shift as the galaxies; these observations provide strong support for the cosmological interpretation.

9.3. THE BIG BANG

The Hubble constant tells us the rate at which the universe is expanding today. It is natural to inquire what happened in the past and what will happen in the future. Cosmological models provide answers to these questions; each model specifies how the scale factor R varies with time. Some model universes are discussed in section 9.5.

A feature common to many of the models is that the scale factor was zero at some time in the past. Since the measured distance between any two galaxies is proportional to R, all such distances must have been zero when R was zero; the density of the universe was infinite at that instant. This highly singular state is called the "big bang"; the term was coined by Fred Hoyle. In any big bang cosmology, the time since the big bang can be considered the age of the universe.

Our two-dimensional spherical analogy is helpful in picturing the big bang. As one looks back in time, the universe consists of smaller and smaller spheres, approaching a single point at the time of the big bang. The same thing happens in a three-dimensional universe whose space is spherical, although it cannot be visualized.

We must resist the temptation to picture all the matter in the early stages of a big bang spherical universe as being concentrated in a tiny sphere surrounded by vacuum. Matter always fills all of space, but *space itself* consists of a tiny sphere at the beginning. In a sense the term "big bang" is misleading because it suggests an explosive event. An explosion is driven by a strong outward pressure gradient, whereas in any model universe the pressure is always the same everywhere, in accord with the cosmological principle. The big bang was not an explosion.

Another common misconception is that the volume of the universe was necessarily zero at the time of the big bang. That is true only for "closed" universes. A universe whose space is Euclidian, for example, always has infinite volume even though the distance between any two galaxies approaches zero and the density approaches infinity as we approach the big bang.

Figure 9.5 shows the scale factor as a function of time for several generic universes; curves (a), (b), and (c) describe big bang universes.

In model (a) the expansion has always proceeded at the same rate as today: the curve of R versus time is a straight line. From equation (9.14) we find that $R = 0$ at $t = t_0 - 1/H_0$. The age of a uniformly expanding universe is therefore $1/H_0$, which is called the Hubble time and is denoted by t_H. With the currently accepted value of H_0, the Hubble time is about 15 billion years. The age of any big bang universe is of this order of magnitude.

Curve (b) illustrates a universe with a decreasing rate of expansion (deceleration). Most cosmological models are of this type; the expansion is slowed by the gravitational attraction of matter. Any decelerating universe must originate from a big bang; its age is less than the Hubble time, as the figure shows.

Curve (c) illustrates a big bang universe with accelerating expansion; its age is more than the Hubble time.

A universe in which the expansion accelerates need not have evolved from a big bang. As one looks back in time, the curve of R versus t could approach zero asymptotically, as in curve (d), or could even turn up, as in curve (e). The steady-state universe, discussed in section 9.6, is of type (d). A universe of either type (d) or type (e) is infinitely old.

With the original value of the Hubble constant, the age of the universe predicted by most of the big bang cosmologies was only a couple of billion years. That value was embarrassingly small because it is considerably less than the ages of some earth rocks and meteorites, determined by radioactive dating. There is clearly something wrong with a model in which the earth is older than the universe. When the value of H_0 was revised downward after the recalibration of the distance scale, the age of the universe increased by a factor between five and ten and the problem disappeared.

Cosmologists define a "deceleration parameter" q, which measures the rate of change of the expansion. Positive q implies deceleration, negative q implies acceleration, and $q = 0$ implies uniform expansion. The deceleration parameter could itself vary with time and could even change its sign;

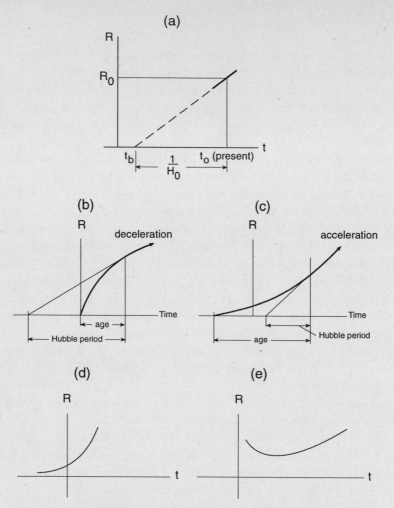

Fig. 9.5. Possible histories of an expanding universe. (a), (b), and (c) are big bang universes. The measured value of H_0 gives the present slope of the curve of the scale factor R as a function of time. This is shown as the solid segment in graph (a). The dashed line in this graph is a linear extrapolation in which the rate of expansion is assumed to be constant. In this model the age of the universe (time since the big bang) is $1/H_0$. Graph (b) shows a decelerating expansion; the age of the universe is less than $1/H_0$. Graph (c) shows an accelerating expansion; in this case the age of the universe is greater than $1/H_0$. The favored models are of type (b). Graphs (d) and (e) show expanding universes without a big bang. In (d) the scale factor increases exponentially with time; in (e) the universe was actually contracting a long time ago, reached a minimum size, and then began to expand. In each case the rate of expansion is increasing (q is negative). Graph (d) is the behavior according to the steady-state theory. The time $t = 0$ in these graphs has no significance.

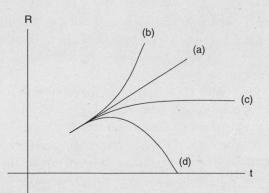

Fig. 9.6. Possible future histories of the universe. In curve (a), the rate of expansion remains constant. In (b), the expansion increases and R approaches infinity. In (c) and (d), the expansion decreases; in (c), R approaches a constant, and in (d), it goes to zero.

its present value is called q_0. The value of q_0 is one of the parameters whose values cosmologists are trying to determine from observations.[15]

Figure 9.6 illustrates possible future courses of the universe. If the rate of expansion remains constant ($q = 0$, curve [a]) or increases (positive q, curve [b]), the scale factor goes to infinity as time increases. If q is negative, the expansion slows down. The rate of expansion either approaches zero asymptotically or becomes negative. In the former case, the scale factor approaches a constant (curve [c]). In the latter case, the expansion turns around and the universe begins to contract, perhaps collapsing to another big bang (curve [d]).

9.4. THE HUBBLE SPHERE AND HORIZONS

The product of the Hubble time and the speed of light is a distance called the Hubble length:

$$L_H = \text{Hubble length} = c/H \qquad (9.26)$$

15. The mathematical definition of q is

$$q = -\ddot{R}\,R/\dot{R}^2 \qquad (9.25)$$

where $\ddot{R}$ stands for the time rate of change of $\dot{R}$. If $\dot{R}$ is constant (uniform expansion), q is zero. The minus sign is included in the definition to make q positive for a universe with slowing expansion.

Since H varies with time in most models, so does L_H; its present value is about 15 billion light-years. A sphere of radius L_H is called the Hubble sphere. In models with slowing expansion, L_H increases with time and the Hubble sphere expands steadily.

Equations (9.12) and (9.26) imply that the recession velocity of a galaxy located on the Hubble sphere is exactly c. Light emitted from such a galaxy remains at a constant distance from earth: the expansion of space just compensates for the velocity of the light moving toward us.[16] A galaxy outside the Hubble sphere is receding at a speed greater than c; light that it emits toward us is losing ground to the expansion of space.[17] If that continues to hold true, the light will never reach us and we will never observe the galaxy in question; such a source is said to be beyond our *event horizon*.

A galaxy outside the Hubble sphere at a given moment is not necessarily beyond our event horizon. If the expansion of the universe were to slow down sufficiently, at some time in the future the recession velocity at the location of the light wave could become less than c. In that case the distance between the light wave and earth would begin to diminish and the light could eventually reach us. That is precisely what happens in the Einstein–de Sitter universe, described in the next section.

An event horizon can arise in another way. If the expansion reverses and the universe eventually collapses to another singularity, light from a particular event may not have enough time to reach us before the collapse has been completed. In that case the event would not be visible to us; it would be outside our event horizon.[18]

A different type of horizon is applicable only to big bang universes. Consider light emitted from earth just after the big bang. That light has now traveled a finite distance, say, d_0, determined by the rate of expansion during its transit time. (As discussed above, d_0 is not just c times the elapsed time.) Observers on a galaxy at a distance greater than d_0 have never seen earth; light emitted from here at the very first instant of time

16. As the Red Queen told Alice in *Through the Looking Glass*, "Here, you see, it takes all the running you can do to keep in the same place."
17. Observers in the vicinity of the galaxy see light traveling at its normal speed c in all directions. But the proper distance of a light wave emitted in the direction of earth increases with time.
18. In a collapsing universe the distance between the earth and an approaching light wave decreases at a rate faster than c. A lot of ground can be covered during the last moments of the collapse. In some models the speed of approach becomes infinite. The rate at which R approaches zero determines whether or not an event horizon exists.

has not yet reached them. By a symmetry argument, it follows that we likewise have never seen that galaxy (nor one at any greater distance). Wolfgang Rindler calls this a *particle horizon*. The particle horizon expands with time, as the light emitted at the earliest times continues to propagate away from each source.

The two types of horizons are quite distinct. The particle horizon depends on how the universe has expanded up to now, while the event horizon depends on the future course of its expansion. A specific model universe can exhibit one type of horizon or the other, or both, or neither.[19] Specific examples of both types of horizons will be analyzed in the following section.

The preceding discussion of horizons tacitly assumes that galaxies live and continue to radiate forever: a galaxy is considered observable if it is kinematically possible for light emitted by the galaxy to reach us. In the real world, galaxies have finite lifetimes; hence a galaxy might be unobservable even if it is not outside our horizon, simply because it has stopped radiating (or has not yet begun to radiate). In any big bang universe, as we shall see, no galaxy with a red shift greater than about 25 can be observed. Light reaching us from such a source would have to have been emitted during a very early epoch, when the density of the universe was so high that galaxies could not yet have formed.

9.5. COSMOLOGICAL MODELS: FRIEDMANN UNIVERSES[†]

Numerous cosmological models have been advanced over the years. In all these models the matter in the universe is assumed to be distributed uniformly, as demanded by the cosmological principle. The mean density of matter is a parameter that can be varied. In nearly all the models the

---

19. For the reader familiar with the calculus, I give the conditions for the existence of the two types of horizons. An event horizon exists if the integral

$$\int_{t_0} dt/R(t)$$

is finite, where t_0 is the present time and the upper limit is either the time of collapse or $+\infty$ if there is no collapse. A particle horizon exists if the integral

$$\int^{t_0} dt/R(t)$$

is finite, where the lower limit is either the time of the big bang or $-\infty$ if the universe is infinitely old.

† Section 9.5 is fairly technical. It may be skimmed without losing the main thread of the exposition.

geometry of space-time is assumed to be governed by general relativity. The available observational evidence eliminates some candidates and sets some constraints but is not sufficient to select a specific model from all the candidates.

The first cosmological model was proposed by Einstein himself in 1917, soon after his paper on general relativity had been published; it is known as the Einstein universe. Since the expansion of the universe had not yet been discovered, Einstein took it for granted that the universe is static; he further assumed that the spatial part of space-time has positive curvature (spherical geometry).

To his dismay, Einstein found that the field equations of general relativity for this model have no static solutions. The Einstein universe in its original form first expands and ultimately collapses; the collapse is due to the gravitational self-attraction of matter.

To salvage the situation, Einstein arbitrarily introduced a "cosmological" term into the field equations. The new term, which is effectively a repulsive interaction, contained a constant Λ, which Einstein named the cosmological constant. If Λ is assigned an appropriate value, a static solution of the equations exists. The solution is, however, unstable: any small perturbation will cause it either to blow up or to collapse.

After the discovery of the expanding universe, a cosmological term was no longer needed; Einstein later told the physicist George Gamow that introducing the Λ term was his "greatest blunder." Models that include such a term nonetheless continue to be studied.

In 1922, Alexander Friedmann investigated the class of homogeneous universes governed by general relativity, without any Λ term. The cosmological principle demands that the curvature be everywhere the same. Hence there are in essence only three possibilities, characterized by a constant, k, that specifies the curvature of the spatial part of space-time:

(i) $k = +1$: positive curvature (spherical space)
(ii) $k = 0$: zero curvature (flat space)
(iii) $k = -1$: negative curvature (hyperbolic space)

All three Friedmann models turn out to be big bang universes: the scale factor goes to zero at a finite time in the past. In case (i) the universe is closed and finite in extent; in the other two cases the universe is open and infinite.

Case (ii) is particularly simple; it is called the Einstein–de Sitter universe. Although it is almost surely not right in every detail, it illustrates many features common to other models.

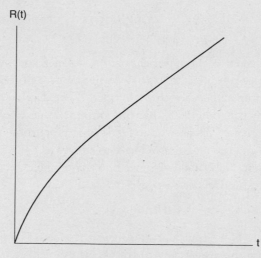

Fig. 9.7. Scale factor $R(t)$ as a function of time for the Einstein-de Sitter universe. $R(t)$ is proportional to the two-thirds power of time. $t = 0$ is the time of the big bang.

The reader may wonder how a universe described by general relativity can be flat. In the neighborhood of any galaxy, to be sure, space must be curved. But cosmology is not concerned with the geometry of small regions. In any cosmological model, all the matter in the universe is spread out uniformly. With this distribution of matter, the field equations admit a solution in which the space at any fixed time is Euclidian. Because space expands, the full space-time metric is not the Lorentzian one of special relativity. Thus the Einstein–de Sitter universe is in fact curved even though its spatial part is flat.

Solution of the Einstein field equations gives the scale factor as a function of time. In the Einstein–de Sitter universe $R(t)$ is proportional to the two-thirds power of the time:

$$R(t) = R_0(t/t_0)^{2/3} \qquad \text{(in Einstein–de Sitter universe)} \quad (9.27)$$

Equation (9.27) is plotted in figure 9.7. The model is a big bang universe; $t = 0$ is the time of the big bang and t_0, the present time, is also the age of the universe.

An elementary application of differential calculus shows that when a variable R is proportional to t^n, its rate of change $\dot{R}$ is proportional to nt^{n-1}. In the present case n has the value $2/3$. Hence the rate of expansion $\dot{R}$ is given by

$$\dot{R}(t) = \frac{2}{3}\left(\frac{R_0}{t_0}\right)\left(\frac{t}{t_0}\right)^{-1/3} \tag{9.28}$$

$\dot{R}$ is always positive but approaches zero as t increases; the Einstein–de Sitter universe continues to expand, but at an ever-decreasing rate. It begins with a bang and ends with a whimper.[20] The deceleration parameter q, defined by equation (9.25), has the constant value $+\frac{1}{2}$.

Since the measured distance of any galaxy is proportional to R, it too increases as $t^{2/3}$ and the recession velocity decreases as $t^{-1/3}$.

From equations (9.27) and (9.28), we find that Hubble's constant in the Einstein–de Sitter model decreases as $1/t$:[21]

$$H(t) = 2/(3t) \tag{9.29}$$

At early times H was much greater than it is today. The present value H_0 is $\frac{2}{3}$ t_0. From the measured value of H_0 we can deduce the age of the Einstein–de Sitter universe:

$$t_0 = \frac{2}{3}\, t_H \approx 10^{10} \text{ y} \tag{9.30}$$

The Hubble length, $L_H = c/H(t)$, increases linearly with time:

$$L_H = \frac{3}{2}\, ct \tag{9.31}$$

The present radius of the Hubble sphere is 22.5 billion light-years; it expands at the constant speed $1.5c$. Since the recession velocities of the galaxies decrease to zero, the Hubble sphere eventually overtakes every galaxy. This implies that the Einstein–de Sitter universe has no event horizon: every galaxy will eventually be observed.

We next calculate the look-back times for the model. According to equations (9.10) and (9.27), light emitted at time t_e is detected as redshifted by the amount

$$1 + z = R_0/R(t_e) = (t_0/t_e)^{2/3} \tag{9.32}$$

Solving this equation for t_e as a function of z, we obtain

$$t_e = t_0(1+z)^{-3/2} = \frac{2}{3}\, t_H\, (1+z)^{-3/2} \tag{9.33}$$

As expected, t_e approaches zero as $z \to \infty$: sources with large red shifts are seen as they looked at very early times. The look-back time to a source with red shift z (the age at which we are seeing it today) is

20. As T. S. Eliot put it in *The Wasteland*, "This is the way the world ends / Not with a bang but with a whimper."
21. This is true of any universe in which R varies as a power of t.

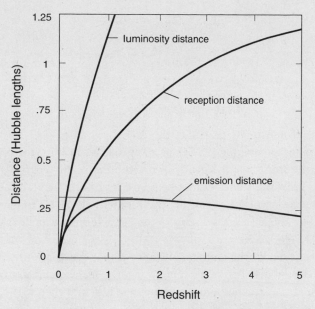

Fig. 9.8. Distance of galaxies as a function of red shift for the Einstein–de Sitter universe. The curve labeled "reception distance" gives the present distance of a galaxy whose red shift is z. The "emission distance" is the distance at the time the light now reaching us was emitted. In the text these are referred to as d_0 and d_e, respectively. For a discussion of luminosity distance, see the text. Adapted from Harrison, *Cosmology: The Science of the Universe*, copyright © 1981 by Cambridge University Press. Reprinted with the permission of Cambridge University Press.

$$t_0 - t_e = \tfrac{2}{3}\, t_H [1 - (1+z)^{-3/2}] \qquad (9.34)$$

For small z, we can approximate

$$(1+z)^{-3/2} \approx 1 - \tfrac{3}{2}\, z \qquad (9.35)$$

and the look-back time is just $z t_H$. All models give this result for small z.

In section 9.2 we defined three different cosmological distances—the present distance d_0, the distance at the time of emission, d_e, and the luminosity distance, d_L. A calculation too complicated to reproduce here yields the following expression for $d_0(z)$:

$$d_0(z) = 2\, L_H [1 - (1+z)^{-1/2}] \qquad (9.36a)$$

Equation (9.19) then gives $d_e(z)$:

$$d_e(z) = \frac{2L_H}{1+z}[1 - (1+z)^{-\frac{1}{2}}] \qquad (9.36b)$$

and equation (9.24) gives $d_L(z)$:

$$d_L(z) = 2L_H(1+z)[1 - (1+z)^{-\frac{1}{2}}] \qquad (9.36c)$$

The three distances are plotted in figure 9.8.

According to equation (9.36a), as z approaches infinity the present distance of the source approaches $2L_H$, or 30 billion light-years. Any galaxy now at a greater distance cannot be seen and has never been seen; $2L_H$ is therefore the particle horizon for the Einstein–de Sitter universe.[22] The particle horizon is twice as distant as the Hubble sphere and expands twice as fast, at velocity $3c$.

The behavior of the emission distance $d_e(z)$ reveals an interesting feature of the model. For small z, d_e is nearly identical to d_0, but as the figure shows, d_e reaches a maximum at $z = 1.25$ and then turns down. Its maximum value is $(8/27)L_H$; this is the greatest emission distance of any galaxy we can see today. The present distance of such a galaxy is $\frac{2}{3}L_H$. Its look-back time is $0.47\, t_H$, about 7 billion years.

Of two sources with z greater than 1.25, the one with the greater red shift was *closer* to us than the other when it emitted the light we detect. As $z \to \infty$, d_e in fact approaches zero. For each emission distance there are two values of z, one less than 1.25 and the other greater than 1.25.

To make sense of all this, consider light emitted at some early time t_1 by a source outside the Hubble sphere, where the expansion velocity is greater than c (fig. 9.9a). That light at first receded from us, at a speed $v - c$, but at a later time, t_2, the Hubble sphere caught up to it and its proper distance began to diminish (fig. 9.9b). At a still later time, t_3, the proper distance of the light was back to its original value (fig. 9.9c). Light emitted at t_3 by another galaxy at that proper distance reaches us simultaneously with the first wave but with a much smaller red shift because it was emitted much later (fig. 9.9d). The red shifts for the two galaxies are quite different even though their emission distances are equal. The behavior illustrated in figure 9.9 is not confined to the Einstein–de Sitter universe but is characteristic of all models in which the expansion decelerates without reversing.

---

22. The existence of a particle horizon but not of an event horizon is consistent with the conditions stated in footnote 19 of this chapter. With $R(t)$ proportional to $t^{2/3}$, the integral $\int_0^{t_0} dt/R(t)$ is finite but $\int_{t_0}^{\infty} dt/R(t)$ is infinite.

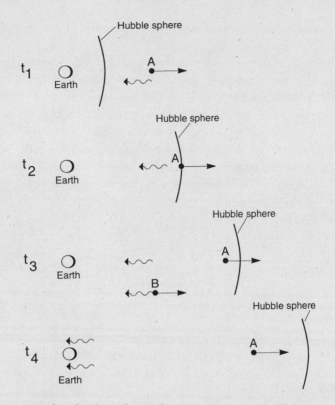

Fig. 9.9. This sketch explains why two galaxies with different red shifts can have the same emission distance in an Einstein–de Sitter universe. (a) At time t_1 galaxy A emitted a light ray toward earth. The expansion velocity at the point of emission was greater than c; hence the distance between earth and the light at first increased. (b) At time t_2 the expansion had slowed enough that the light from galaxy A began to approach earth. (c) At time t_3 its distance from earth was the same as when it was first emitted. (d) Light from a second galaxy, B, at this distance at t_3 reaches earth at t_4, together with the light from A. The red shift of the light from B is much smaller than that of the light from A, since it was emitted when the scale factor of the universe was much greater.

One may be tempted to conclude from the plot of $d_e(z)$ in figure 9.8 that a galaxy with a very large red shift could appear brighter than one with a smaller red shift, because its emission distance is smaller. As argued in section 9.2, however, d_L is the distance that is related to apparent brightness by an inverse-square law, and $d_L(z)$ increases monotonically with z. Hence the sources in fact grow progressively dimmer with increasing z.

The angular size of a galaxy is the ratio of its diameter to the emission distance. In the Einstein–de Sitter universe the angular size at first decreases with increasing z, reaches a minimum at $z = 1.25$, which corresponds to the maximum emission distance, and thereafter increases gradually.

In section 9.2 we defined two recession velocities for every source: the present recession velocity, v_0, and the velocity when it emitted the light we now detect, v_e. Equations (9.20a,b) give the relation between each velocity and the corresponding distance:

$$v_0(z) = H_0 \, d_0(z) \qquad (9.20a)$$

and

$$v_e(z) = H(t_e) \, d_e(z) \qquad (9.20b)$$

The relation between $H(t_e)$ and H_0 is given by equations (9.29) and (9.32):

$$H(t_e) = \left(\frac{t_0}{t_e}\right) H_0 = (1+z)^{3/2} \, H_0 \qquad (9.37)$$

The two recession velocities in the Einstein–de Sitter universe are therefore

$$v_0(z) = 2c \, [1 - (1+z)^{-1/2}] \qquad (9.38a)$$

and

$$v_e(z) = 2c \, [(1+z)^{1/2} - 1] \qquad (9.38b)$$

Both velocities are plotted in figure 9.10. As expected, v_0 is less than v_e for all z. For small z, each velocity is approximately cz, in accord with the relation (9.2). As z approaches infinity, however, v_e becomes infinite whereas v_0 approaches the finite value $2c$. Infinite z represents the particle horizon, which is at twice the Hubble length.

As shown above, the particle horizon is receding at three times the speed of light, faster than the recession velocity of a galaxy located there, which is only $2c$. The particle horizon therefore eventually overtakes every galaxy; this confirms our earlier conclusion that the model has no event horizon.

Figure 9.11 shows the history (measured distance as a function of time) of several typical galaxies in an Einstein–de Sitter universe. According to

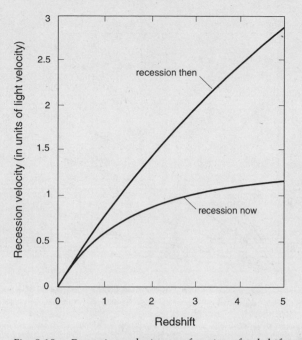

Fig. 9.10. Recession velocity as a function of red shift for the Einstein–de Sitter universe. "Recession then" refers to the velocity when the light was emitted; "recession now" refers to the present velocity. (See fig. 9.8.) From Harrison, *Cosmology: The Science of the Universe,* copyright © 1981 by Cambridge University Press. Reprinted with the permission of Cambridge University Press.

equation (9.9), the time dependence of d is the same as that of R; each curve therefore has a $t^{2/3}$ variation. The solid circles mark the emission time and emission distance for each galaxy.

Figure 9.12 presents the same information in a different format. Shown are the Hubble sphere, the particle horizon, and the galaxies from figure 9.11. Each observable galaxy is labeled with its red shift, present distance, and present recession velocity; a dotted line leads back to the point of emission.

The behavior described in the preceding discussion is apparent in this figure. For very small z, d_0 and d_e are nearly the same; for larger z, d_e is substantially smaller. G_3 $(z=1.25)$ has the largest emission distance, 0.29 L_H. G_2 $(z=0.53)$ and G_4 $(z=3.0)$ have the same emission distance, even though t_e is much smaller for G_4. G_4 is on the Hubble sphere; its

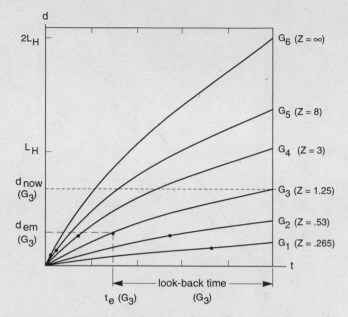

Fig. 9.11. Distance as a function of time for representative galaxies in an Einstein–de Sitter universe. The emission time for light received from each galaxy is indicated by a filled circle. The look-back time and emission distance are indicated for one galaxy (G_3). G_4 is on the Hubble sphere; G_6 is on the particle horizon.

present recession velocity is c. G_6 is on the particle horizon; its recession velocity is $2c$ and its red shift is infinite. Finally, G_7 is beyond the particle horizon; it is not visible to us today.

A short time afterward, the galaxies, the Hubble sphere, and the particle horizon will all have receded, at the velocities indicated. G_4 will be inside the Hubble sphere and G_7 inside the particle horizon. No matter how distant a galaxy is today it will eventually be overtaken, first by the particle horizon and later by the Hubble sphere.

The other two Friedmann universes (with $k = \pm 1$) can be similarly analyzed; we summarize the results here. The time dependence of the scale factor for all three Friedmann models is shown in figure 9.13. In each case, R is proportional to $t^{2/3}$ for small t.

In the case $k = +1$ (spherical geometry, closed universe), the deceleration parameter q starts with the value $+\frac{1}{2}$ and increases with time. The deceleration is sufficient to make the expansion turn around, as the figure shows: the universe reaches a maximum size and then collapses, ending

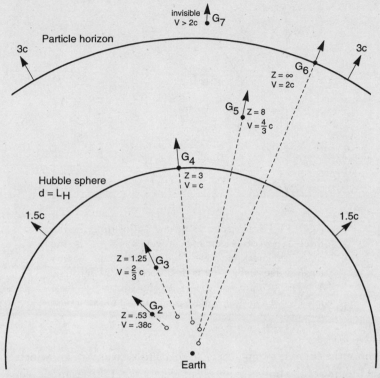

Fig. 9.12. Another picture of the Einstein–de Sitter universe. Shown are the Hubble sphere, at a distance $L_H = c/H_0$, and the particle horizon at a distance $2L_H$. Several galaxies are shown, each labeled by its red shift and its present expansion velocity. G_4, on the Hubble sphere, has expansion velocity c. The dashed line from each galaxy leads back to its emission distance (the position of the galaxy when it emitted the light we now detect). Note that the maximum emission distance is for G_3, with $z = 1.25$. Galaxies with greater red shifts have smaller emission distances, as explained in the text.

its life with a second big bang. The dependence of R on time is a cycloid (the curve generated by a fixed point on the circumference of a circle that rolls on a flat surface). The maximum value of the scale factor, R_m, and the time at which the maximum is reached, t_m, can both be expressed in terms of R_0, H_0, and q_0. The relations are

$$t_m = \frac{\pi q_0}{H_0(2q_0 - 1)^{3/2}} \tag{9.39}$$

$$R_m = R(t_m) = \frac{2q_0}{2q_0 - 1} R_0 \tag{9.40}$$

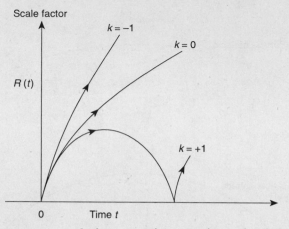

Fig. 9.13. Scale factor as a function of time for the three types of Friedmann universe. $k = +1$ is the closed (cycloidal) universe, $k = 0$ is the Einstein–de Sitter universe, and $k = -1$ is the hyperboloidal universe. The last two are both open. At short times all three models are identical, with R proportional to the square root of t.

Since the curve is symmetric, R returns to zero at $t = 2t_m$, which is the total lifetime of the universe in this model. If $q_0 = 1$, for example, equation (9.39) gives $t_m = \pi/H_0$, or about 50 billion years. The lifetime of the universe is 100 billion years. t_0, the present age of the universe, can also be expressed in terms of H_0 and q_0. With $q_0 = 1$ the present age is about 8.5 billion years; the universe is still young. Nonetheless, according to equation (9.40), the scale factor has already reached half its maximum value.

The cycloidal universe has both an event horizon and a particle horizon. Light from a galaxy now at the event horizon will reach us at the completion of the collapse, just before the second big bang. During the contracting phase the spectra of some galaxies will be detected as blue-shifted, since R_0/R in equation (9.8) will be less than unity. The largest blue shift is associated with the galaxy whose light was emitted when R had its maximum value. (Under a conventional Doppler interpretation, that source would exhibit no frequency shift at all, since it would have been at rest with respect to us at the moment the light was emitted.)

The closed Friedmann model is typical of universes that begin and end with a bang. Some have speculated that if such a model is correct, the universe could "bounce" after the second bang and begin expanding again. The history of the universe could consist of a succession of expansions and contractions, with a big bang at the end of each cycle. Because of the singular nature of the conditions at the time of a big bang, there is no way to predict whether such bounces would actually occur.

In the case $k = -1$, the deceleration parameter begins at $+\frac{1}{2}$ and diminishes steadily to zero. For large t, the rate of expansion is constant; R is proportional to t and H is proportional to $1/t$. The present age depends on the value of q_0. If $q_0 = 0.014$, for example, $t_0 = 0.96\ t_H \approx 14.5$ billion years. As figure 9.13 indicates, the general behavior is quite similar to that for $k = 0$. The hyperbolic universe is open; the model has a particle horizon but no event horizon.

9.6. THE STEADY-STATE UNIVERSE

A model that enjoyed great popularity for some time was the steady-state universe, proposed by Hermann Bondi, Thomas Gold, and Fred Hoyle in 1948. The steady-state universe obeys the perfect cosmological principle: all observable properties are independent of time, a feature that appealed to many. The characteristics of this model are in marked contrast with those of Einstein–de Sitter and of other big bang universes.

The appearance of a given galaxy does *not* remain unchanged in a steady-state universe. Galaxies can be formed, evolve, and die. Only the appearance of the sky as a whole remains unchanged.

The expansion of the universe poses a serious problem for any steady-state cosmology. As space expands, the distance between any two galaxies increases; concurrently, the mean density of the universe decreases. These changes, which are observable, would seem to rule out a steady-state model.

To overcome this difficulty, the proponents of the steady-state universe postulated that matter is continuously being created, thus compensating for the expansion. The newly created matter eventually forms new galaxies that fill the "holes" left by the expansion and maintain a constant density.

Continuous creation is a radical hypothesis. It violates the conservation of energy: the new matter is being created out of nothing. However, the required rate of mass creation is only about one hydrogen atom per 6 cubic kilometers per year. No experiment could detect (or exclude) so minute an effect. To advocates of the steady-state universe, continuous creation is a less bizarre proposal than is the creation of the entire universe in one cataclysmic event.

The steady-state model does not fit into the conventional framework of general relativity; a term must be added to the field equations to obtain a steady-state solution. This is not the same as the Λ term of the Einstein universe. A steady-state universe is not to be confused with a static universe; the latter cannot expand.

In a steady-state universe, Hubble's constant is a true constant and we have from equation (9.13):

$$H(t) = \dot{R}/R = H_0 \qquad (9.41)$$

The Hubble length is likewise constant; the Hubble sphere has a fixed radius c/H_0, about 15 billion light-years.

The solution of equation (9.41) is

$$R(t) = R_0 \, e^{H_0(t - t_0)} \qquad (9.42)$$

where t_0, as before, denotes the present time. The scale factor increases exponentially with time, as in figure 9.5d. Space is flat and the deceleration parameter has the constant value -1.

In the steady-state model, R was never zero; there was no big bang and the universe is infinitely old. The measured distance of each galaxy and its recession velocity both increase exponentially with time.

The steady-state model had particular appeal before the distance scale had been recalibrated, when the Hubble time was believed to be less than the estimated age of the earth. Since in the steady-state model the Hubble time has nothing to do with the age of the universe, the discrepancy posed no problem for the model.

The relation between red shift and time of emission for the steady-state universe is

$$1 + z = e^{H_0(t_0 - t_e)} \qquad (9.43)$$

and the look-back time for a source with red shift z is

$$t_0 - t_e = \frac{1}{H_0} \ln(1 + z) \qquad (9.44)$$

where ln stands for the natural logarithm. For small z,

$$\ln(1 + z) \approx z$$

and the look-back time is approximately $z/H_0 = zt_H$, just as in the Einstein–de Sitter model. For large z, however, the two models give quite different results: in the steady-state universe the look-back time approaches infinity as z increases.

Once again we distinguish between the three types of distance defined in section 9.2. The results for the steady-state universe are

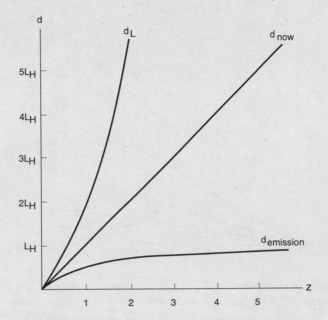

Fig. 9.14. Distance as a function of red shift for the steady-state universe. Plotted are the present distance (d_{now}) and the distance at the time of emission. (Cf. fig. 9.8, which shows the same quantities in an Einstein–de Sitter universe). With a change of scale, the same curves give the velocities (present and emission) as functions of z.

$$d_e(z) = L_H z / 1 + z \qquad\qquad (9.45a)$$
$$d_0(z) = L_H z \qquad\qquad (9.45b)$$
$$d_L(z) = L_H z (1 + z) \qquad\qquad (9.45c)$$

The recession velocities v_e and v_0 are related to the corresponding distances by equations (9.20a,b). Since $H(t)$ is constant, each velocity is just H_0 times the corresponding distance:

$$v_e(z) = cz / 1 + z \qquad\qquad (9.46a)$$
$$v_0(z) = cz \qquad\qquad (9.46b)$$

A galaxy with $z = 1$ is now on the Hubble sphere; its present recession velocity is c and its velocity at the time of emission was $c/2$.

Relations (9.45) and (9.46) are plotted in figure 9.14, which may be compared to figures 9.8 and 9.9. For the steady-state model a single plot suffices; the ordinate measures both distance and (with a change of scale) recession velocity.

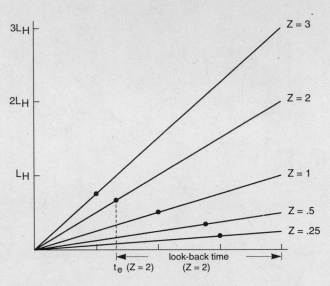

Fig. 9.15. Distance as a function of time for representative galaxies in the steady-state universe. (See the caption for fig. 9.11.)

In the steady-state universe, v_0 is greater than v_e for every galaxy because the expansion is accelerating. The present recession velocity increases without bound as z increases; v_e, however is always less than c (eq. 9.46a). There is no particle horizon, as was to be expected inasmuch as the universe is infinitely old.

Unlike the case of Einstein–de Sitter, the curve of d_e vs. z, equation (9.45a), does not turn over but continues to increase, approaching L_H asymptotically as $z \to \infty$. The emission distances of high-z sources are all clustered around the Hubble sphere.

Since the Hubble sphere is fixed in the steady-state universe, every galaxy must eventually cross it. (In the Einstein–de Sitter universe, the opposite happens: the Hubble sphere overtakes the galaxies.) According to equation (9.46b), once a galaxy has crossed the Hubble sphere its recession velocity exceeds c and thereafter continues to increase. Hence light emitted from a source outside the Hubble sphere can never reach us and the Hubble sphere is also an event horizon.[23]

23. The integrals in footnote 19 above confirm that the model has an event horizon but not a particle horizon: with $R(t)$ given by eq. (9.42), $\int_{t_0}^{\infty} dt/R(t)$ is finite but $\int_{-\infty}^{t_0} dt/R(t)$ is infinite.

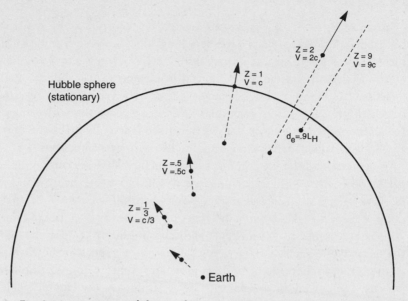

Fig. 9.16. A picture of the steady-state universe. The Hubble sphere is stationary and constitutes the event horizon for this model. The dashed lines lead back to the emission distance of each galaxy, as in fig. 9.12. Notice that for any value of z, the emission distance is less than the Hubble radius and the galaxy is visible even though its present position may be far outside the Hubble sphere. The galaxy with $z = 9$ is outside the range of the picture.

Figures 9.15 and 9.16 show the histories of typical galaxies in the steady-state universe. They are to be compared to figures 9.10 and 9.11, which describe Einstein–de Sitter; the contrast between the two universes is apparent.

One property of the steady-state model seems paradoxical. As we have just seen, every galaxy eventually crosses the Hubble sphere/event horizon. Light emitted from anywhere outside the horizon will never reach us. One might expect, therefore, that as we follow a particular galaxy we should see it cross the event horizon and disappear from view, never to be seen again. That is not at all what happens, however. On the contrary, we continue to see every galaxy forever, in spite of the fact that they spend most of their lives outside our event horizon.

Figure 9.16 explains this paradoxical behavior. All galaxies with z greater than 1 are outside the Hubble sphere; we shall never see them as they look *now*. Yet we see them all, each at an emission distance less than L_H.

After some time, each galaxy has receded; concurrently, its red shift has increased. High-z sources recede a long way. Suppose G_5 recedes to the current distance of G_6; its red shift increases from 2 to 3. Its new emission distance will be the old emission distance of G_6, only a small increase. The emission distance increases but only very slowly. As time passes, therefore, we "see" each galaxy slowly approach the horizon as its red shift approaches infinity, but we never see it actually reach the horizon. This behavior is characteristic of models with event horizons.

A somewhat different argument leads to the same conclusion. As a galaxy nears the event horizon and its red shift increases, the wavelength of every light wave increases and so does its period $T = 1/f = \lambda/c$. A single cycle of the wave takes longer and longer to pass us. In effect, the red shift stretches out time, by a factor that approaches infinity as z does; the light that records the small fraction of a galaxy's history spent in the vicinity of the Hubble sphere/event horizon takes an infinite amount of time to pass us. The portion of the galaxy's life spent outside the Hubble sphere is not accessible to us.

The steady-state model has been pretty much demolished by observational evidence discussed in the following sections. Its demise has been a disappointment to its proponents, many of whom had great affection for the theory. In the words of Dennis Sciama, "For me the loss of the steady-state theory has been a cause of great sadness. The theory has a sweep and beauty that for some unaccountable reason the architect of the universe appears to have overlooked. The universe in fact is a botched job, but I suppose we shall have to make the best of it."[24]

9.7. Observational Tests

The measured value of H_0 tells us the present rate of expansion of the universe. Cosmologists are interested in its entire history: What happened in the past? Did it all begin with a big bang? What will happen in the future? Will the universe continue to expand or will it eventually turn around and collapse?

In section 9.8, I present the evidence that supports the big bang hypothesis and reconstruct the past history of the universe. Here I address the question of the future.

24. Dennis Sciama, "Cosmology before and after Quasars," in *Cosmology + One* (San Francisco: W. H. Freeman, 1977), 31–33. (Originally published in *Scientific American*, September 1967.)

On the theoretical side, if the universe is governed by general relativity and the cosmological principle is valid, the choice is among the three Friedmann models described in section 9.5.[25] Each model makes a definite prediction concerning the future course of the expansion.

Let us review the properties of the Friedmann models. With $k = +1$, the deceleration parameter q is always greater than $\frac{1}{2}$; the universe is closed (spherical space, finite volume) and ultimately collapses. With $k = 0$ (Einstein–de Sitter universe), q is always exactly $\frac{1}{2}$. The universe is open (flat space, infinite volume) and continues to expand but ever more slowly; the expansion velocity approaches zero. Finally, with $k = -1$, q is always less than $\frac{1}{2}$. The universe is again open (hyperbolic space) and expands forever; the expansion velocity approaches a constant.

If the universe is of the Friedmann type, then, determination of the value of q_0 would decide whether the expansion will continue forever or whether it will eventually turn around. Several types of observations can provide information concerning the value of q_0. The available data are meager and somewhat contradictory; the jury is still out on whether the universe is open or closed and on the future of the expansion.

The Density of the Universe

One approach to the determination of the deceleration parameter is through the mean density of matter in the universe. The mutual attraction between every bit of matter and every other bit slows down the expansion. The more densely packed the matter, the greater the deceleration. If the density is high enough, the expansion will eventually turn around and the universe will begin to collapse.

Although the argument of the preceding paragraph has been couched in old-fashioned Newtonian language, general relativity leads to the same conclusion. In Friedmann cosmology, the future course of the expansion is determined by the ratio of the mean density of matter ρ_0, to a critical density, ρ_c, defined as

$$\rho_c = 3H^2/8\pi G \qquad (9.47)$$

where G is the gravitational constant. The dimensionless ratio ρ_0/ρ_c is assigned the symbol Ω.

25. Inclusion of Einstein's Λ term extends the range of possibilities but does not change the qualitative conclusions we shall draw concerning the course of the expansion.

It was shown by Friedmann and later by Lemaitre that in a Friedmann universe the deceleration parameter is directly related to the value of Ω:

$$q_0 = \tfrac{1}{2}\,\Omega = \rho_0/2\rho_c \qquad (9.48)$$

According to equation (9.48), if the mean density is equal to or less than ρ_c ($\Omega \leqslant 1$) the universe is open and will continue to expand. If ρ_0 is greater than ρ_c ($\Omega > 1$), the universe is closed and will ultimately collapse. With the currently accepted value of H_0, the critical density is about 10^{-29} g/cm^3, which corresponds to about five hydrogen atoms per cubic meter.

A lower bound on the value of ρ_0 can be obtained by estimating the amount of mass contained in galaxies. The average mass of a galaxy and the number of galaxies per unit volume are known with fairly high confidence. The mass contained in galaxies, spread uniformly throughout the universe, corresponds to a mean density about 3×10^{-31} g/cm^3, only about one-thirtieth of ρ_c.

If galaxies account for nearly all the mass in the universe, then, k must be negative and, according to equation (9.48), q_0 is about 0.015. Space is hyperbolic and will continue to expand forever. As we shall see, the red shift data are not consistent with this estimate; they suggest a considerably larger value of q_0.

A closed universe would have to contain at least 30 times as much mass as the amount accounted for in galaxies. Many possible sources for the "missing" mass have been suggested. They include intergalactic matter (in the form of stars or tenuous gas), dwarf galaxies too dim to be observed, black holes, and relativistic particles such as cosmic rays, neutrinos, gravitons, or even more exotic particles. Although the existence of such matter cannot be excluded, there is no evidence to suggest that it is present in the large amount required to close the universe. The cosmic abundance of deuterium, discussed in the following section, supports the conclusion that ρ_0 is substantially less than ρ_c.

A cosmological term would not help to close the universe if the value of Λ is positive, as is generally assumed in models that contain such a term. Any positive Λ would make q_0 even smaller and would cause an even more rapid expansion.

Determination of q_0 from Red Shift Data

In principle, one could determine the value of the deceleration parameter by measuring how the slope of the Hubble diagram varies with time.

However, the predicted fractional change in H is only about one part in 15 billion per year, far too small to be detectable.[26]

A more promising approach is to examine the departure from linearity at the high-z end of the Hubble diagram. Each cosmological model predicts a specific relation between red shift and distance; the results for the Einstein–de Sitter and steady-state universes are shown in figures 9.8 and 9.14.[27]

The departure from linearity in the Hubble diagram can be attributed to two causes. First, the slope of the high-z end of the curve reflects the value of the Hubble constant a long time ago, when the light from those sources was emitted. A decrease with time in the value of H, predicted by most models, causes the diagram to steepen at the high-z end.

The curvature of space also affects the shape of the Hubble plot for distant sources, by modifying the look-back times. Even if H were constant, a plot of d_0 vs. z in a universe with curved space would depart from linearity at the high-z end. Only in the steady-state universe is the relation linear for all values of z (eq. [9.45b]).

Comparison of the predicted red shift–distance relations with observational data could provide a sensitive test of the models. Unfortunately, distance estimates are available only for galaxies with red shifts up to about 0.5; within this range, the predictions of the models differ very little and it is hard to discriminate among them.

What the observers actually plot is red shift against visual magnitude for the brightest galaxy in a cluster.[28] The red shift–distance relation predicted by a model translates into a relation between red shift and visual magnitude if one assumes that the brightest galaxy in a cluster can be used as a standard candle, that is, its absolute luminosity is the same for all clusters.

In addition to random fluctuations, two other effects make that assumption questionable for very distant galaxies. First, the luminosity of a

26. Eq. (9.12) states that $H = \dot{R}/R$. On differentiating this equation with respect to time, one finds directly that

$$\dot{H} = -(1+q)H^2 \tag{9.49}$$

where $\dot{H}$ is the rate of change of H and q is defined by eq. (9.25). Thus measurement of $\dot{H}$ would determine the value of q_0. If H decreases with time, q_0 must be greater than -1.

27. Notice that in those figures the ordinate and abscissa are reversed as compared to the conventional Hubble plot, fig. 9.1.

28. The visual magnitude is a measure of the energy that reaches the telescope in the particular spectral band to which the photographic plate is sensitive.

galaxy varies over its lifetime because young stars are generally brighter than old ones. Hence it is to be expected that after an early period during which stars are being formed, the luminosity of a typical galaxy should diminish with time. Since galaxies with high z are being observed as they looked a long time ago, the stars in such galaxies are on the average younger than those in galaxies with smaller red shifts. This effect tends to make the high-z galaxies more luminous.

The "galactic evolution" effect has been studied by Beatrice Tinsley, who concluded that for the models she examined, the true value of q_0 is higher by between 0.6 and 1.5 than the estimates obtained by ignoring the effect. Galactic evolution can evidently be an important effect.

Another correction arises from the possibility that a massive galaxy can capture matter from a smaller neighbor and perhaps even swallow it entirely. This "galactic cannibalism" effect, which has been studied by James Gunn and Tinsley, tends to make older galaxies more luminous and therefore works in the opposite direction from galactic evolution. The magnitude of the change in q_0 is hard to estimate quantitatively but could be comparable to that of galactic evolution.

Yet another problem faced by the observer is caused by the fact that the red shift brings a different part of a galaxy's spectrum into the visible band. In comparing the luminosities of galaxies with different red shifts, a correction must be made. This is called the K correction.

Estimates of the value of q_0 based on analysis of red shift data have been made by Gunn and J. B. Oke, by J. Kristian, A. Sandage, and J. A. Westphal, and by others. The results vary over a wide range, depending on the method of analysis employed, the sample galaxies included, and the corrections applied to the raw data.

Kristian et al. concluded that the best fit is obtained with $q_0 = 1.6 \pm 0.4$. (They made no correction for the effect of galactic evolution or cannibalism.) This value for q_0 is substantially greater than the one based on mass density; it suggests a closed universe. Gunn and Oke, however, obtained much smaller values of q_0; with galactic evolution included, their analysis suggested that q_0 is probably negative.

Still another way to estimate q_0 is to compare the angular sizes of distant galaxies with their red shifts, assuming that all galaxies of a given type have about the same linear diameter. In Euclidean geometry, the angle θ (in radians) subtended by an object of diameter D at a distance d is just $\theta = D/d$. (In an expanding universe, d_e is the distance that must be used in this formula.)

As we saw in section 9.5, the Einstein–de Sitter universe, with $q_0 = 0.5$, predicts that the angular size should reach a minimum at $z = 1.25$ and then turn up. Similar behavior is predicted by models with higher values of q. A plot of angular size vs. z, made by W. A. Baum, is inconclusive but suggests a value of q_0 lower than those inferred from the luminosity data. According to Baum, the value of q_0 that best matches his data is 0.15; this value again implies an open universe.

Finally, indirect information on the deceleration is provided by estimates of the ages of very old objects in the universe. As we saw in section 9.3, the age of the universe in any big bang model depends on q_0. The greater the value of q_0, the younger the universe. According to fairly reliable estimates, some stars in globular clusters are at least 8 billion years old. This then constitutes a lower bound on the age of the universe. The Einstein–de Sitter universe, with $q_0 = 0.5$, is only some 10 billion years old. Hence the age data suggest that q_0 cannot be much greater than 0.5.

The principal conclusion that emerges from this analysis is that the deceleration parameter is quite sensitive to effects whose magnitude is hard to estimate. We are still far from having a reliable estimate of q_0, and the question of whether the universe is open or closed is far from settled. The data are compatible with any value between 0 and 2; values greater than 2 are probably (though not definitely) excluded, as is the value -1 predicted by the steady-state model.

9.8 The History of the Universe

I conclude my survey of cosmology by reconstructing the history of the universe from the point of view of big bang cosmology, working backward in time from the present toward the singularity. The qualitative features of the picture that emerges do not depend on any specific cosmological model.

The major piece of evidence that lends credence to the entire picture is the microwave background radiation, discovered serendipitously in 1965. Using a horn antenna designed for satellite communication, Arno Penzias and Robert Wilson detected a weak radio-frequency noise, at a wavelength of 7 cm, whose intensity was independent of the direction in which their antenna was pointed.

The isotropy of the radiation, now confirmed to better than one part in a thousand, rules out any possible source within the solar system or the galaxy and strongly suggests a cosmic origin. Dicke and his coworkers immediately identified the radiation as the "thermal background," whose

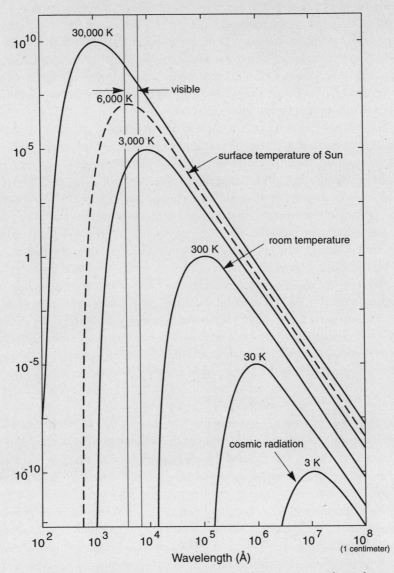

Fig. 9.17. Blackbody spectra at a number of temperatures. The scales are logarithmic. The peak of the spectrum at each temperature is given by eq. (9.50). Notice that the spectra fall off very steeply on the low-wavelength side of the peak. The visible portion of the spectrum is indicated by the narrow vertical band. From Harrison, *Cosmology: The Science of the Universe,* copyright © 1981 by Cambridge University Press. Reprinted with the permission of Cambridge University Press.

existence had been predicted nearly twenty years earlier by Gamow on the basis of big bang cosmology.

I digress briefly to summarize the properties of thermal radiation. Every body at a given temperature emits a continuous spectrum of radiation, called the thermal or blackbody spectrum. The curve of intensity vs. wavelength, called a Planck curve, is plotted in figure 9.17 for several different temperatures.

The shape of the thermal spectrum is the same at all temperatures, but as the temperature increases, two things happen: the amount of energy radiated increases sharply (note that the scale is logarithmic), and the emission shifts toward lower wavelengths. Theory predicts (and experiments verify) that the total energy radiated is proportional to the fourth power of the absolute temperature,[29] T, and the maximum emission is at a wavelength λ_m which is inversely proportional to T:

$$\lambda_m = b/T \qquad (9.50)$$

The constant b has the value 0.3 cm (deg). The distribution of intensity drops off steeply on the short-wavelength side of the peak; very little energy is emitted at wavelengths below about $\lambda_m/4$. Since the energy of a photon is hc/λ, we see that thermal photons become more and more energetic as the temperature increases.

Gamow had predicted that the universe should now be filled with thermal radiation at a very low temperature, emitted when the universe was young and hot but cooled enormously as a result of the subsequent expansion. He estimated that the present temperature of the radiation should be roughly 10 K; a later calculation by Ralph Alpher and Robert Herman gave $T \approx 5$ K.

Penzias and Wilson concluded that if the radiation they had detected is part of a thermal spectrum, its temperature is 3.5 ± 1 K. A measurement at a single wavelength says nothing, of course, about the shape of a spectrum. However, subsequent measurements detected the radiation at other wavelengths and confirmed that the spectrum indeed has a thermal shape. Figure 9.18 shows the data; the solid line is a Planck curve with $T = 2.7$ K.[30] The fit is excellent. The peak of the spectrum is at a wavelength

29. Absolute temperature (denoted by the symbol K) is 273 + the Celsius temperature.
30. The point labeled CN in fig. 9.18 is based not on microwave data but on the absorption spectrum of the cyanogen molecule, found in interstellar space. This independent measurement supports the interpretation of the microwave data.

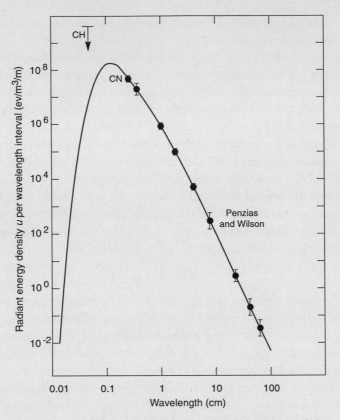

Fig. 9.18.   Blackbody curve fitted to the observations of the microwave background. The temperature is 2.7 K. The point labeled CN comes from the relative intensities of interstellar cyanogen, as explained in the text.

about 1 mm, in the microwave band. Penzias and Wilson's point is far below the peak; the intensity at their wavelength, $\lambda = 7$ cm, is only about a millionth of the peak intensity.

Figure 9.18 is convincing proof that the universe is now filled with thermal radiation at a very low temperature. According to the generally accepted interpretation, this radiation is a relic of the "primeval fireball" and constitutes our only direct link to the early universe.

The energy density of the thermal background can be converted to a mass density by dividing by c^2; the result,

$$\rho_{\rm rad} \approx 10^{-33} \ {\rm g/cm^3} \tag{9.51}$$

is at least a thousand times less than the mean density of matter, discussed in the preceding section. At present, then, matter dominates radiation. During the very early universe, as we shall see, the opposite was true.

If the matter in the universe were in thermal equilibrium with the radiation, it too would be at 3 K. That however is *not* the case; the 3 K temperature is characteristic only of the radiation. When matter and radiation are in thermal equilibrium, photons are being copiously emitted and absorbed at equal rates. At present the universe is largely transparent to the thermal background. The microwave photons roam freely in space; their "mean free path" between collisions is very long. Radiation and matter are said to be decoupled.

We are now prepared to embark on our journey backward in time. I take 10^{10} years as the age of the universe and 10^{-30} g/cm^3 as the present density of matter.[31] The value of z can be used to specify the scale factor: the two are related by equation (9.10). For example, "$z = 20$" refers to the epoch when R was $1/21$ of its present value. In any specific cosmological model, the time is also determined by the value of z. For illustrative purposes I use the Einstein–de Sitter model, in which the relation between z and t is given by equation (9.32). Other models give similar results.

As we go back in time we see the universe contract; the density of matter, the density of the thermal radiation, and its temperature all increase, each at a different rate. Consider first the radiation. As the scale factor diminishes, both the energy of each photon and the density of photons increase. When the scale factor was R, the energy of a photon whose present energy is E_0 was

$$E = E_0(R_0/R) = E_0(1 + z) \qquad (9.52)$$

Equation (9.52) is simply the red shift expressed in terms of photon energy.

The number of photons in the cosmic background remains unchanged as the size of the universe changes; hence the number per unit volume varies inversely as the volume, which is proportional to R^3 or to $(1 + z)^{-3}$. The energy density (the product of the number density and the energy of each photon) varies as R^{-4} or as $(1 + z)^4$:

31. As noted in the preceding section, this number represents the mass in galaxies and is only a lower bound on the mass density. The actual density might be higher because of unseen "dark" matter. If there is enough mass to close the universe, some of the numerical estimates that follow are altered, but not a great deal. The general nature of the history is unchanged.

$$\rho_{rad} \sim \frac{1}{R^4} \sim (1+z)^4 \sim t^{-8/3} \qquad (9.53)$$

The dependence of ρ_{rad} on R and z is the same for all models. However, the time dependence, in this and all subsequent relations, applies only to the Einstein–de Sitter model.

Since the energy density of thermal radiation varies as T^4, we may infer from equation (9.53) that the temperature of the radiation varies as $1/R$ or, equivalently, as $1+z$:

$$T = T_0\left(\frac{R_0}{R}\right) = T_0(1+z) = T_0\left(\frac{t}{t_0}\right)^{-2/3} \qquad (9.54)$$

where T_0 is the present temperature, 2.7 K. The result depends on the assumption that the radiation does not undergo any interactions; the change in the spectrum is due entirely to the expansion of space. That assumption is justified today and for some time in the past. However, the $1/R$ dependence of temperature turns out to be valid even when radiation and matter interacted strongly. (See eq. [9.57c].)

I turn next to the matter. Since the total amount of matter remains constant, its density varies inversely as the volume, that is, as $1/R^3$:

Table 9.1. Temperature and density of the universe as a function of red shift (z) and of time since the big bang. The time dependence is based on the Einstein–de Sitter universe; the other quantities are model-independent.

$1+z = \dfrac{R_0}{R}$	T (K)	ρ_{matter} (g/cm^3)	ρ_{rad} (g/cm^3)	t	Remarks
Present	3	10^{-30}	10^{-33}	10^{10} years	
100	300	10^{-24}	10^{-25}	10^7 years	galaxy formation
1000	3000	10^{-21}	10^{-21}	3×10^5 years	decoupling; radiation era ends
10^4	3×10^4	10^{-18}	10^{-17}	3000 years	
10^5	3×10^5	10^{-15}	10^{-13}	30 years	
10^6	3×10^6	10^{-12}	10^{-9}	100 days	radiation era
10^7	3×10^7	10^{-9}	10^{-5}	1 day	
10^8	3×10^8	10^{-6}	10^{-1}	15 min	
10^9	3×10^9	10^{-3}	10^3	10 sec	cosmic nucleosynthesis
3×10^9	10^{10}	10^{-1}	10^5	1 sec	lepton era ends; neutrinos decouple
3×10^{10}	10^{11}		10^9	0.01 sec	
3×10^{11}	10^{12}		10^{13}	10^{-4} sec	
3×10^{12}	10^{13}			10^{-6} sec	hadron era ends

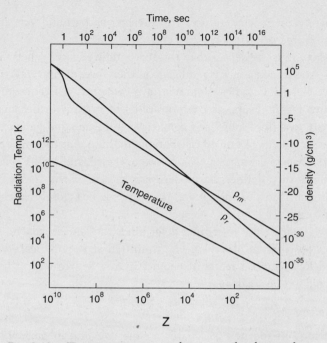

Fig. 9.19. Temperature, matter density, and radiation density according to the standard big bang model. Temperature scale is at the left, density scale at right. Red shift scale is at the bottom, time scale at top. (N.B. time scale is not regular.)

$$\rho_{\text{matter}} \sim R^{-3} \sim (1+z)^3 \sim t^{-2} \tag{9.55}$$

The mass density decreases with R more slowly than does the radiation density, which contains one additional factor $1/R$ because of the red shift. As we go back in time, therefore, the ratio $\rho_{\text{rad}}/\rho_{\text{matter}}$ increases steadily.

Table 9.1 lists the relevant quantities ρ_{rad}, ρ_{matter}, T, and t, starting with the present and going back toward the big bang; all quantities except the time are model-independent. The same quantities are plotted in figure 9.19, on logarithmic scales.

Galaxy Formation

At $z = 100$, when the universe was about 10 million years old, the density of matter was 10^{-24} g/cm^3, a million times greater than today. That density corresponds to about one hydrogen atom per cubic centimeter, roughly the mean density in a galaxy. (The sizes and densities of galaxies are determined by local physical laws and do not change as the universe expands.) The temperature of the radiation at this epoch was 300 K, approximately room temperature. The thermal background was mostly in the far infrared ($\lambda \sim 10^{-3}$ cm).

A galaxy must be denser than the surrounding space. When the entire universe was as dense as a typical galaxy, therefore, distinct galaxies could not have existed. Galaxies could not have come into existence until some time after $t = 10^7$ years, as concentrations of matter formed out of the previously existing "soup" and condensed into stars and galaxies, drawn together by gravity. Detailed calculations indicate that the most plausible epoch for galaxy formation was between $t = 10^8$ and $t = 10^9$ years. The period since then defines the modern era, during which matter has been the dominant form of energy and the universe has closely resembled its present form.

The time of galaxy formation determines the largest galactic red shift that can possibly be observed. Light emitted at $t \sim 10^8$ years would be detected today with a red shift about 25. Any source with a larger red shift could not be a galaxy.

Decoupling

Continuing our journey backward in time, we find in table 9.1 that at $z = 1,000$ ($t = 300,000$ years), the radiation temperature was 3,000 K. At that temperature the thermal spectrum extends into the visible range; the universe was bathed by a sea of visible and infrared light.

This epoch marks a significant change in the universe. At temperatures above 3,000 K, an appreciable fraction of thermal photons have enough energy to ionize a hydrogen atom; the fraction increases rapidly with increasing temperature. The density of photons also increases. As a result, the hydrogen atoms in space become ionized. The transition between the two regimes is fairly abrupt: at 3,000 K the ionized fraction is near zero, whereas at 5,000 K it is nearly 100 percent. At somewhat higher temperatures, helium also becomes ionized.

Ionization is important because when an atom is ionized, an electron is released; free electrons are more effective in scattering radiation than are bound ones. At temperatures above 5,000 K, when all the hydrogen is ionized and free electrons abound, the background radiation interacts strongly with matter and the two quickly reach thermal equilibrium; the universe is opaque to the background radiation instead of being transparent as it is later on.

According to this picture, matter and radiation became decoupled when the universe was about 300,000 years old; that epoch can be regarded as the end of the big bang. The thermal radiation detected today was emitted then at a temperature between 3,000 and 5,000 K and has been roaming freely ever since, influenced only by the expansion of space, which has cooled it by a factor of a thousand. During the first 300,000 years, radiation and matter were strongly coupled and had the same temperature.

The Radiation Era

By coincidence, the time of decoupling, $t = 300,000$ years, is also the time when the densities of radiation and of matter became equal; each was about 10^{-21} g/cm^3. At earlier times the density of radiation exceeded that of matter. That era, which began a few seconds after the big bang, is called the radiation-dominant era.

The densities of matter and of radiation during the radiation era vary as R^{-3} and R^{-4}, respectively, just as they do afterward; the temperature therefore varies as $1/R$. The time dependence of R, however, is different from that during the matter-dominated era. When the density of radiation is much greater than that of matter, all three Friedmann models have the same solution: R is proportional to the square root of the time. The time dependence of density and temperature during the radiation-dominant era are therefore given by

$$\rho_{matter} \sim R^{-3} \sim t^{-3/2} \tag{9.56a}$$

$$\rho_{rad} \sim R^{-4} \sim t^{-2} \tag{9.56b}$$

and

$$T \sim R^{-1} \sim t^{-1/2} \tag{9.56c}$$

Both ρ_{rad} and ρ_{matter} fall off more slowly with time than they do during the matter-dominated era, as shown in figure 9.19.

Cosmic Nucleosynthesis

During most of the radiation era, nothing very exciting happens. The universe simply gets denser and hotter as we go back in time. Near the beginning of this era, however, an important sequence of events took place.

According to generally accepted theory, fusion reactions in stellar interiors result in the formation of all the elements up to iron, starting from hydrogen. This process, which requires a temperature of around 10^8 K, is called nucleosynthesis.

During the early part of the radiation era, the density and temperature of the universe resembled those in stellar interiors; conditions were therefore conducive to the occurrence of fusion reactions. Gamow realized this and suggested that most of the elements were in fact formed in the first few minutes after the big bang. His idea turned out to be only partially right.

How could the early universe accomplish in only a few minutes what stars take hundreds of millions of years to do? Neutrons provide the key to the answer: they are major players in cosmic nucleosynthesis, whereas they contribute almost nothing to element formation in stars. Stellar nucleosynthesis begins with the fusion of two protons to form a deuteron:

$$p + p \rightarrow d + e^+ + \nu \tag{9.57}$$

Reaction (9.57), like any in which a neutrino is involved, proceeds very slowly because it proceeds via the so-called weak interaction. Only a tiny fraction of proton-proton collisions results in the formation of a deuteron. That explains why stellar nucleosynthesis is so slow. In the early universe, however, deuterons can be formed as well by the fusion of a proton and a neutron:

$$n + p \rightarrow d + \gamma \tag{9.58}$$

Reaction (9.58), which is mediated by strong and electromagnetic interactions, is much more probable than is (9.57). Neutron-proton fusion does not occur in stellar interiors because practically no neutrons are present there; the cores of young stars consist almost entirely of protons and electrons. But during the early universe, neutrons were present in abundance, having been produced during an earlier era when the temperature was still higher. (See below.)

The binding energy of the deuteron is about 2.2 MeV. A photon with energy greater than that can dissociate a deuteron by initiating reaction (9.58) in the reverse direction. At temperatures above 10^{10} K, photons with the required energy are abundant; hence any deuteron formed is promptly dissociated. Under such conditions very little deuterium is formed and nucleosynthesis cannot proceed. When the temperature falls below 10^{10} K, however, the rate of dissociation drops and deuterium begins to be created at a rapid rate. This change marks the onset of cosmic nucleosynthesis; it occurred when the universe was between 1 and 10 seconds old.

The synthesis of deuterium, reaction (9.58), is quickly followed by several other fusion reactions that ultimately lead to the formation of ^{4}He, as well as a small amount of ^{3}He. The relevant reactions are the following:

$$d + d \rightarrow {}^3\text{H} + p \tag{9.59a}$$
$$d + d \rightarrow {}^3\text{He} + n \tag{9.59b}$$

$$d + {}^3H \rightarrow {}^4He + n \tag{9.59c}$$
$${}^3He + {}^3He \rightarrow {}^4He + p + p \tag{9.59d}$$

All these reactions involve only strong interactions and proceed very rapidly at temperatures above 10^9 K. The radiative reactions

$$d + n \rightarrow {}^3H + \gamma \tag{9.60a}$$
$$d + p \rightarrow {}^3He + \gamma \tag{9.60b}$$
$$d + d \rightarrow {}^4He + \gamma \tag{9.60c}$$
$${}^3H + p \rightarrow {}^4He + \gamma \tag{9.60d}$$
$${}^3He + n \rightarrow {}^4He + \gamma \tag{9.60e}$$

also contribute to the formation of helium.

Calculations show that within a few minutes, about a quarter of the mass in the universe is transformed into helium as the result of reactions (9.58) through (9.60). The outcome is insensitive to the details of the cosmological model. Because reactions (9.59a–d) are so fast, the process continues until virtually all the available neutrons have been exhausted. The limiting factor is the neutron/proton ratio at the onset of nucleosynthesis; for a broad range of values of the relevant parameters, that ratio is about 1:6. (See below.)

Consider a group of twelve protons and two neutrons. After the fusion reactions have run their course, one ^4He nucleus will have been formed and ten protons will be left over. (The amounts of deuterium, ^3H, and ^3He formed are very small.) Hence the helium fraction at the conclusion of nucleosynthesis is about 9 percent by number or 28 percent by mass. This is just about the measured abundance of helium; the agreement constitutes important evidence in support of the big bang picture. Moreover, the abundance of helium (unlike that of heavy elements) is nearly the same in different parts of galaxies, as well as in intergalactic space; this observation further supports the hypothesis that most of the helium in the universe is of primordial origin.

To be sure, some helium is generated also in stellar nucleosynthesis, as reaction (9.57) is followed by (9.59) and (9.60). However, the amount of hydrogen converted to helium in the core of a typical star is only 1 or 2 percent of a stellar mass, not nearly enough to account for the measured abundance of helium.

Cosmic nucleosynthesis must also produce deuterium. Most of the deuterons formed, however, are quickly reprocessed and end up as helium. The theory predicts that very little deuterium should be left over. Unlike the case of helium, the amount of deuterium produced is quite sensitive

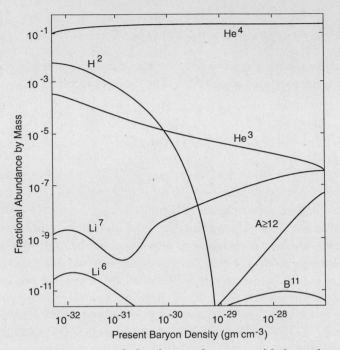

Fig. 9.20. Fractional abundances of isotopes of light nuclei produced in cosmic nucleosynthesis, as functions of the present density of matter. Calculated by Wagoner, Fowler, and Hoyle. Note that the abundance of ^{4}He is practically independent of the assumed present density, whereas the abundance of deuterium (^{2}H) is very sensitive to the present density.

to the density of matter while nucleosynthesis is taking place. The higher the density, the more frequently a deuteron collides with a neutron or with another deuteron and the less deuterium survives.

The mass density during nucleosynthesis is a known multiple of the present density. Hence if the deuterium in the universe is of primordial origin, its measured abundance provides indirect evidence concerning the present density of matter. Robert Wagoner, William Fowler, and Fred Hoyle have calculated the abundance of various isotopes produced in the early universe as a function of the present mass density; their results are shown in figure 9.20. Notice that the calculated abundance of ^{4}He increases very little as ρ_0 varies between 4×10^{-32} and 4×10^{-28} g/cm^3, whereas the abundance of deuterium falls steeply with increasing ρ_0.

The measured cosmic abundance of deuterium is about one part in 10^5 by mass. According to the figure, the value of ρ_0 which gives that abundance is 10^{-30} g/cm^3 or a little less. This is close to the lower limit on the present mass density and is substantially less than the amount needed to close the universe. The data on deuterium therefore support the hypothesis that the universe is open. The argument is not conclusive because some deuterium might have been produced in other ways.[32]

Tiny amounts of lithium, beryllium, and boron are also formed in cosmic nucleosynthesis, as shown in figure 9.20. The measured cosmic abundances of these elements are all very low.

Gamow thought that all the elements could be synthesized during the early universe, but his hope was not fulfilled. The absence of any stable isotope with mass number 5 or 8 acts as an effective barrier to the production of heavier elements. If ^{4}He captures a neutron or a proton it forms ^{5}He or ^{9}Li, both of which are unstable. The fusion of two alpha particles would produce ^{8}Be, but that isotope is unstable too. Cosmic nucleosynthesis was therefore a dud: it came to an end with helium.[33]

The Lepton Era

At $t \approx 1$ sec, the beginning of the radiation era, the temperature of the universe stood at 10^{10} K. At this temperature, the mean energy of thermal photons is more than enough to create electron-positron pairs. Prior to 1 second, the universe was therefore filled with such pairs, which outnumbered the protons by nearly a hundred million to one. Earlier still, when the temperature was even higher, muon-antimuon pairs were similarly created. (The rest mass of a muon is about 200 times that of an electron, so a correspondingly higher temperature is required.) The period between about 10^{-4} seconds and 1 second is called the lepton era. (Leptons are light particles that interact only through weak and electromagnetic interactions; they comprise electrons, muons, and neutrinos.)

During the lepton era, the creation of particle-antiparticle pairs made the density of matter much higher than the estimate (9.57a); in fact, with pair production and annihilation taking place in rapid succession, the density of matter and radiation must be about equal. This is shown schematically in figure 9.19.

32. Stellar nucleosynthesis produces almost no deuterium. Hence the small amount measured is either a product of the early universe or has some other (unknown) origin.
33. The stars manage to break through the barrier at mass number 8 by means of a very rare reaction—the fusion of three alpha particles to form ^{12}C. Although a three-body collision is highly improbable, over time enough ^{12}C is formed to enable nucleosynthesis to proceed. In the early universe, both density and tempera-

The lepton era is an important link in the chain of events leading to cosmic nucleosynthesis because it provides the required neutrons. A collision between a proton and an electron can produce a neutron through the reaction

$$p + e \leftrightarrow n + \nu \tag{9.61a}$$

The two-headed arrow indicates that the reaction can occur in either direction.

In a single collision between an electron and a proton, the formation of a neutron is highly improbable because reaction (9.61a), like (9.58), proceeds via the weak interaction. During the lepton era, however, the electron density was so high that collisions took place at a prodigious rate. Note from table 9.1 that at $t = 1$ sec, the density was 10^5 g/cm^3, a hundred thousand times the density of water and much greater than the density in the cores of stars. At 10^{-4} seconds, the density was 10^{13} g/cm^3, about 1 percent of nuclear density. Consequently, a large number of neutrons were produced in a very short time, enabling nucleosynthesis to proceed.

Two other reactions compete with (9.61a) to establish the balance between neutrons and protons during the lepton era:

$$p + \bar{\nu} \leftrightarrow n + e^+ \tag{9.61b}$$

and the neutron beta decay

$$n \to p + e^- + \bar{\nu} \tag{9.61c}$$

where $\bar{\nu}$ stands for the antineutrino.

The neutron is heavier than the proton by 1.29 MeV/c^2; hence reactions (9.61a,b) are endoergic when proceeding from left to right: the colliding particles must have enough kinetic energy to provide the difference in rest mass. At temperatures well above 10^{10} K, which are characteristic of the early part of the lepton era, the thermal energies of neutrons, protons, and electrons are much greater than 1.29 MeV;[34] hence the neutron-proton mass difference has a negligible effect. Under those conditions the numbers of neutrons and protons are about equal.

---

ture are falling rapidly. By the time helium has been formed, the three-alpha reaction cannot occur to any appreciable extent.

34. The energy distribution of protons and neutrons at a given temperature differs from that of photons. Particles are described by the Maxwell-Boltzmann distribution. Their mean energy at absolute temperature T is $(3/2)kT$, only slightly smaller than the mean photon energy at the same temperature.

At lower temperature, the colliding particles have less kinetic energy and the neutron-proton mass difference becomes more important. The reactions that change a proton to a neutron take place at a slower rate than those that do the reverse, and the neutron fraction drops. Finally, when the temperature reaches a few times 10^9 K, the electrons and positrons have almost entirely recombined and reactions (9.59a,b) can no longer take place. The lepton era has come to an end. Thereafter, the neutron fraction diminishes slowly as neutrons decay.[35] Nucleosynthesis begins at about this time. James Peebles has done the calculations and found that the neutron fraction at the onset of nucleosynthesis is about ⅙. As we have noted, this fraction determines the amount of helium produced.

A very large number of neutrinos is left over at the end of the lepton era. These cosmic neutrinos, degraded in energy by the expansion of space, should still be around, forming a thermal background similar to the microwave background. Their density is nearly as great as that of the photons. Detection of such a neutrino background would provide additional evidence in support of the big bang hypothesis. Because low-energy neutrinos interact so weakly, however, there is no way to confirm their presence.

The First Ten Thousandth of a Second

We have traced the history of the universe back to 10^{-4} seconds. If we are daring enough to push still farther back in time, we enter a truly uncharted domain. According to the models, both the density and the temperature of the universe during this epoch were higher than anything of which physicists have any experience; no experimental evidence is available to compare with theoretical predictions. This has not discouraged theorists from carrying out elaborate calculations. The study of the first ten thousandth of a second is particularly interesting because it involves an interplay between astrophysics and high-energy particle physics. But the results must be viewed with a healthy dose of skepticism.

Above 10^{12} K, thermal photons are energetic enough to create pions, strongly interacting particles whose rest energy is 140 MeV; above 10^{13} K, proton-antiproton pairs can be created. At still higher temperatures, pairs of heavier strongly interacting particles known collectively as hadrons can be created; these particles are currently being intensively

---

35. The lifetime of the neutron, about 12 minutes, is long on the time scale of the early universe.

studied by high-energy physicists. The universe during that fleeting "hadron era" was a maelstrom of particles, antiparticles, neutrinos, and photons, all in thermal equilibrium at immense density and temperature. At some earlier time free quarks might have made an appearance.

At $t = 10^{-23}$ sec, whimsically referred to as a jiffy, the Hubble radius was 10^{-13} cm: the entire observable universe was the size of one proton! The density was some 10^{55} times that of water. Undaunted, cosmologists have carried the history back still farther, to the incredibly short age of 10^{-43} seconds, known as the Planck time. Before that time, quantum gravity governed.

Most theorists believe that general relativity is valid as far back as the Planck time and that the laws of physics as we know them still apply under the extreme conditions that then prevailed. This, however, is an article of faith. The assumption of isotropy and homogeneity might also not be justified. Small anisotropies could have profound effects when extrapolated back to such infinitesimal times.

9.9. DID THE BIG BANG REALLY HAPPEN?

I recapitulate here the evidence in support of the big bang hypothesis.

1. *The microwave background.* This is the most persuasive single piece of evidence. As we have seen, the existence of a thermal background is an unambiguous consequence of big bang cosmology, independent of the details of the model adopted. The temperature at the time of decoupling (3,000–5,000 K) is fixed by accepted physical laws. After decoupling, the radiation must cool as a result of the expansion of space. The radiation has been detected just as the theory predicts, and no alternative origin for this radiation has been suggested.

2. *The age of the universe.* According to the most plausible cosmological models, the time elapsed since the big bang is somewhat less than the Hubble time—10 billion years or so. Before the distance scale had been recalibrated, the cosmological age of the universe was shorter by a factor of about ten; this created a serious problem for the big bang theory, as it would have made the universe younger than some of its constituents. After the recalibration, however, the geological and cosmological age estimates mesh together quite comfortably. The oldest earth rocks, as well as lunar rocks and meteorite fragments, are estimated on the basis of radioactive dating to be some 4 to 6 billion years old. This is quite consistent with the age of 10 billion years for the universe inferred from the cosmological models and the measured value of the Hubble constant. It seems a

priori unlikely that the agreement between the ages should be fortuitous. One should keep in mind, however, that the value of the Hubble constant depends on distance determinations that are not totally reliable. The "measured" value of H has changed before; it could change again.

3. *The abundance of helium.* As we have seen, big bang cosmology accounts quite naturally for the high cosmic abundance of helium, independently of the details of the cosmological model adopted. There are no free parameters whose value can be adjusted to give the measured abundance. (The result does, however, depend on the cross sections for reactions like [9.59] and [9.60], which have not been directly measured.) Without cosmic nucleosynthesis, one would be hard put to understand where all the helium in the universe comes from.

4. *The abundance of deuterium.* This supports the helium data. In this case, the data do discriminate among the models. Matching the measured abundance of deuterium requires a specific choice of the present mass density. The density required is low, suggesting that the universe is open.

5. *Counts of radio sources.* During the 1950s, many powerful extragalactic radio sources were discovered, some of them associated with galaxies whose red shifts could be measured. Study of the distribution of these sources in intensity and in red shift provides some evidence of evolution in the universe. If the sources were uniformly distributed in space and their intrinsic luminosity were independent of time, the number of sources at a distance less than R should vary as R^3. Since the apparent brightness decreases as $1/R^2$, the number of sources with flux greater than F should vary as $F^{-3/2}$. A number of surveys have shown that the actual distribution is different: there are many more weak sources than would be expected according to the $-3/2$ power law. Since the weak sources are, on the average, the most distant and therefore the youngest, the result implies that radio sources are brighter when they are young. This is evidence of evolution on a cosmological time scale and therefore weakly supports the big bang picture, although it says nothing about the big bang itself. An oscillatory universe without singularities would be consistent with the data.

The evidence is strong, but is it convincing? The reader must make up his or her own mind on that. One should surely be skeptical about the very early part of the picture drawn by the theory. The microwave background supports only the portion of the story since the age of 300,000 years or so, and the age evidence has nothing to say about the very early period. The helium data go back farther and are the strongest evidence for the early part of the history. If the predicted neutrino background could be detected (see the discussion on p. 351), we would have direct evidence supporting the big bang picture back to the age of about 1 second. The required measurements, unfortunately, are far beyond the capabilities of present-day technology.

Finally, we address a question that must intrigue anyone who thinks at all about cosmology: what was there before the big bang? One possibility, already mentioned, is the oscillating universe model according to which the universe has undergone an endless sequence of expansions and contractions, separated by big bangs. In this case, there is no conceptual problem, although we can say nothing about conditions before the last big bang.

Another possibility is that there was nothing before the big bang—that time began with the singularity some 10 billion years ago. This notion depends on Einstein's conception of curved space-time. If space-time were flat, time would have to extend infinitely in both directions. In curved space-time, however, the total extent of time can be finite, just as the area of a sphere is finite. If the geometry of space-time is of this kind, the question, What was there before the big bang? is equivalent to asking, What is north of the north pole? It is not a scientifically meaningful question.

Problems

9.1. (a) Using the data found in section 9.1, estimate the mean distance between a star in our galaxy and its nearest neighbor. If the radius of a typical star is 10^9 m, what fraction of the volume of the galaxy is occupied by stars?

(b) Assume the universe is a Euclidean sphere of radius 10^{10} light-years. If the universe contains 10^{11} galaxies, what is the mean distance between a galaxy and its nearest neighbor? Using the dimensions of our galaxy as typical, what fraction of the volume of the universe is occupied by galaxies?

9.2. A good telescope can resolve two point sources whose lines of sight form an angle of about 1 sec. Estimate the largest distance at which individual stars in a galaxy identical to ours can be resolved. Compare your answer to the distance of the Magellanic Clouds.

9.3. A galaxy has a red shift of 0.05.

(a) How long ago was light from this galaxy emitted if it is now reaching earth?

(b) If the expansion of the universe has been uniform, what was the distance of this galaxy when it emitted the light now reaching us? What is the present distance of the galaxy? What is its present recession velocity? Assume Euclidian geometry.

9.4. A galaxy has a red shift of 1. Answer the questions posed in problem 9.3, assuming (a) an Einstein–de Sitter universe and (b) a steady-state universe.

9.5. Consider a model universe that is static (no expansion) and infinite. Suppose all the galaxies were created simultaneously T years ago. What is the distance of the farthest galaxy we can observe? Does this constitute an event horizon or a particle horizon? Explain.

Index

Designer: U.C. Press Staff
Text: 10/13 Aldus
Display: Aldus
Compositor/Printer/Binder: Maple-Vail Book Manufacturing Group